Information Processing
in Cells and Tissues

Information Processing
in Cells and Tissues

Edited by

Mike Holcombe

University of Sheffield
Sheffield, United Kingdom

and

Ray Paton

University of Liverpool
Liverpool, United Kingdom

Plenum Press • New York and London

Library of Congress Cataloging-in-Publication Data

Information processing in cells and tissues / edited by Mike Holcombe
and Ray Paton.
       p.   cm.
    Includes bibliographical references and index.
    ISBN 0-306-45839-X
    1. Cell interaction--Congresses. 2. Cellular signal transduction-
-Congresses. 3. Information theory in biology--Congresses.
4. Second messengers (Biochemistry)--Congresses.   I. Holcombe, W.
M. L. (William Michael Lloyd), 1944-   . II. Paton, Ray.
QH604.2.I52   1998
571.6--dc21   98-18265
                                                           CIP

Proceedings of the International Workshop on Information Processing in Cells and Tissues,
held September 1–4, 1997, in Sheffield, United Kingdom

ISBN 0-306-45839-X

© 1998 Plenum Press, New York
A Division of Plenum Publishing Corporation
233 Spring Street, New York, N.Y. 10013

http://www.plenum.com

10 9 8 7 6 5 4 3 2 1

Printed in the United States of America

PREFACE

The Second International Workshop on Information Processing in Cells and Tissues (IPCAT) was held in Sheffield from 1st to 4th of September, 1997. The meeting took place in Halifax Hall, a former steel baron's mansion located in Broomhill, a leafy suburb of Sheffield once described by John Betjeman as the "prettiest suburb in England". As with the First International IPCAT in 1995 the purpose of the workshop was to bring together a group of scientists working in the general area of modelling cells and tissues. The result was a multidisciplinary event involving over 50 biologists, physicists, computer scientists and mathematicians from Europe, Japan and the USA.

A central theme underlying the workshop was the nature of biological information and the ways it is processed in cells and tissues. The workshop sought to provide a forum to report research, discuss emerging topics and gain new insights into information processing systems. As a result many areas were considered including enzyme and gene networks, second messenger systems and signal transduction, automata models, PDP models, cellular automata models, molecular computing, single neuron computation, information processing in developmental systems, information processing in neural and non-neural systems and new insights into non-linear aspects of physiological behaviour.

This post-conference proceedings has been organised into three sections. The first two sections consist of papers and a selection of posters that were presented during the three and a half days of the meeting. Section 1 deals with the topic of signalling and communication and covers a number of subjects ranging from gap junctions and calcium signals to neurones, the pituitary and the heart. The emphasis of section 2 is computation and information. The range of biological systems covered here is quite wide including embryos, gene expression, immune system, enzymes, neurones and signalling. The organisation of the first two sections has not been straightforward and there is considerable overlap in content. Such is the nature of this area of research. However, it is hoped that the incorporation of some structure will facilitate the readability of the whole. The final section contains abridged notes based on discussions that were held during the workshop.

A key motivation of the IPCAT workshops is to provide a common ground for dialogue and reporting research without emphasising one particular research constituency or way of modelling or singular issue in this area. IPCAT '97 sought to further the meaningful dialogue and exchange of ideas started at IPCAT '95. We hope this book will reflect some of the valuable work that is going on.

Our thanks to the programme review committee:

Guenter Albrecht-Buehler (Cell Biology) - Northwestern University
Dennis Bray (Zoology) - University of Cambridge
David Brown (Neurobiology) - Babraham Institute, Cambridge
Robin Callard (Immunology) - University of London, London
Chris Cannings (Statistics) - University of Sheffield
Teresa Chay (Biological Sciences) - University of Pittsburgh

Michael Conrad (Computer Science) - Wayne State University, Detroit
John Easterby (Biochemistry) - University of Liverpool
Bard Ermentrout (Mathematics) - University of Pittsburgh
Leon Glass (Biological Sciences) - McGill University
Albert Goldbeter (Nonlinear Systems) - University Libre, Brussels
Uwe an der Heiden (Mathematics) - University of Witten/Herdecke
Mike Holcombe (Computer Science) - University of Sheffield
Felix Hong (Physiology) - Wayne State University, Detroit
Tsuguchika Kaminuma (Biosciences) - NIHS, Tokyo
George Kampis (Philosophy of Science) - Budapest
Douglas Kell (Biochemistry) - University of Wales, Aberystwyth
Rolf Koetter (Neurobiology) - University of Dusseldorf
Gareth Leng (Physiology) - University of Edinburgh
Philip Maini (Mathematics) - University of Oxford
Pedro Marijuan (Electronics/BioInformatics) - University of Zaragoza
Koichiro Matsuno (BioEngineering) - University of Nagaoka
Hiroshi Okamoto (Neurobiology) - Fuji Xerox, Kanagawa
Ray Paton (Computer Science) - University of Liverpool
Peter Schuster (Molecular Biotechnology) - University of Jena
Gordon Shepherd (Neurobiology) - Yale University
Richard Stark (Mathematics) - University of South Florida, Tampa
Rene Thomas (Molecular Biology) - University Libre, Brussels
Chris Tofts (Computer Science) - University of Leeds
Rickey Welch (Biochemistry) - University of Maryland
Hans Westerhoff (Math. Biochemistry) - Free University, Amsterdam
Gershom Zajicek (Medicine) - Hebrew University of Jerusalem

We also wish to thank Matt Fairtlough and Marie Willett who were involved in local arrangements for the workshop and helped a great deal in ensuring that the event ran smoothly. Finally, our thanks to J. Lawrence of Plenum for her encouragement and advice during the production of this book. The editors also acknowledge the help and support of a number of organisations: SmithKline Beecham, GPT, Unilever, Hewlett-Packard, Geomica and EPSRC.

The IPCAT WWW HomePage is located
at:http://www.csc.liv.ac.uk/~biocomp/ipcat/iphome.html

The IPCAT email list was set up after IPCAT '95 in order to promote dialogue among the delegates. Since that time it has grown in size. To subscribe to the IPCAT email list send an email message to the address: ipcat-request@csc.liv.ac.uk containing the simple message: subscribe. Further information about IPCAT can be obtained from Ray Paton.
Further reading Cuthbertson, R., Holcombe, M. & Paton, R. (eds) (1996) Computation in Cellular and Molecular Biological Systems, Singapore: World Scientific.

CONTENTS

Information Processing
in Cells and Tissues

SIGNALLING AND COMMUNICATION
AN INTRODUCTION TO SECTION 1

Ray Paton

Department of Computer Science
University of Liverpool
Liverpool L69 3BX

The first section of the book deals with papers which can be described as focusing on signalling and communication. As noted in the preface there is a degree of arbitrariness in making this separation and so some papers which deal with this subject may be found in the second section. The paper by Steve Baigent, Jaroslav Stark and Anne Warner discusses a model built to explore the possible role of gap junctions in mediating developmentally important signals in a developing *Xenopus* embryo. Biophysical models for both the cells and the gap junctions are outlined and used to build a model embryo in the form of an arbitrary network of cells linked together by gap junctions through which various molecular species are communicated. They outline the dynamics of the network model, which may be viewed as a singularly perturbed system, and discuss the model from a thermodynamic viewpoint. Finally, they outline how perturbation techniques can be used to study the steady state properties of the network and comment upon the relevance of our results to pattern formation.

Following on from their IPCAT '95 contribution, Martin Bogdan and Wolfgang Rosenstiel present a paper concerned with the real time processing of nerve signals for controlling a limb prostheses. Microfabricated neural interfaces promise to become a powerful tool for applications like the control of motor/sensory limb prostheses for amputees and the direct stimulation of spinal cord injuries (FNS). Such interfaces, which have been fabricated within the INTER-project (Intelligent Neural InTERface), have been successfully implanted and signals of the peripheral nerve system (PNS) have been recorded. The group has proposed a modus operandi to process recorded nerve signals using Artificial Neural Nets (ANNs). In the present paper they show that it is possible to process nerve signals and to direct a limb prostheses in real time using Artificial Neural Nets (ANN). After a short introduction to the INTER-project, they discuss the signal processing using recordings from the stomatogastric nervous system (STNS) of the crab *Cancer parugus*. The recording have been classified by Kohonen's self-organizing map (SOM). The obtained clusters have been used to assign an action to limb prostheses. After the training of the SOM and the assignment of the clusters to an action of the prostheses, the system is able to identify and to classify incoming nerve signals. Due to this classification of the incoming nerve signals, a limb prostheses is able to react with movements defined by the frequency of occurrence (speed) and the class of the nerve signal (direction). Both tasks were evaluated in real time.

Teresa Chay presents a paper on stimulus-secretion coupling in pancreatic beta-cells explained by her store-operated model. The burst pattern of pancreatic beta-cells is influenced by various external stimuli such as glucose, extracellular Na^+, K^+ and Ca^{2+}, and the neurotransmitters and hormones involved in the phosphatidylinositol signalling and adenylate-cyclase transduction pathways. By constructing a mathematical model she explains how these stimuli may influence the burst pattern.

The short paper by Laurence Clark and Ray Paton reviews some of the progress made so far towards a computational model of the cellular signalling system used in *E. coli* chemotaxis. Information is received through receptors (MCPs) on the cell surface. Signals are then passed through a network of proteins via phosphorylation, in order to process the input. The work reported here seeks to create a computational model of the system using several different methods and will hopefully extend existing models. A number of alternative strategies are appraised, including a parallel distributed processing network with nodes representing proteins, a distributed processing package (such as SWARM) where communicating, individual agents act like proteins, boolean nets, cellular automata, and algebraic machines.

The next three papers deal with various aspects of neuronal information processing. David Friel's paper looks at three modes of calcium-induced calcium release in neurones are discussed. Ca^{2+} release channels are present in a wide variety of cells and are thought to mediate Ca^{2+}-triggered Ca^{2+} release from internal stores by a process known as Ca^{2+}-induced Ca^{2+} release (CICR). Ryanodine receptors, a particular class of Ca^{2+} release channels found in many excitable cells, open in response to elevations in the cytosolic free Ca^{2+} concentration ($[Ca^{2+}]_i$), speeding stimulus-evoked elevations in $[Ca^{2+}]_i$. It is usually assumed that CICR causes net Ca^{2+} release from internal stores. This point has been re-examined in sympathetic neurones, where previous studies implicate CICR as an important factor in defining the kinetics of depolarisation-induced elevations in $[Ca^{2+}]$. It appears that during weak depolarisation, while Ca^{2+} release channel opening leads to an acceleration of the rise in [Ca2+]i, internal stores do not undergo net loss of Ca^{2+}. A simple model is presented which illustrates this as one of three modes of CICR, in which the internal store act as a Ca^{2+} buffer whose strength is modulated by Ca^{2+} release channel activity.

The paper by Boris Gutkin and Bard Ermentrout presents a 1-d spiking model that results from a formal reduction of conductance-based neural models and can account for a number of both *in vitro* spiking characteristics and *in vivo* spike train statistics of non-bursting cortical pyramidal neurones. The theta-neurone is a phase model that can be thought of as a non-linear extension of the integrate-and-fire model. Inherent in the model is the type I membrane excitability characteristic of a number of compartmental cortical neurone models, discussed in detail in [Rinzel 1989]. They outline the reduction method by which the model is derived. They further show that the saddle node bifurcation dynamics accounts for the *in vitro* input/output properties found in non-bursting cortical pyramidal neurones, in particular the appearance of low frequency oscillations. The model can also account for the high level of randomness observed in *in vivo* spike trains. They suggest that the theta-neurone can replace the integrate and fire neurone as a basic computational unit in large scale spiking neural network models.

Tim A. Hely, Arjen van Ooyen and David J. Willshaw present a new theoretical model to explain growth cone filopodia dynamics. Filopodia behaviour is usually attributed to random, and spontaneous events in the growth cone. However, recent experimental work has linked the presence of a high concentration of intra-cellular calcium ions to filopodial outgrowth. Both influx of calcium from the external medium and calcium induced calcium release from internal calcium stores affect filopodia behaviour. They propose that the calcium ions act as a diffusible morphogen - a chemical which directly affects the development of the cell shape. A Turing reaction-diffusion system with calcium as an activator is modelled, and cAMP as an inhibitor. The growth cone dynamically changes shape in response to the local calcium concentration, and the "random and spontaneous" creation of filopodia occurs as a direct result of the underlying calcium pattern. This model provides a new computational perspective for understanding the dynamics of filopodia cone growth.

The paper by Arun Holden, G Kremmydas and A Bezerianos presents an anatomically accurate model of the diastolic dog ventricle, in which muscle fibre orientation at each point in the ventricle is specified, and coupled with excitable maps to construct a coupled map lattice (S) model of propagation of electrical activity in the whole canine ventricle. This anatomical model of propagation within the mammalian heart is then used to drive local, ordinary differential equation models of guinea pig ventricular cell excitability at different points within the ventricle to simulate the effects of disturbances in the pattern of ventricular propagation on the electrical activity behaviour. This hybrid construction of a chimaeric model of propagation and excitation in a mammalian heart allows the efficient simulation of detailed cellular electrical activity at different parts of the ventricle during normal and abnormal propagation. It also illustrates a general approach to inter-relating tissue and organ behaviour (on coarse time and space scales, with models that have few dynamic variables) to high order, stiff models of local cell behaviour.

Marc Keulers and Hiroshi Kuriyama discuss extracellular signalling in an oscillatory yeast culture. When the yeast *Saccharomyces cerevisiae* was grown under aerobic condition a metabolic oscillation appeared. This oscillation was observed due to synchronisation of oscillatory metabolism in a population. The synchronised metabolic oscillation was dependent on the aeration rate, high aeration stopped the oscillation suggesting that synchronisation was caused by a volatile compound in the culture. Ethanol, acetate, acetaldehyde and oxygen were found not to be the synchroniser of the oscillation. Stepwise increase in carbon dioxide concentration of the gas flow rate ceased synchronisation, but the oscillation continued in each individual cell. Stepwise increase of the aeration rate keeping carbon dioxide at oscillatory condition did not cease the oscillation. Based on these facts it is postulated that carbon dioxide, through the influence of its dissociation to bicarbonate or through its direct effect on major metabolic pathways, could be the synchronisation affecter of the metabolic oscillation of *S. cerevisiae*.

Rolf Kötter, Dirk Schirok and Karl Zilles present a quantitative kinetic model of signal transduction pathways linking stimulation of metabotropic receptors to the phosphorylation state of membrane channels in striatal principal neurones. The model is based on an experimental study by Surmeier *et al.* (*Neuron*, 14: 385, 1995) and replicates the reported effects of a variety of pharmacological agents on the reduction of N/P-type calcium currents by D1 dopamine receptor stimulation. They show that an adequate model of calcium channel regulation requires, in addition to protein kinases A and G and protein phosphatase 1, a protein phosphatase 2B-like activity, which has not been considered previously. Further exploration of differential equations describing cAMP-dependent signalling pathways predicts kinetic characteristics of its components that can be tested experimentally. This model extends previous work on dopaminergic effects on striatal principal neurones and provides further insights into the complex role of dopamine in striatal processes.

Xiao Mang and Siobhan North describe a WWW cytokine database project to explore methods of constructing a cytokine database on the Web which will provide on-line access to a full scale compendium of cytokines and their receptors across the Internet. It will offer various representations of the data available including text, signal pathway diagrams and 3D images. By constructing the database on the Web, true cross-reference links can be made between different cytokines, cells, receptors and pathways. The management and maintenance of the database will also be made easier. Furthermore, by using the Web as the platform, the Web cytokine database will be able to present some of the cytokine attributes like the amino acid sequence and signal transduction in more intuitive ways, such as images and diagrams, rather than static text. The Web cytokine database features dynamic information generation, dynamic information collection and dynamic information display. It could also provide information on cytokine related research projects, discussions and vendors. It aims to build a wide ranging home for cytokine related information and research from which anyone who is interested in cytokines can benefit.

The next three papers return us to various aspects of molecular information processing. Juergen Nauroschat and Uwe an der Heiden report on a mathematical model of the action of G-protein-coupled receptor-kinase upstream from cAMP synthesis in cells. Triggering of cellular processes often requires the availability of the cytosolic cyclic nucleotide cAMP. They discuss a differential system modelling the availability of cAMP as a second messenger, mediated by stimulatory G-protein coupling receptor. Action of cognate receptor-kinase upstream from cAMP synthesis is considered in the model at the level of agonist-liganded receptor.

Hiroshi Okamoto and Kazuhisa Ichikawa look at a model for biochemical-reaction networks describing autophosphorylation versus dephosphorylation of Ca^{2+}/calmodulin-dependent protein kinase II (CaMKII) exhibits switching characteristics with respect to the intensity of the Ca^{2+} signals (Okamoto & Ichikawa, IPCAT '95). Recent experimental studies have demonstrated that autophosphorylation of CaMKII results in i) drastic increase in the calmodulin-binding affinity and ii) full activation of total activity. In the present study, they show that the switching characteristics are amplified if either of these two effects is taken into the model. They further demonstrate that the amplification of the switching characteristics can provide a possible molecular-mechanical explanation for why a single burst, if it is given at a peak but not at a trough of the theta oscillation in the hippocampus provoked by cholinergic agonist, can induce significant long-term synaptic potentiation.

Stefan Schuster, Marko Marhl, Milan Brumen and Reinhart Heinrich present a refined model for intracellular calcium oscillations, which are of importance for cell signalling in many living cells. Special attention is drawn to the concomitant oscillations of the potential difference across the membrane of intracellular calcium stores and to the buffering of calcium by proteins. The fast binding to specific proteins such as calmodulin is taken into account using a rapid-equilibrium approximation. Moreover, a slower association to another class of calcium-binding sites is included. The quasi-electroneutrality condition allows us to reduce the number of independent variables to two. The subcritical and supercritical Hopf bifurcations are analysed. It is pointed out that frequency encoding is particularly well feasible near the subcritical Hopf bifurcation. The oscillations obtained by numerical integration show the typical spike-like shape and reasonable values for the amplitudes of calcium and potential oscillations which are in better agreement with experiment than the values obtained when only one class of proteins is included.

In the final paper of this section, Sinead Scullion, David Brown and Gareth Leng discuss a model of LHRH "self-priming" at the pituitary. Luteinising hormones (LH) is released from the anterior pituitary in response to the secretion of luteinising hormone-releasing hormone (LHRH) from the hypothalamus. The pattern of LH release varies during the reproductive cycle and a large surge of LH triggers ovulation in all female mammals. However, the pre-ovulatory LHRH surge alone is too small to trigger the LH surge. The LH surge is also dependent on a rapidly increasing pituitary responsiveness to LHRH. This is due to the "self-priming effect" of LHRH; initial exposure to LHRH potentiates subsequent LHRH-induced LH release in pituitaries exposed to oestrogen. Self-priming varies with the duration of LHRH exposure and the time between exposures. A non-linear differential equation model of LHRH induced LH release is described. This includes components related to self-priming to account for the changes in pituitary responsiveness that occur before the LH surge. The results of simulations are compared with *in vivo* findings.

SOME ASPECTS OF GAP JUNCTION DYNAMICS

IN EMBRYONIC SYSTEMS

Stephen Baigent[1], Jaroslav Stark[1], and Anne Warner[2]

[1]Centre for Nonlinear Dynamics and Its Applications
[2]Department of Anatomy & Developmental Biology
University College London, Gower Street, London WC1E 6BT

INTRODUCTION

Effective intercellular communication is essential for the proper integration of any multicellular system into a functioning syncytium. In many tissues an intercellular link is provided by arrays of aqueous protein channels known as *gap junctions*. Individual cells communicate with their neighbours by the exchange of ions and small molecules through gap junctions. In this way a biological signal may be relayed from one cell to a distant neighbour via a chain of cells.

Gap junctions are ubiquitous in both the developing embryo and the mature animal. Their role is very diverse, varying from tissue to tissue and species to species. In some tissues they provide electrical coupling between excitable cells. For example, in heart myocardial tissue gap junctions permit the rapid spread of an action potential (De Mello, 1987), whereas between pacemaker cells in the sinoatrial node they facilitate sychronisation of beating (Dongming *et al*, 1994). Gap junctions also provide electrical coupling between non-excitable cells, such as between photoreceptors in the retina, where it is thought that light dependent changes in the degree of signal averaging may help to optimise resolution in the presence of noise (Attwell *et al.*, 1885).

Gap junctions are also found in embryos of virtually all animals, where their properties can vary with the developmental stage and actual location within the embryo. Their precise role in development is not yet understood, but a variety of experiments suggest that they mediate developmental signals. We briefly review the experimental evidence for this in the next section.

The aim of our project is to explore the communicative role of the gap junction during early development of the *Xenopus* embryo. For this purpose we are building a *biologically realistic* network model of a developing *Xenopus* embryo. The model is based upon actual electrophysiological data recently obtained from voltage clamp experiments carried out on *Xenopus* embryos. Thus, for the present, our model focuses on the electrical properties of the cells and gap junctions. Later, we hope to include intracellular concentrations and their modulative affect on both the surface membrane

Information Processing in Cells and Tissues
Edited by Holcombe and Paton, Plenum Press, New York, 1998

7

and the gap junctions. The development of the model is achieved by testing predictions of the model experimentally and feeding the conclusions back into the model. This will ensure that our model remains biologically realistic. We feel that this approach will offer an insight, both mathematical and biological, into the difficult problem of embryonic pattern formation. Moreover, the ubiquitous appearance of the gap junction as a communicating junction means that the theory and methods developed during the project may be applicable to multicellular systems in mature organisms.

GAP JUNCTIONS IN *XENOPUS* DEVELOPMENT

In this section, we will briefly review some of the experimental evidence that suggests gap junctions may be important for mediating developmentally important signals during embryogenesis. Direct evidence is difficult to obtain, partly because multicellular systems are so hard to analyse experimentally.

1. When antibodies blocking gap junction channels were micro-injected into one cell of an early *Xenopus* embryo (Warner *et al.*, 1984) dye transfer and electrical coupling between the progeny of the injected cell were found to be selectively disrupted and the embryo later developed deformities. It is not clear, however, how the blocking of gap junction communication lead to the observed developmental deformities.

2. A number of water soluble dyes were injected into cells of a *Xenopus* embryo between the 16-cell and early blastula stages (Guthrie *et al.*, 1988). As early as the 32-cell stage, cells lying on the future dorsal side exchanged Lucifer Yellow through gap junctions more frequently and more extensively than cells in the future ventral region. How these nonuniformities might arise, and their physiological significance is not yet known.

3. The dorso-ventral axis of a *Xenopus* embryo was experimentally perturbed by u.v. irradiation or Lithium treatment (Nagajski *et al.*, 1989). After u.v. treatment the transfer of Lucifer Yellow was found to be the same throughout the animal hemisphere and at the low level characteristic of future ventral regions of normal embryos. The embryos developed with massive reductions in dorsal axial structures. Lithium had the opposite effect, converting all gap junctions into dorsal type junctions. This suggests that the properties of gap junctions at the 32-cell stage indicate the future developmental fate of animal cells.

4. The permeability of some gap junctions has been shown to be a time-dependent function of the electrochemical potential at each terminal of the junction (Mulrine & Warner, in preparation). The passage of ions and small molecules through a gap junction changes these electrochemical potentials and the gap junction permeability responds by relaxing exponentially to a new steady state. This suggests that various forms of electrochemical feedback could control gap junction permeability.

GAP JUNCTION DYNAMICS

Gap junctions are not passive linear passages for intercellular communication. They are dynamic structures whose permeability changes with both the local physicochemical state and time. For example, the permeability of some gap junctions are affected by the presence of chemical modulators, such as Ca^{2+}, H^+, cyclic AMP, or an electric field.

Permeability changes are not instantaneous, but generally occur with an exponential transient.

A popular approach to modelling permeability changes is to envisage entry to, and exit from, a channel as being controlled by a number of gates. Each gate is either entirely open or entirely closed. A channel then has a certain set of possible gating states, each with a certain permeability, and random fluctuations cause gates to open or close, so that channels continually switch between states. The transition rates between gating states may be modulated by chemicals or an electric field. The total permeability of a gap junction is the weighted sum of all fractional populations of channels in the available gating states.

Other processes, such as phosphorylation and dephosphorylation of gap junction proteins, binding of ligands, or the physical blocking of channels also change gap junction permeability. Our present research, however, focuses on the gating properties of the gap junction.

These ideas lead to the following simple model for gap junction dynamics. We assume that a gap junction channel has N gating states and that the fraction of the total number of channels in the gap junction in state k is s_k, so that $\sum_{k=1}^{N} s_k = 1$, and that the permeability of state k is ρ_k. Next we suppose that the probability per unit time of a channel transition from state k to state l is α_{kl}. These transition rates depend upon the physicochemical conditions surrounding the gap junction. (Note that, in general, $\alpha_{kl} \neq \alpha_{lk}$). Then the permeability of the gap junction is

$$\rho = \sum_{k=1}^{N} \rho_k s_k,$$

where

$$\frac{ds_k}{dt} = \sum_{l=1}^{N} \alpha_{lk} s_l - \left(\sum_{l=1}^{N} \alpha_{kl} \right) s_k, \quad k = 1, \ldots, N. \tag{1}$$

When the transition rates are considered constant, the system (1) has a unique, globally attracting steady state s^* with permeability $\rho(s^*)$. This steady state is a function of the physicochemical state. It is useful to use the constraint $\sum_l s_l = 1$ to eliminate one gating state population, say s_1 and rewrite (1) in matrix form

$$\frac{ds}{dt} = b - As, \tag{2}$$

where $s = (s_2, \ldots, s_N)$ and b are $(N-1) \times 1$ column vectors and A is a $N-1$ positive definite square matrix. In this notation, the globally attracting steady state is $s^* = A^{-1}b$. For a gap junction from a *Xenopus* embryo, and many others, the dominant determinant of gap junction permeability is the trans-junctional voltage. Here the steady state voltage-conductance curve is bell-shaped and the corresponding voltage-current relation is N-shaped (see Figure 1).

MULTICELLULAR TRANSPORT DYNAMICS

Before considering network models for *Xenopus* embryos, let us first consider a general system of cells linked by *time-independent* gap junctions. (The extension to time-dependent gap junctions will be made later. The time-independence means that we need not concern ourselves yet with the gap junction dynamics.) We assume that the cells lie in an infinite bath of constant chemical composition. The cells are able to

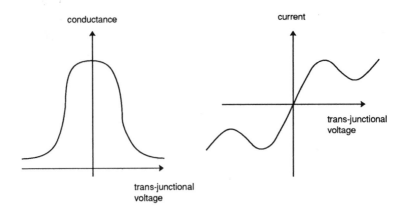

Figure 1: Typical steady state voltage-conductance and voltage-current characteristics for a *Xenopus* gap junction. Note that for large trans-junctional voltages a small residual conductance remains.

communicate with each other via the gap junctions and with the outside world via membrane transport mechanisms.

As a first step, our model is concerned only with transport of species and not with chemical reactions. The key assumption we make is that

Assumption A: all transport fluxes, both for the cells and the gap junctions, depend upon differences, and not absolute values, of electrochemical states.

According to the theory of nonequilibrium *network thermodynamics* [see, for example, Oster *et al.* (1971) and references therein], this assumption ensures that the evolution of such a system can be understood solely in terms of energy storage elements and a thermodynamic dissipation function. For small concentration changes, this assumption reduces to the experimentally established dependence of the junctional conductance on the trans-junctional voltage. The validity of the assumption for the general case needs experimental verification, for which data is not yet available.

Mathematically, these facts can be expressed as follows. We suppose that in and between each of the n cells there are N chemical species transported. For cell k let μ_l^k be the chemical potential thermodynamically conjugate to the molar quantity n_l^k of a chemical species l, $C_l^k = \partial \mu_l^k / \partial n_l > 0$ be the incremental capacitance of species l, and $C^k = \text{diag}\{C_l^k\}$. Next we group all vector electrochemical potentials into one large vector $\mu = (\mu^1, \dots, \mu^n)$. Then the dynamics of the species transport takes the concise form

$$C(\mu)\frac{d\mu}{dt} = -\frac{\partial \Phi}{\partial \mu}, \tag{3}$$

where C is the $n \times n$ diagonal matrix $\text{diag}\{C^k\}$.

Although at first sight these equations may appear daunting, it is easily shown (Brayton & Moser, 1964; Oster *et al.*, 1971) that their dynamics is trivial: the gradient structure of the righthand side of (3) and the positive definiteness of C are enough to ensure that all initial distributions of the chemical species asymptotically approach a steady state. This particular steady state, if there is more than one, depends on the initial state of the system.

MULTICELLULAR TRANSPORT WITH DYNAMIC GAP JUNCTIONS

When the gap junctions are no longer time-independent, we have to augment the system (3) with equations for the gap junction dynamics. The resulting equations are

$$\left. \begin{array}{rcl} C(\mu)\dfrac{d\mu}{dt} & = & -\dfrac{\partial \Phi}{\partial \mu}(\mu, s) \\[2mm] \dfrac{ds}{dt} & = & b(\mu) - A(\mu)s. \end{array} \right\} \tag{4}$$

This is a major change from the gradient form of the system (3). Now the transport of species is coupled to the dynamics of the gap junctions and vice versa. This is the price paid for introducing the biological fact that gap junctions are dynamically coupled to the cell electrochemical state; the system is no longer *reciprocal* in the thermodynamic sense (Oster *et al.*, 1971).

The dynamics of a system of the form (4) is generally difficult to handle unless there are additional properties, such as differences in time scales. For example, if each of the diagonal elements of incremental capacitance $C(\mu)$ is small (in comparison to $\partial \Phi / \partial \mu$), then we may rescale time $t \mapsto \epsilon \tau$, where $\epsilon \ll 1$, to give a perturbation problem

$$\left. \begin{array}{rcl} \bar{C}(\mu)\dfrac{d\mu}{dt} & = & -\dfrac{\partial \Phi}{\partial \mu}(\mu, s) \\[2mm] \dfrac{ds}{dt} & = & \epsilon\left(b(\mu) - A(\mu)s\right), \end{array} \right\} \tag{5}$$

where $\bar{C}(\mu) = C(\mu)/\epsilon$.

When $\epsilon = 0$ the gating states of each gap junction is constant, say s_0, and system (5) admits a globally stable steady state $\mu(s_0)$. As we range over all the possible gating states s_0, we map out a manifold (i.e. a surface) I_0 given by the set of points $\{(s, \mu(s)) \mid s \in \mathbb{R}^N\}$. The dynamics for $\epsilon = 0$ ensures that each initial state ends up on this manifold and that there is no dynamics on this manifold. For $\epsilon > 0$ sufficiently small, it can be shown using the results in Sakamoto (1990) that, for strictly convex Φ and ϵ sufficiently small, there exists a new manifold, I_ϵ, near to I_0, within which the now finite number of steady states lie, and onto which all initial states are exponentially attracted. This result has also been established (Iannelli *et al.*, 1997 [submitted]) using a method which yields explicit estimates for the smallness of ϵ, which enables comparison with experiment. However, the dynamics restricted to this manifold are no longer trivial, since $\dot{s} \neq 0$. For two *Xenopus* cells coupled by a gap junction we have shown that the model fitted with recent experimental data (Mulrine & Warner, in preparation) cannot exhibit oscillations: all initial states tend asymptotically to a steady state. Essentially this is because, for the experimentally determined parameters used in the model, the transport and gap junction dynamics are weakly coupled, and that in the completely uncoupled case the transport and gap junction dynamics are globally stable. For larger networks, we have not yet studied the dynamics restricted to the manifold, and oscillations may well be possible beyond two cells. We can determine, however, that the asymptotic dynamics are slow, with the velocity vector $\dot{s} = O(\epsilon)$. This means that any oscillations will have a long period.

NETWORK MODELS FOR *XENOPUS* EMBRYOS

Now consider a *Xenopus* embryo consisting of a group of cells linked by gap junctions. We represent this as a linear graph $\mathcal{G}$ whose nodes are cells and whose branches are gap junctions. First we need to specify which cells are connected to which cells. For this purpose we define the set N_k to be the set of nodes (cells) which are connected by a branch (gap junction) to node k (cell k). In order to be able to define local and nonlocal effects on the network, it is convenient to have a notion of distance on the graph. The distance between two nodes a and b on $\mathcal{G}$ is simply the least number of branches in a path joining a to b. Next, for simplicity, let us also assume that only one species is transported. (The extension to many species can be easily made by using matrices of permeability coefficients.) Let $\rho^{lk} = \rho^{lk}(s^{lk})$ be the permeability of the gap junction linking cell l to cell k, where s^{lk} is the gating state of the gap junction, and ρ^k the permeability of the membrane of cell k. We assume that ρ^k is a constant, so that the gating of the membrane channels is ignored. While this is biologically unrealistic, it allows us to focus on the nonlinearities introduced by the gap junctions. If μ^k is the electrochemical potential of the species in cell k, then the transport dynamics is given by

$$C^k(\mu)\frac{d\mu^k}{dt} = -\rho^k(\mu^k - \bar{\mu}) - \sum_{l \in N_k} \rho^{lk}(s^{lk})\left(\mu^k - \mu^l\right) + J^k, \tag{6}$$

where J^k is a *forcing* flux representing transport of the species to and from the extra-cellular fluid, and $\bar{\mu}$ the constant and uniform extracellular electrochemical potential of the chosen species. The potential $\bar{\mu}$ is a convenient reference potential, and we consider the μ^k to be deviations from this potential. This is equivalent to setting $\bar{\mu} = 0$ in (6).

In addition to (6), we also have the dynamics of the gap junctions. However, in the sequel, since we will be concerned with spatial patterns of electrochemical potentials, we shall only be concerned with the existence and stability of steady state solutions of (6). We therefore make the assumption that for any *fixed* trans-junctional potential difference $\mu^l - \mu^k$, the equations for the gap junction dynamics have a unique stable fixed point $g^{lk}(\mu^l - \mu^k)$.

SPATIAL PATTERNS OF ELECTROCHEMICAL POTENTIAL

The emergence of spatial inhomogeneities from an initially homogenous state is an important event in the context of developmental biology. Typically, this will involve the switching from one steady state to another, following a perturbation. Thus the first investigation we have carried out is to locate the steady states of the transport dynamics coupled with the gap junction dynamics. Since both the transport and the gap junction dynamics must be simultaneously in steady state, the steady state electrochemical potentials are solutions of

$$-\rho^k \mu^k - \sum_{l \in N_k} f^{lk}(\mu^k - \mu^l) + J^k = 0, \tag{7}$$

where $f^{lk}(x) = x g^{lk}(x)$. Here we make a second assumption

Assumption B: f^{lk} has the general property $x f^{lk}(x) > 0$ for all $x \neq 0$.

This simply asserts that the flow is down a concentration gradient. In general classifying the solutions of (7) is a complex problem. The number and stability of steady states depends upon the parameters in both the cells and the gap junctions. Some progress, however, can be made if assumptions regarding symmetry are made; for example, one can assume that some aspects of the cells and gap junctions are identical and then consider small perturbations from this symmetric state.

12

STEADY STATES AND THEIR PERSISTENCE UNDER PERTURBATIONS

The starting point is to investigate under what conditions (7) has a unique (globally stable) steady state. First, let us rewrite (7) in the form

$$F_k(\mu; \rho; J; \delta) = -\rho_k \mu^k + \sum_{l \in N_k} \delta_{lk} \hat{f}^{lk}(\mu^k - \mu^l) + J_k = 0, \tag{8}$$

where δ_{lk} is the maximum permeability of the junction and $\hat{f}_{lk}$ is the normalised steady state junctional permeability. Let us list two extreme conditions under which (8) has a unique steady state:

C1 $J_k = J$ for all $k = 1, \ldots, n$, for some J; i.e. the flux due to membrane pumps is the same for each cell. In this case the unique solution is $\mu_k = -J$ for $k = 1, \ldots, n$.

C2 $\delta_{lk} = 0$ for all $l, k = 1, \ldots, n$; i.e. the cells are completely uncoupled and the solution is $\mu_k = -J_k/\rho_k$ for $k = 1, \ldots, n$.

A third case of biological interest is more complex:

C3 $\rho_k = 0$ for $k = 1, \ldots, n$; i.e. the cell membranes are perfectly impermeable to the species. Here solutions may or may not be unique. When $J_k = 0$ for $k = 1, \ldots, n$ there is a continuum of steady states.

Note that these facts are true regardless of the network size and topology, and do not depend upon the particular steady state properties of the gap junctions (other than the satisfaction of assumption B).

Does uniqueness of the steady states persist for small perturbations from the above conditions C1,C2? Mathematically, we can check this using the Inverse Function Theorem (IFT) [see, for example, Golubitsky & Schaeffer (1988)]. We find that for C1 above the uniqueness of the steady state persists when the J_k are sufficiently similar (i.e. when $\min |J_l - J_k|$ is sufficiently small). Similarly, if the cells are weakly coupled, i.e. $0 < \delta_{lk} < \epsilon$ and $\epsilon > 0$ is sufficiently small, the uniqueness persists. In the case C3, when $J_k = 0$ for $k = 1, \ldots, n$ an adaptation of the IFT can be used to show that a small perturbation from $\rho = 0$ yields a unique steady state, and the general case is more complex.

Biologically, these results tell us that when the cells are either weakly or strongly coupled, they have a unique (globally stable) steady state, provided also that the membrane pumping fluxes are sufficiently different. Techniques are available (Mackay, 1996) to estimate how strong the coupling can become before new steady states are possible (and similarly for the reverse process of weakening initially strong coupling). However, in general, for large perturbations the analysis breaks down; other methods, such as numerical computation, are necessary to make further progress. We have, however, completely classified the steady states for two cells connected by a single gap junction (Baigent et al., 1997a; Baigent et al., 1997b [submitted]). Our analysis for the two cell model confirms the above analysis and shows that there can be either a unique stable steady state, or two stable and one unstable steady states. Further, the two-cell analysis also demonstrates that hysteretic behaviour is possible under certain parameter variations, provided that the coupling is moderate. In future work we intend to investigate the onset of hysteresis in larger networks and its implications for pattern formation.

STRUCTURE OF THE PERTURBED STEADY STATES

We can also use perturbation analysis to investigate quantitatively the changes in the electrochemical potentials of each cell in the network following a perturbation. Consider solutions of

$$F_k(x; \lambda) = 0, \quad k = 1, \ldots, n, \tag{9}$$

where λ is a set of parameters. Suppose that for $\lambda = 0$, (9) has a unique solution x^*. Then provided that the Jacobian matrix $D_x F$ at $x = x^*$ and $\lambda = 0$ is invertible, then for small λ the solution of (9) is unique and is given by

$$x^*(\lambda) = x^* - D_x F(x^*; 0)^{-1} D_\lambda(x^*; 0)\lambda + o(\|\lambda\|). \tag{10}$$

For example, consider the situation C1 above. Here we have $x^* = 0$ and

$$J = \frac{\partial F_i}{\partial \mu_j} = \begin{cases} -\rho_i - \sum_{l \in N_i} f'_{li}(0) & i = j \\ f'_{ij}(0) & i \neq j. \end{cases} \tag{11}$$

Now $f'_{lk}(0) > 0$ for all l, k and $\rho_k > 0$ for each k, so that J is diagonally dominant and thus invertible. According to Mackay (1996) the elements of the matrix J^{-1} are exponentially decaying (roughly speaking, this means that its elements decay exponentially with the distance on the graph). This means that the effect of some perturbation in pumping $J_k \mapsto J + z$ with z small and $J_l = J$ for $l \neq k$ decays exponentially from cell k. Thus perturbations in the network model only have a local effect, unless parameters are close to values where steady states are created or destroyed.

CONCLUSIONS AND OUTLOOK

In this article we have discussed dynamical aspects of gap junctional communication between nonexcitable cells with linear membrane properties. The dynamics of these multicellular systems are typically different from those of multicellular systems consisting of coupled excitable cells whose membranes are nonlinear and have an autocatalytic component. In the case of two *Xenopus* nonexcitable cells connected by a gap junction, the cells and gap junction tend to a steady state from all initial states. For larger networks, such as those modelling *Xenopus* embryos, we cannot rule out the possibility of oscillations, although the difference in membrane and junction time constants ensures that any oscillations would be slow.

One planned extension of the model is to include the nonlinear properties of the cell membrane and this could introduce additional dynamical features. A more drastic extension of the model will be to consider transport and gap junction properties that depend upon the absolute value of the electrochemical potential and not merely the electrochemical gradient. Such an extension would not yield a potential function representation and therefore our approach would have to be modified.

We have also outlined how perturbation analysis can help to understand the local and global effects that gap junction nonlinearities have on communication. Our analysis suggests that, typically, the effect of a local perturbation tails off exponentially with distance from its source. This may not be the case when membrane nonlinearities are included. We intend to investigate this further both analytically and computationally.

ACKNOWLEDGEMENTS

This work was funded by a grant from EPSRC to J. Stark and A. Warner (grant no GR/J37904). We thank J.E. Hall for useful discussion in the initial stages of the project.

REFERENCES

ATTWELL, D., WILSON, M. & WU, S. M. (1985): The Effect of Light on the Spread of Signals through the rod network of the Salamander Retina, *Brain Research*, **343**, No. 1, 79-88.

BAIGENT, S. A., STARK, J. & WARNER, A. E. (1997a): Modelling The Effect Of Gap Junction Nonlinearities In Systems of Coupled Cells'. *J. theor. Biol.*, **186**, 223-239.

BAIGENT, S. A., STARK, J. & WARNER, A. E. (1997b): The Convergent Dynamics of Two Cells Coupled by a Nonlinear Gap Junction. (Submitted to *IMA J. Math. Appl. Med. Biol.*).

BRAYTON, R. K. & MOSER, J. K. (1964), *A theory of Nonlinear Networks, I*, Q. Appl. Math., Vol XXII, No. 1, 1-33 .

DE MELLO, W. C. (1987): Modulation of Junctional Permeability. In: Cell-to-cell communication (ed. Walmor C. de Mello), Plenum Press, New York.

DONGMING, C., WINSLOW, R. L. & NOBLE, D. (1994): Effects of Gap Junction Conductance on Dynamics of Sinoatrial Node Cells: Two-Cell and Large-Scale Network Models. *IEEE Trans. Biomed. Eng.*, **41**, No. 3, 217-231.

GOLUBITSKY, M. & SCHAEFFER, D. G. *Singularities and Groups in Bifurcation Theory*, Vol. 1, Springer-Verlag, 1988.

GUTHRIE, S. C., TURIN, L. & WARNER, A. E. (1988): Patterns of junctional communication during development of the early amphibian embryo. *Development*, **103**, 769-783.

IANNELLI, P., BAIGENT, S. & STARK, J. (1997): Inertial Manifolds for Dynamics of Cells Coupled by Gap Junctions. (Submitted to *Dynamics and Stability of Systems*).

MacKAY, R. S. (1996). DYNAMICS OF NETWORKS: Features which Persist from the Uncoupled Limit. (Lectures Notes, From Finite to Infinite Dimensional Systems, Newton Institute, Cambridge).

NAGAJSKI, D. K., GUTHRIE, S. C., FORD, C. C. & WARNER, A. E. (1989): The Correlation Between Patterns of Dye Transfer Through Gap Junctions and Future Developmental fate in *Xenopus*: the Consequences of U. V. Irradiation and Lithium Treatment, *Development*, **105**, 747-752.

OSTER, G., PERELSON, A. & KATCHALSKY, A. (1971): Network Thermodynamics, *Nature (Lond.)* **234**, 393-399.

SAKAMOTO, K. (1990): Invariant manifolds in singular perturbation problems for ordinary differential equations. *Proc. Royal Soc. Edin.* **A**, Vol. 116, 45-78.

WARNER, A. E., GUTHRIE, S. C. & GILULA, N. B. (1984): Antibodies to Gap-Junctional Protein Selectively Disrupt Junctional Communication in the Early Amphibian Embryo, *Nature*, **311**, 127-131.

Real Time Processing of Nerve Signals for Controlling a Limb Prostheses

Martin Bogdan and Wolfgang Rosenstiel
Lehrstuhl für Technische Informatik, Universität Tübingen,
Sand 13, D-72076 Tübingen, Germany
E-mail: <bogdan|rosenstiel>@informatik.uni-tuebingen.de

INTRODUCTION

The aim of the INTER[1]-project (*I*ntelligent *N*eural In*TER*face) is to investigate fundamental issues related to the design and fabrication of a new generation of microsystems applicable as neural prostheses. A global overview for a PNS-remoted limb prostheses is given in [Dar93] and is shown in figure 1.

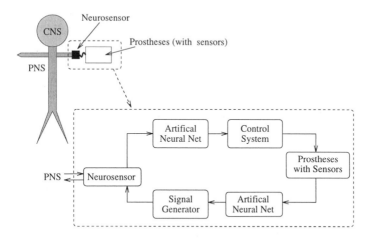

Figure 1: Scheme Configuration of a bio-neural controlled prostheses

Nerve signals will be recorded and amplified by a neurosensor. The neurosensor is a regeneration-type sensor. The principle of the neurosensor is explained below. Then, an

---

[1] The INTER-project is granted by the European Community under ESPRIT BR project #8897.

artificial neural net (ANN) is applied which classifies the resulting signals in order to assign certain limb movements to the signal classes. A control unit uses the resulting information to regulate the movement of the prostheses.

Ideally, the prostheses is equipped with sensors. Signals from the sensors will be processed by an ANN and transmitted via a signal generator and the neurosensor to the peripheral nervous system (PNS). This means, the prostheses will be completely controlled from the PNS like a natural limb.

After presenting the neurosensor used within the INTER-project and a short presentation of the proposed signal processing unit, we are concentrating in this paper on the signal processing of the recorded nerve signals using Kohonen's self-organizing map (SOM). In this paper, we show the possibility to process nerve signals and to control a limb prostheses in real time using recordings from the stomatogastric nervous system (STNS) of the crab *Cancer parugus*.

THE NEUROSENSOR OF THE INTER-PROJECT

The principle of the implementation [Coc93] and the neurosensor which is used in the INTER-project is shown in figure 2.

Peripheral nerves of vertebrates will regenerate if severed. For this reason, the peripheral nerve can be surgically severed in order to insert the proximal and the distal stump into the guidance channel. The guidance channel encloses the chip. The main functions of the

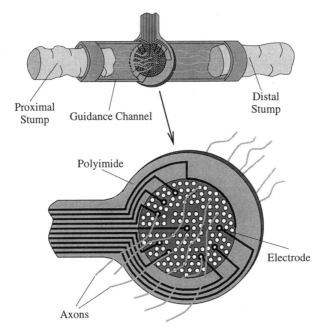

Figure 2: Implementation scheme for the regeneration-type neurosensor

guidance channel are to provide a stable physical connection between the chip and the stumps of the nerve, and to promote nerve generation and guide axons growth from the proximal nerve stump towards the distal nerve stump.

The sensor is fabricated of polyimide perforated by multiple 'via holes'. The axons regenerate through the via holes from the proximal stump towards the distal stump of the nerve. Nerve signals can be recorded by electrodes, which are enclosing some of the via holes. A circuitry amplifies and preprocesses the nerve signals. The amplified signals are transferred to the units which are controlling the prostheses as shown in figure 1. For more details about the chip, please refer to [Sti96].

PROPOSED SIGNAL PROCESSING USING ANN

A global overview of our proposal for the signal processing is shown in figure 3. For separating the different nerve signals from the recorded mixture of different axons grown through an electrode, we are proposing to use INCA, presented by Jutten and Hérault [Jut91]. The

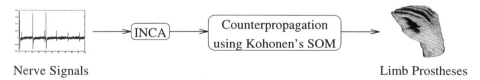

Nerve Signals   Limb Prostheses

Figure 3: Proposal for signal processing

algorithm allows to separate different signals from several independent sources (in our case a source is an axon) without any knowledge on the sources. It is a so called blind separation of sources.

Since the recordings will be obtained by In Vivo tests with animals, there will be no defined target pattern corresponding clearly to the recorded nerve signal [Bog96]. Thus the classification of the nerve signals must be done by unsupervised learning algorithms. We are going to apply Kohonen's self-organizing map (SOM) [Koh82, Koh89]. The classification of the nerve signals in real time using the SOM will be described below.

A more detailed description of the proposal is given in [Bog94, Bog96].

DATA SET

The data set, which is used for the classification, has been recorded by the Institut für Biomedizinische Technik (IBMT) using a hawk electrode. IBMT has chosen the stomatogastric nervous system (STNS) of the crab *Cancer pagurus* as described in [Hei93].

The STNS of the crabs is probably the only nervous system which contains a reasonably small number of neurons and in which almost all neurons have been identified. It is composed of the commisural ganglia, the oephageal ganglion and the stomatogastric ganglion. The stomatogastric ganglion contains about 30 nerve cell bodies, 24 of which are motorneurons and 6 of which are interneurons. Among the 24 identified motorneurons 2 innervate the cardiac sac muscles (CD2 and AM), 9 innervate the musculature of the gastric mill (LG, MG, 2 LPGs, 4 GMs and DG), 2 innervate the cardio-pyloric valve (VD and IC) and 11 innervate the pyloric muscles (2 PDs, LP and 8 PYs). The axons of all of the motorneurons except AM, DG and CD2 leave the ganglion via the dorsal ventricular nerve (dvn).

A typical recorded sequence of the signals of these cells are shown in figure 4. The action potentials corresponding to the PD, LP and PY motorneurons can be easily identified. The durations of the recordings are 24 respectively 40 seconds. The data set were recorded using a sample frequency of 5 kHz. This system has been chosen since it is very well known. Thus, we are able to verify the results obtained by the classification of the SOM for their correctness.

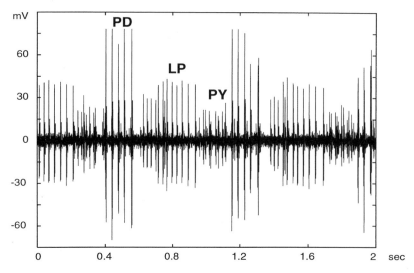

Figure 4: One detail out of the recordings from the gastric nerve of a crab. PD, LP and PY cells can be easily identified.

DATA PREPROCESSING

In the primary step of the data preprocessing we are using a low pass filter in order to reject the drift effects which are introduced by the recording unit. In fact, this filter eliminates the deviation of the signal in order to obtain a signal with a defined reference voltage of 0 V.

In the upper illustration of figure 5 the typical waveform of a nerve signal after the rejection of the drift is shown. The shape of the recorded nerve signal can be divided into three parts: one positive peak followed by one negative peak and again one positive peak. This (+|-|+)-sequence appears everytime a spike occurs, except if there is a superposition of two or several spikes. The third peak is not physiological and is due to the filtering during the recording. In addition to the occurance of a (+|-|+)-sequence a time criterion must be fulfilled: the (+|-|+)-sequence must be finished within 6 msec. If both criteria are fulfilled, we assume that this (+|-|+)-sequence was due to a true nerve signal.

Another problem is the noise within the signal. As the upper illustration of figure 5 shows, there are also (+|-|+)-sequences within the part of the signal which must be considered as noise. For this, a spike cannot be judged as a simple (+|-|+)-sequence within the recording.

In order to detect the nerve signals, we have combined the criteria of the (+|-|+)-sequence with a threshold. So we judge a sequence of the signal as a nerve signal, if the first peak exceeds the threshold, and a (+|-|+)-sequence follows within 6 msec. A critical point of this method is the threshold. Because we want to detect as many nerve signals as possible, we have chosen a threshold of 10 mV, which is in fact very close to the noise. Using this method to judge the recorded data, all nerve signals within the noise are rejected. To avoid this problem, we are currently working on another method, which will judge the signal by a fuzzy rule [Cec96].

The next generation of amplifiers and filters are expected to avoid the third peak within the (+|-|+)-sequence. So, we cut the third peak from the detected signals. The result of the nerve signal detection using this method is shown in the lower illustration of figure 5. Compared with the upper illustration, only the sequences of the recorded data judged as nerve signals are available.

The waveform of these signals are characterized using the area, the width of base and the maximum respectively the minimum value of the two peaks. Using this characterization we

20

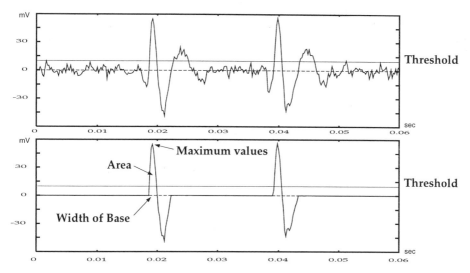

Figure 5: Upper illustration: Typical waveform of a recorded nerve signal.
Lower illustration: The signal after filtering and spike detection. To characterize the waveform, the area, the width of the base and the maximum respectively minimum values are detected.

have computed for each spike a vector containing six components (area, width of base and maximum (minimum) value of the two peaks). The obtained 2667 vectors has been used for the training of the SOM.

CLASSIFICATION WITH KOHONEN'S SOM

For the classification of the data set a two dimensional SOM with 10 neurons in both dimension have been applied. The training data set consists of 2667 vectors with six components. The computation of the training vectors has been described above.

The trained SOM is presented in figure 6. The map is very well disposed and has no topological defects. After obtaining a well ordered map, we have identified the clusters within

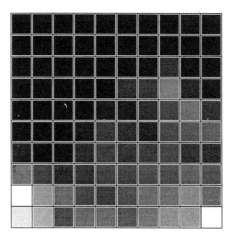

Figure 6: Representation of the trained SOM (first component, area of the first peak). Obviously the map is very well disposed. Black corresponds to highest value, white corresponds to lowest value within the range.

this map. Each of the obtained clusters will represent one specific signal from an axon respectively from a group of axons (e.g. PD or PY cells). This means, if we are able to identify clusters within the trained SOM, we can assign an action to each cluster conditioned by a signal of an axon. The obtained clusters are presented in figure 7.

Each square of the trained SOM shown in figure 7 represents a neuron. The greyscaled rectangles are corresponding to the euclidean distances between the neurons. A black rectangle stands for the longest distance and a white one for the shortest distance. The curves within the squares are the codebook vectors.

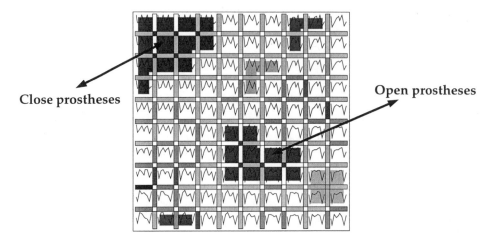

Figure 7: The trained SOM with obtained clusters. Two of the clusters has been chosen to assign an action of the prostheses.

The SOM also contains the clusters we have obtained after training. As mentioned above, each cluster represents the signals from an axon respectively from a group of axons (e.g. PD or PY cells). Every time a nerve signals occurs, it will be classified to its corresponding cluster. Because of this we are able to recognize the signals from certain axons in order to control the movement of the limb prostheses. In our case we control an artificial hand which has two degrees of freedom: open/close and speed/power. The first degree of freedom has been directly coded to the SOM as indicated in figure 7. The second degree of freedom is coded within the frequency of occurance of the signal.

CONTROL UNIT OF THE PROSTHESES

After classifying the nerve signals to their corresponding clusters the occurance of a nerve signal of a certain cluster must be assigned to its corresponding action. Since the information of the nerve signals are pulse-frequency encoded we have to change into the time domain. The problem within this case is that the occurance of one single nerve signal does not carry useful information. This might be a spontaneous or hazardous signal. For these reasons we have decided to build an integration based signal interpreter to remote the control unit of the prostheses. The principle of the system is shown in figure 8.

In fact we use only two of the six clusters to remote the prostheses (refer to figure 7). One corresponds to the information 'close hand', the other one to the information 'open hand'. Using these two clusters we encode the first degree of freedom of the artificial hand, the direction of the movement.

The second degree of freedom is directly connected to the speed respectively power of the hand. The speed respectively the power of movements (if the hand runs against an obstacle

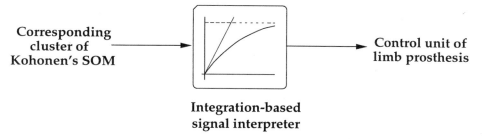

Corresponding
cluster of
Kohonen's SOM

Integration-based
signal interpreter

Control unit of
limb prosthesis

Figure 8: The principle of the control unit. The output for the control unit is integration based, but not linear.

it changes automaticly into the power mode) is encoded through the frequency of occurance of nerve signals in one direction. The interpretation of the frequency has been done by an integration based signal interpreter.

Assuming the occurance of a spike train has been classified by the SOM. The spike train will be integrated and will assign an action to the prostheses if the integrated signal exceeds a minimum value. This avoids that a hazardous or spontaneous single nerve signals leads to an action of the prostheses. The higher the frequency of the spike train the higher the integration value. In order to stop an action as fast as possible the integration value will be decreased much faster than increased. For this reason the function of the control system is nonlinear.

Since the integration value is directly coupled with the speed of the movement respectively the power, the second degree of freedom can be controlled through the frequency of the occurance of the nerve signals. In fact, the processing system consisting of the SOM and the integration based signal interpreter realizes a pulse-frequency decoder. This version of the integration based signal interpreter is still a very simple model. Thus, we have to validate this approach using a more complex biologically inspired function in the future.

DISCUSSION

One important question must be asked: How realistic is it to use data from the stomatogastric nerve in order to control a limb prostheses?

Obviously, there will be no information within the nerve signals which are dedicated to control extremities. The goal of the work is to prove the applicability of artificial neural nets (ANN) to control limb prostheses in real time in principle.

Following requirements must be achieved in real time:

- Detection of nerve signals

- Classification of detected nerve signals

- Interpretation of the classified nerve signals

The advantage of the used data is that the signals of the stomato-gastric nerve are well known, what provides the facility to verify if the classification done by the ANN is correct or not. For other data sets such as recordings from the sciatic nerve of a rat, the verification cannot be done as precise as for the recordings from the STNS of the crab *Cancer pagurus*. Thus, in a first attempt we have used the data from the STNS of the crab *Cancer pagurus*.

Using the proposed system we are able to identify axons respectively functional groups of axons out of the data set. The experimental system is shown in figure 9. After extracting the characteristics for each detected nerve signal providing the recognition, Kohonen's self-organizing map (SOM) is able to classify the nerve signals. Finally, the classified signals will

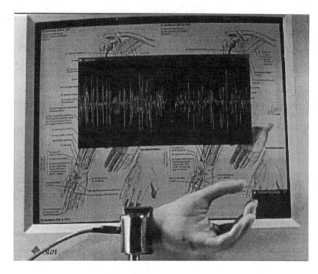

Figure 9: The experimental system of the processing unit. The prostheses in front of the monitor is controlled by the displayed nervesignals.

be interpreted by an integration-based signal interpreter in order to perform the encoding of the frequency modulated signal. Using this system of signal processing, the requirement of a real time processing unit is fulfilled.

CONCLUSION AND FUTURE WORK

In this paper, we have presented a signal processing unit basing on artificial neural networks which is able to interprete real nerve signals and to control a limb prostheses in real time. The nerve signal processing unit will be applied within the INTER-project. It consists of Kohonen's self-organizing map (SOM) and an integration based signal interpreter. The SOM classifies the nerve signal corresponding to its origin. The integration based signal interpreter controls the movement of the limb prostheses based on the frequency of occurance of the nerve signals. The two parts of the processing unit realize a pulse-frequency decoding of nerve signals. In a further step, the function of the integration based signal interpreter will be changed to a more complex and biologically inspired function.

Currently, the processing system is implemented on a suitable microprocessor system in order to speed up the processing time to realize multichannel processing in real time. This system will be based on the PowerPC processor.

To conclude, we have presented a system which is able to remote a limb prostheses due to incoming nerve signals. The whole system works in real time. We have shown that it is possible to interprete nerve signals from recordings of a sum of axons in real time.

ACKNOWLEDGEMENT

The authors like to thank J.-Uwe Meyer, Cornelia Blau and their team from the Fraunhofer Institut für Biomedizinische Technik (IBMT) at St. Ingbert, for giving the data at our disposal.

The INTER project is supported by the European Community under ESPRIT BR project #8897. The INTER consortium consist of 6 partners:

- *Scuola Superiore S. Anna*, Pisa, Italy

- *Hahn-Schickard-Gesellschaft, Institut für Mikro- und Informationstechnik*, Villingen-Schwenningen, Germany

- *Centro Nacional de Microelectronica*, Barcelona, Spain

- *Centre Hospitalier Universitaire Vaudois*, Lausanne, Switzerland

- *Fraunhofer Institut, Biomedizinische Technik*, St. Ingbert, Germany

- *Universität Tübingen, Institut für Physikalische und Theoretische Chemie* and *Lehrstuhl für Technische Informatik*, Tübingen, Germany

Coordinator of the project is *Scuola Superiore S.Anna* at Pisa. The *Lehrstuhl für Technische Informatik* is junior partner of the *Institut für Physikalische und Theoretische Chemie*.

References

[Bog94]   M. Bogdan and W. Rosenstiel. Artificial Neural Nets for Peripheral Nervous System - remoted Limb Prostheses. In *Neural Networks & their Applications, Nanterre*, pages 193–202, 1994.

[Bog96]   M. Bogdan, A. Babanine, J. Kaniecki, and W. Rosenstiel. Nerve Signal Processing using Artificial Neural Nets. In *Computation in Cellular and Molecular Biological Systems*, pages 121–133. World Scientific, ISBN 981-02-2878-3, 1996.

[Cec96]   A. Cechin, U. Epperlein, W. Rosenstiel, and B. Koppenhoefer. The Extraction of Sugeno Fuzzy Rules from Neural Networks. In *ESANN '96, Bruges*, 1996.

[Coc93]   M. Cocco, P. Dario, M. Toro, P. Pastacaldi, and R. Sacchetti. An Implantable Neural Connector Incorporating Microfabricated Components. In *Micro Mechanics Europe 1993, Neuchâtel*, 1993.

[Dar93]   P. Dario and M. Cocco. Technologies and Applications of Microfabricated Implantable Neural Prostheses. In *IARP Workshop on Micromachine & Systems 1993, Tokyo*, 1993.

[Hei93]   H.G. Heinzel, Weimann J.M., and E. Marder. The Behavioral Repertoire of the Gastric Mill in the Crab, Cancer pagurus: An in situ Endoscopic and Electrophysiological Examination. *The Journal of Neuroscience*, pages 1793–1803, April 1993.

[Jut91]   C. Jutten and J. Hérault. Blind separation of sources, Part I: An adaptive algorithm based on neuromimetic architecture. *Signal Processing, Elsevier*, 24:1–10, 1991.

[Koh82]   T. Kohonen. Self-organized formation of topologically correct feature maps. In *Biological Cybernetics 43*, pages 59–69, 1982.

[Koh89]   T. Kohonen. Self-Organization and Associative Memory. In *Springer Series in Information Sciences*. Springer-Verlag, 1989.

[Sti96]   T. Stieglitz, H. Beutel, and J.-U. Meyer. A flexible, light-weighted, multichannel sieve electrode with integrated cables for inter-facing regenerating peripheral nerves. In *Proceedings of 10th Eurosensors Conference*, 1996.

STIMULUS-SECRETION COUPLING IN PANCREATIC B-CELLS EXPLAINED BY CHAY'S STORE-OPERATED MODEL

TERESA REE CHAY

Department of Biological Sciences, University of Pittsburgh
Pittsburgh, Pennsylvania 15260USA
E-mail: TRC1@VMS.CIS.PITT.EDU

INTRODUCTION

The cells in the pancreas that secrete insulin are called b-cells. They are situated in tightly bound cellular aggregates known as the islets of Langerhans. In an islet, the b-cells are surrounded by at least three other cell types (Notkins, 1979). These are the alpha (a-) cells which secret the hormone glucagon, the delta (d-) cells which secrete the hormone somatostatin, and the PP cells which secrete pancreatic polypeptide hormone. Both glucagon and somatostatin affect insulin release such that glucagon enhances the secretion of insulin whereas somatostatin inhibits the secretion of both insulin and glucagon. In addition, insulin secretion is also affected by neurotransmitters released from both sympathetic and parasympathetic fibers, which enter pancreatic islets. The main role of insulin is to facilitate the storage of glucose in the liver and muscle. Without insulin the blood glucose level rises dangerously high. Glucagon, on the other hand, mobilizes glucose from the liver to the blood when needed. This glucose may then be used as an energy source by the entire body. The phenomena associated with electrical activity of b-cells (which will be described below) are very interesting to nonlinear dynamists because they are due to nonlinearity embedded in the ion channels in the plasma membrane and the intracellular events that take place in the cytosol.

One of the most interesting properties of the b-cell is that it exhibits a characteristic electrical activity known as bursting when glucose is added in the perifusion medium. Bursting was discovered in mouse pancreatic b-cells (Dean and Mathews 1970). As shown in Fig. 1A, the voltage across the plasma membrane of b-cells remains at the resting level of about -70 mV as long as glucose is maintained at the subthreshold concentration of 2.8 mM. When the glucose concentration is then raised to 11.1 mM, the b-cells exhibit a trihasic response. First, the intracelluar Ca^{2+} concentration, $[Ca^{2+}]_i$, falls after the addition of

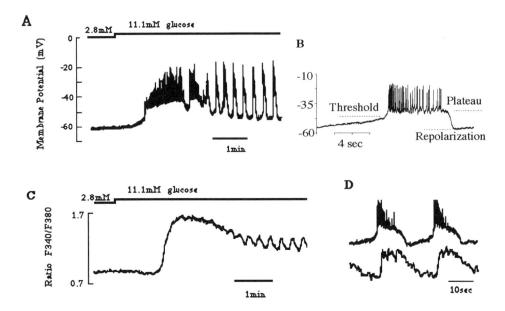

Figure 1. Left: Simultaneous measurements of membrane potential (A) and the intracellular Ca^{2+} concentration (C) induced by glucose when the glucose concentration is raised from 2.8 mM to 11.1 mM. Right: One burst in the limit cycle in an expanded scale, showing the locations of the repolarization, threshold, and plateau potentials (B); Two bursts in the limited cycle regime in an expanded scale (D), which reveal the relation between electrical bursts and intracellular Ca^{2+} oscillation. Frames A and C are retouched from Ref. 2, and Frames D is retouched from Ref. 3.

glucose (see Fig. 1C). In the transient state, the potential climbs to the plateau potential of about -37 mV and remains there for several minutes. Superimposed on this plateau potential are rapid action potentials known as spikes. The spikes oscillate between -20 mV and -37 mV. The plateau potential ends with repolarization to about -45 mV. Shortly afterwards, the membrane depolarizes again to the plateau potential. In the next several minutes after the rich transient state, short repetitive bursts occur, which gradually stabilize to limit-cycle bursts (see Fig. 1B&D). These limit cycle bursts continue with the same duration and frequency as long as glucose remains at the elevated level. Each burst in the limit cycle has four distinct phases: i) slow depolarization, ii) sudden rise to the plateau potential, iii) fast spikes that appear on the top of the plateau potential, and iv) "gentle" hyperpolarization.

Fluorescent measurements showed that [Ca^{2+}]$_i$ oscillates concomitantly with electrical bursting (see Fig. 1C&D). In the limit cycle regime, [Ca^{2+}]$_i$ rises rapidly during the upstroke of the electrical burst and decays slowly during the repolarization until another electrical burst is initiated. Note that the shape of the [Ca^{2+}]$_i$ oscillation is quite different from electrical bursting. Glucose can lengthen the plateau fraction (i.e., the ratio between the active phase and the silent phase) of electrical bursting (Meissner & Schmelz, 1974) without affecting the amplitude of [Ca^{2+}]$_i$ oscillation.

When mouse b-cells are dissociated from the islet of Langerhans, isolated b-cells burst very differently from b-cells embedded in islets. First, the frequency of bursting in isolated cells (Smith, Aschcroft & Rorsman, 1990) is at least 10 times lower than that in intact b-cells (Hattori et al., 1994). Likewise, the frequency of [Ca^{2+}]$_i$ oscillations in isolated b-cells is 10 times lower (Smith, Aschcroft & Rorsman, 1990) (Grapengiesser, Gylfe & Hellman,

1988) than that in intact b-cells [Hattori, Kai & Kitasato, 1994) (Worley, McIntyre, Spencer & Dukes, 1994). Second, both $[Ca^{2+}]_i$ oscillation and electrical bursting in intact cells occur in much more regular intervals than in isolated b-cells. The amplitude of $[Ca^{2+}]_i$ oscillation in isolated b-cells (Grapengiesser, Gylfe & Hellman, 1988), however, changes little from that in intact b-cells (Hattori, Kai & Kitasato, 1994) (Worley, McIntyre, Spencer & Dukes, 1994). Third, each isolated b-cell shows a heterogeneous response to glucose (Herchuelz et al, 1991), while intact b-cells show homogenous response (Meissner & Schmelz, 1974). As in intact b-cells, bursting of isolated cells transformed to continuous spiking when the glucose concentration was raised from 8 mM to 20 mM. Why do isolated b-cells burst so slowly while intact b-cells burst very fast? What information do b-cells try to convey by changing the mode of bursting?

The bursting frequency is also influenced by hormones secreted from the intra-islet cells and by neurotransmitters released from both sympathetic and parasympathetic fibers. Glucagon (a receptor-mediated adenylate cyclase activator) and forskolin (an activator of the catalytic subunit of adenylate cyclase) can increase both the plateau fraction and frequency of electrical bursting (Ikeuchi & Cook, 1984). Epinephrine and somatostatin, on the other hand, enhance hyperpolarization and decreases the frequency of burst activity (Cook & Perara, 1982).

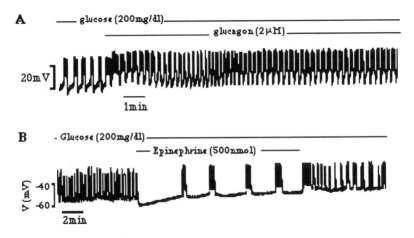

Figure 2. A. Effect of glucagon (2mM) on glucose-induced electrical activity. **B.** Continuous recording of membrane potential of b-cell exposed to a steady flow of epinephrine (500 nmol) for 15 min. Frames A and B are retouched from Refs. 11 and 13, respectively.

This intriguing b-cell bursting behavior is shown in Fig, 2. Note that in the presence of glucagon a period of burst lasts for only several seconds (the top trace), while in the presence of epinephrine it lasts for several minutes (the bottom trace). Insulin production is increased by several-fold in the presence of glucagon, and the opposite is observed in the presence of epinephrine.

My interest in b-cells grew after reading a review volume of the Journal of Experimental Biology (Berridge, Rap & Treherne (Eds.), 1979). Over the past decade our group have worked on mathematical modelling of electrical bursting and $[Ca^{2+}]_i$ oscillations (for a review see Refs. 12-14). The present paper explains how the intracellular calcium stores communicate with the ion channels in the plasma membrane to generate these interesting bursting patterns. My explanation is based on Chay's store-operated model (Chay, 1995, Chay, 1997], which assumes that the calcium concentration in the lumen of the calcium store, $[Ca^{2+}]_{lum}$, drives electrical bursting and the $[Ca^{2+}]_i$ oscillation. The dynamic change of $[Ca^{2+}]_{lum}$, in turn, is controlled by the neurotransmitters and hormones that influence the PI- and AC-signalling pathways.

THE MODEL

As shown in Figure 3, three signalling-pathways exist in b-cells, which may influence the frequency of bursting, $[Ca^{2+}]_i$, and insulin secretion (Prenki & Matschinsky, 1987): i) the phosphatidylinositol (PI-) signalling pathway (left) that can activate protein kinase C (PKC), ii) the glucose-sensing pathway (lower right) that can activate calcium-calmodulin kinase (Ca-CAM K), and iii) the adenylate-cyclase (AC-) transduction pathway (right) that can activate cyclic AMP-dependent kinase (PKA). These kinases can release insulin from its granules via phosphorylation.

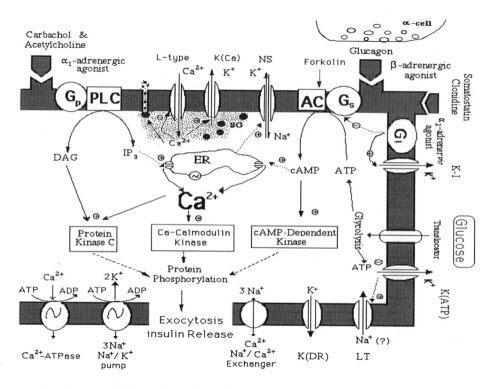

Figure 3. Three signalling pathways involved in secretion of insulin from pancreatic b-cells: The phosphatidylinositol-signalling pathway, glucose-sensing pathway, and adenylate-cyclase transduction pathway. Here, the symbols + and - indicate activation and inhibition, respectively.

Intact b-cells are equipped with hormones (e.g., cholecystokinin) and neurotransmitters (e.g., acetylcholine) secreted from axons passing the b-cell. The agonist (e.g., neurotransmitters) activates the GTP-bound G-protein (G_p) when it is bound to its receptor (see the left side). This agonist produce *myo*-inositol 1,4,5-triphosphate (IP_3) and diacylglycerol (DAG) by activating phospholipase C (PLC) (left). The IP_3 thus produced by PLC releases Ca^{2+} from the IP_3-sensitive Ca^{2+} releasing channel (CRC) in the endoplasmic reticulum (ER). The second product, DAG, along with elevated Ca^{2+}, enhances the activity of PKC, which in turn can release insulin from b-cells. The ER contains a Ca²⁺-ATPase pump which pumps intracellular Ca²⁺ into the store.

Glucagon secreted from neighboring a-cells, on the other hand, enhances the AC-transduction pathway, which in turn raises the concentration of cyclic adenosine monophosphate (cAMP) by activating G_s-proteins. (See the right side.) How a-cells influence b-cell is further evidenced by secretory response which is markedly amplified when single b-cells are incubated in the presence of a-cells or glucagon (Gorus, Malaisse & Pipeleers, 1984). Somatostatin secreted from d-cells, on the other hand, inhibits activity of adenylate cyclase via the G_i-proteins. The cAMP raised by glucagon can enhance a release of luminal Ca²⁺ from the ER via another type of CRC.

In the glucose-sensing pathway (lower right), the ATP-sensitive K⁺ channel (K-ATP) has been regarded as an initiator of depolarization [Cook & Hales, 1984, (Himmel & Chay, 1987), which is required to activate a low-threshold transient channel (LT). This channel is assumed (Chay, 1997) to activate and inactivate in the manner similar to the Hodgkin-Huxley Na⁺ channel (Hodgkin & Huxley, 1952). Activation of I_{LT} leads to activation of a high-threshold voltage-dependent Ca²⁺ channel (L-type) which opens when the membrane depolarizes, permitting extracellular Ca²⁺ ions to come into the cell, and inactivates when [Ca²⁺]ᵢ is undesirably high. The Ca²⁺ entered from the L-type Ca²⁺ channel can influence a voltage-independent Ca²⁺-sensitive K⁺ (K-Ca) channel, which activates when [Ca²⁺]ᵢ becomes high (see top middle). Underneath the hot spot where L-type Ca²⁺ channel clusters lie, secretory granules (SGs) sequester intracellular Ca²⁺ via its Ca²⁺-ATPase pump and then release insulin to the external medium during the exocytosis.

In addition, the plasma membrane contains i) a voltage-independent cationic non-selective channel (NS) which is activated when [Ca²⁺]lum becomes low and ii) a voltage-dependent delayed-rectifying K⁺ (K-DR) channel which activates and inactivates similar to the H-H K⁺ channel. The K⁺ channel (K-I) regulated by the G_i-protein and sensitive to somatosatin (Rorsman et al., 1991) is not considered in this paper.

In this model, a rise of [Ca²⁺]ᵢ is due to two sources — influx of Ca²⁺ ions through the Ca²⁺ channels and a release of luminal Ca²⁺ from the ER (the first and third terms in Eq. 1). The fall of [Ca²⁺]ᵢ is due to two effluxes — sequestration of intracellular Ca²⁺ into the SGs (second term) and the ER (fourth). Accordingly, the [Ca²⁺]ᵢ dynamic is expressed as

$$\frac{d[Ca^{2+}]_i}{dt} = -\phi I_{Ca} - k_{Ca}[Ca^{2+}]_i + k_{rel}\left([Ca^{2+}]_{lum} - [Ca^{2+}]_i\right) - k_{pump}[Ca^{2+}]_i$$
(1)

where f measures the surface-volume ratio, and k_{Ca} is the sequestration rate of intracellular calcium by SGs. The last two terms in Eq. 1 are due to the events taking place in the ER, i.e., the dynamic change of the calcium concentration in the ER, [Ca²⁺]lum,

$$\frac{d[Ca^{2+}]_{lum}}{dt} = -k_{rel}\left([Ca^{2+}]_{lum} - [Ca^{2+}]_i\right) + k_{pump}[Ca^{2+}]_i$$
(2)

where k_{rel} measures the activity of the calcium releasing channel, and k_{pump} is pump activity of Ca²⁺-ATPase in the ER.

The membrane potential necessary to compute I_{Ca} in Eq. 1 can be found from the charge neutrality condition (Hodgkin & Huxley, 1952):

$$-C_m \frac{dV}{dt} = \sum I_{ionic}$$
(3)

where C_m is the membrane capacitance, and I_{ionic} is an ionic current component. The expression of each ionic component and the basic parametric values associated with this component are given in the Appendix.

RESULTS

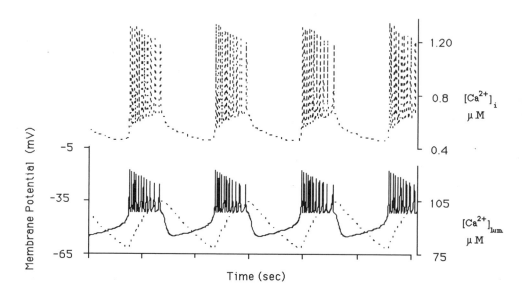

Figure 4. Pancreatic b-cells bursting. Here, the lower traces show the time course of membrane potential (solid) and $[Ca^{2+}]_{lum}$ (dash), and the upper trace shows that of $[Ca^{2+}]_i$.

Figure 4 is the result obtained in the limit cycle by solving the differential equations in Sec. 2. Here, the upper frame shows the time course of $[Ca^{2+}]_i$. (dash), membrane potential (solid), and $[Ca^{2+}]_{lum}$ (dots). Note that $[Ca^{2+}]_{lum}$ oscillates slowly between 78.2 mM and 105.0 mM. In conjunction with this slow oscillation, the membrane potential (V) bursts between -55 mV (the repolarization potential) and -41 mV (the plateau potential). On the top of the plateau, fast electrical spikes appear. The plateau terminates when $[Ca^{2+}]_{lum}$ reaches the maximum amplitude, followed by the silent period. After the termination of the plateau, $[Ca^{2+}]_{lum}$ decreases slowly until V reaches the threshold potential of about -47 mV. Then, a rapid upstroke of V follows. Note that this slow luminal Ca^{2+} oscillation is what drives electrical bursting and $[Ca^{2+}]_i$ oscillation.

Typical bursts start with rapid membrane depolarization, which is accompanied by a rapid rise in $[Ca^{2+}]_i$. A rise of $[Ca^{2+}]_i$ is due to I_{Ca}, which was activated by the lower-threshold I_{LT}. The upstroke of the electrical spike during the plateau phase is due to a combined effect of I_{LT} and I_{Ca} and the down-stroke is due to the combined effect of I_{K-DR} and the inactivating component of I_{LT}. The termination of the plateau results when I_{K-Ca} exceeds $I_{Ca} + I_{NS}$.

That the decrease in $[Ca^{2+}]_i$ during the silent phase is gradual is due to luminal Ca^{2+}, which is released from the CRC during this period. This release prevents $[Ca^{2+}]_i$ from decreasing rapidly even after the Ca^{2+} channel closes. The slow decrease in $[Ca^{2+}]_i$, in turn, releases the bound Ca^{2+} from the K-Ca channel. The release brings about a decrease in I_{K-Ca}. Likewise, a decrease of $[Ca^{2+}]_{lum}$ activates I_{NS}. When I_{NS} exceeds I_{K-Ca}, the burst is initiated.

In the limit cycle regime, glucose tends to increase the plateau fraction. In the presence of a supra-threshold glucose concentration, the burst disappears completely and only repetitive spiking remains (Dean & Mathews, 1970).

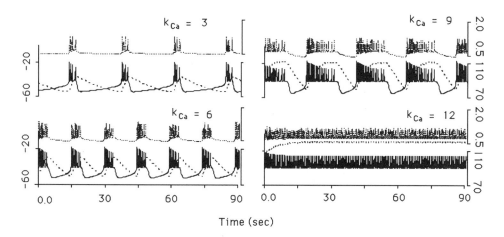

Figure 5. The effect of sequestration by secretory granules. Here, the lower traces show the time course of membrane potential (solid) and $[Ca^{2+}]_{lum}$ (dash), and the upper trace shows that of $[Ca^{2+}]_i$.

This glucose effect can be modelled by varying the SG sequestration rate (i.e., by varying k_{Ca}). How k_{Ca} affects the plateau fraction is demonstrated in Fig. 5. Note that k_{Ca} has little effect on the repolarization potential or the plateau potential. Note also that k_{Ca} affects little on the amplitude of $[Ca^{2+}]_i$ oscillation. On the other hand, k_{Ca} lifts the level of $[Ca^{2+}]_{lum}$ and its amplitude significantly. It is interesting to note that Fig. 5 mimics the glucose effect (Meissner & Schmelz, 1974) and thus suggests that one of the metabolites involved in the glycolysis pathway may enhance the sequestration rate of SGs.

Figure 6 reveals how extracellular calcium affects electrical bursting, $[Ca^{2+}]_i$, and $[Ca^{2+}]_{lum}$. Consistent with experiments of Worley et al. (1994), when extracellular Ca^{2+} is depleted, electrical bursting transforms to repetitive spiking (top trace). During the spiking, $[Ca^{2+}]_i$ decreases to the lowest level (top trace). The spikes seen here are not Ca^{2+} spikes since I_{Ca} is absent when external Ca^{2+} is depleted. The current involved in the spiking cannot be I_{K-Ca}, since it is very weak due to low $[Ca^{2+}]_i$. The top trace thus reveals that the depolarization that accompanies depletion of extracellular Ca^{2+} is due to I_{NS} which is fully activated by depletion of $[Ca^{2+}]_{lum}$.

When $[Ca^{2+}]_o$ is high (see the bottom trace), $[Ca^{2+}]_i$ oscillates with a high amplitude between 0.54 mM and 2.61 mM. $[Ca^{2+}]_{lum}$ also oscillates at a high level between 88.6 mM and 166.9 mM. One may ask: why does the burst amplitude increases as $[Ca^{2+}]_o$ is raised? This can be explained by the following sequences: i) since I_{NS} is weak (due to high $[Ca^{2+}]_{lum}$), the repolarization potential is rather low (i.e., -62.9 mV); ii) during the silent period I_{K-Ca} becomes maximally activated (since $[Ca^{2+}]_i$ is so high) and the inactivation

33

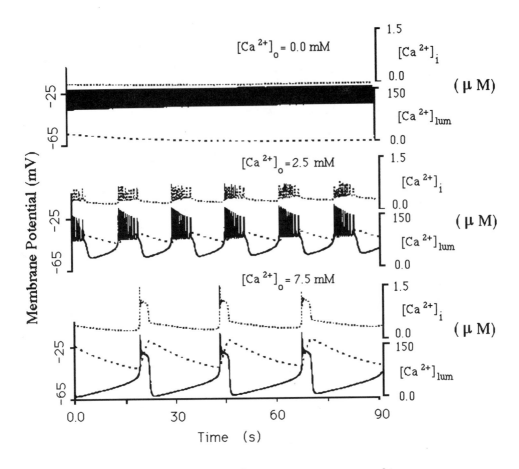

Figure 6. The role of extracellular Ca^{2+} in electrical bursting and [Ca^{2+}]$_i$ oscillation.

gating variable h becomes at its maximal value (near unity); iii) since I_{K-Ca} is in its full strength, it takes a long time before the cell depolarizes; iv) during the fast upstroke of depolarization, m becomes fully activated (i.e., m close to unity) while h is already close to unity; v) as a consequence, I_{LT} gains its full strength, and this leads to the high plateau potential. These five events bring about the high burst amplitude. Then, why is the plateau period so short? The higher plateau potential induces fuller activation of I_{Ca}, and this in turn raises [Ca^{2+}]$_i$. _When [Ca^{2+}]$_i$ is high, I_{K-Ca} gains its maximal strength, and this in turn shortens the plateau length.

This model is capable of explaining how glucagon and ephinephrine affect the period and amplitude of bursting (see Fig. 2). Accroding to the model shown in Fig. 3, the hormones involved in the AC-transduction pathway affect the parameter, k_{rel}. Note in Fig. 7 that a seventeen-fold increase in k_{rel} brings about more than a seventeen-fold increase in the burst frequency. Also note that the increase in k_{rel} leads to a decrease in the amplitude of the [Ca^{2+}]$_{lum}$ oscillation (see the large dots). The repolarization potential is also lifted upward from -56.4mV to -50.0 mV due to an increase in I_{NS}. Note that this simulation is consistent with electrical bursting observed in Fig. 2.

In addition, Fig. 7 is capable of explaining why single b-cells burst with a periodicity lasting for several minutes (Smith, Aschcroft & Rorsman, 1990), while intact b-cells in the

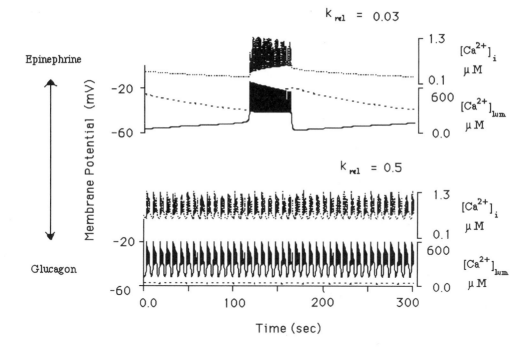

Figure 7. Effect of hormones involved in the AC-transduction pathway.

islet burst much faster (a few tens of seconds). In the islet, glucagon released form the a-cells raises [cAMP], which in turn releases luminal calcium from the store. Since [cAMP] is much lower in isolated single cells than that in intact cells, CRC activity is very idle, i.e., the k_{rel} value is small. This is why single b-cells burst very slowly (Smith, Aschcroft & Rorsman, 1990).

DISCUSSION

The central assumption in the store-operated model is that a Hodgkin-Huxley type mechanism is inappropriate to describe slowly bursting b-cells. In this cell, the slow cyclic variation of $[Ca^{2+}]_{lum}$ drives electrical bursting and $[Ca^{2+}]_i$ oscillation. Using this model, I showed how the intracellular calcium store participates encoding the frequency of bursting (see Fig. 7). What is the significance of frequency encoding in physiology? My hypothesis is that the insulin production is a frequency-encoded phenomenon, and *the calcium store is an internal clock that controls a release of insulin via frequency-encoding*. In fact, isolated b-cells which burst with a periodicity of several minutes secrete little insulin, while b-cells in the islet which burst with a periodicity of few tens of seconds can secrete much more insulin. Likewise, b-cells in the presence of glucagon (top trace of Fig. 2) secrete much more insulin than those in the presence of epinephrine (bottom trace of Fig. 2).

Previous models which did not incorporate I_{LT} are unable to simulate the extracellular Ca^{2+} effect i.e., Fig. 6. The question is then, does I_{LT} exist? This channel fits the description of a TTX-insensitive Na^+ channel, which is coupled to the muscarinic receptors and mediated by guanine nucleotide-binding proteins (Gilon, Nenquin & Henquin, 1995).

The possibility that this channel is a TTX-sensitive Na^+ channel (Plant, 1988) whose activation threshold is raised by glucose should not be eliminated.

I have previously shown that three dynamic variables with some feedback mechanisms are all that are necessary to generate electrical bursting (Chay, 1985). Therefore, the question should not be directed to whether the proposed model can generate the bursting, but the proper question to ask is: does the model generate crucial experiments observed in pancreatic b-cells? Figures 4-7 indeed demonstrate that the present model is capable of simulating most of the important experimental observations.

The benefit of our modelling approach is that it helps clarify the main events that take place in the various interacting processes as well as predicting the areas requiring further experimental work. The model predicts that i) $[Ca^{2+}]_{lum}$ in the ER oscillates in pancreatic b-cells and ii) intra-granular Ca^{2+} in SGs is necessary for exocytosis. Experiments using fluorescent dyes such as mag-fura-2-AM or aequorin could provide relevant information. The model also predicts that a transient low-threshold inward current I_{LT} is essential to explain the extracellular Ca^{2+} effect. The possibility that I_{LT} is a carbachol-sensitive, TTX-insensitive Na^+ current (hypothesis 1) should be elucidated experimentally. Whether this current could be a TTX-sensitive Na^+ current (hypothesis 2) or a transient low-threshold Ca^{2+} current (hypothesis 3) should also be pursued. Hypothesis 1 can be test by demonstrating whether repetitive firing arises in the presence of TTX when extracellular Ca^{2+} depletes. Hypothesis 2 can be tested by demonstrating that glucose can raise the activation threshold of the TTX-sensitive Na^+ current. The work presented in this paper is a clear demonstration of the power and usefulness of mathematical modelling.

This work was supported by National Science Foundation MCB-9411244.

REFERENCES

BERRIDGE M. J., RAPP P. E. & TREHERNE J.E. (Eds). Cellular Oscillators, *J. Experimental Biol.*, **81** (1979).

CHAY T. R., Effects of extracellular calcium on electrical bursting and intracellular and luminal calcium oscillations in insulin secreting pancreatic b-cells. *Biophys. J.* **73**, 1673-1688 (1997).

CHAY T. R., Mathematical modeling for the bursting mechanism of insulin secreting b-cell, *Comments Mol. Cell Biophys.* **4**, 349-368 (1988).

CHAY T. R., The Mechanism of intracellular Ca^{2+} oscillation and electrical bursting in pancreatic b-cells, *Adv. Biophys.* **29**, 75-103 (1993).

CHAY T. R., Modelling for nonlinear dynamical processes in biology, In *Patterns, Information and Chaos in Neuronal Systems*, Editor-in-Chief WEST BJ. World Scientific Publishing. River Edge, N.J. pp 73-122 (1993).

CHAY T.R. Chaos in a three-variable excitable cell model. *Physica* **16D**: 233-242 (1985).

CHAY T.R. Bursting, spiking, and chaos in an excitable cell model: The role of an intracellular calcium store. The 1995 International Symposium on Nonlinear Theory and its Application (NOLTA'95). Las Vegas, Calif. pp 1049-1052 (1995).

COOK D. L. & HALES C. N., Intracellular ATP directly blocks K^+ channels in pancratic B-cells. *Nature* **311**, 271-273 (1984).

COOK D. L. & PERARA E., Islet electrical pacemaker response to alpha-adrenergic stimulation. *Diabetes* **31**, 985-990, (1982).

DEAN P.M. & MATHEWS E. K., Glucose-induced electrical activity in pancreatic islet cells, *J. Physiol.* **210**, 255-264 (1970).

GILON P., NENQUIN M., HENQUIN J-C. Muscarinic stimulation exerts both stimulatory and inhibitory effects on the concentration of cytoplasmic Ca^{2+} in the electrically excitable pancreartic B-cell. *Biochem J.* **311**: 259-267 (1995).

GORUS FK, MALAISSE WJ & PIPELEERS DG (1984). Differences in glucose handling by pancreatic A- and B-cells. *J. Biol. Chem.* **25**, 1196-1200.

GRAPENGIESSER E., GYLFE E. & HELLMAN B., Glucose-induced oscillations of cytoplamic Ca^{2+} in the pancreatic b-cells, *Biochem Biophys. Res. Comm.* **151**, 1299-1304 (1988).

HATTORI M., KAI R. & KITASATO H., Effects of lowering external Na^+ concentration on cytoplasmic pH and Ca^{2+} concentration in mouse pancretic b-cells: Mechanism of periodicity of spike-bursts. *Jap. J. Physiol.* **44**, 283-293 (1994).

HERCHUELZ A., POCHET R., PASTIELS C. H. & PRAET A. V., Heterogenous changes in $[Ca^{2+}]_i$ induced by glucose, tolbutaminde and K^+ in single rat pancreatic B cells, *Cell Calcium* **12**, 577-586 (1991).

HIMMEL D. M. & CHAY T. R., Theoretical studies on the electrical activity of pancreatic ß-cells as a function of glucose, *Biophys. J.* **51**, 89-107 (1987).

HODGKIN A. & HUXLEY A.F. A quantitative description of membrane current and application to conduction and excitation in nerve. *J. Physiol.* (London) **117**: 500-544 (1952).

IKEUCHI M. & COOK D. L., Glucagon and forkolin have dual effects upon islet cell electrical activity, *Life Sciences* **35**, 685-691 (1984).

MEISSNER H. P. & SCHMELZ H., Membrane potential of beta-cells in pancreatic islets, *Pflugers Arch.* **351**, 195-206 (1974).

NOTKINS A. L., The causes of diabetes, Sci. Amer. **241**: 62-73 (1979).

PLANT T.D. Na^+ currents in cultured mouse pancreatic B-cells. *Pfluegeres Arch.* **411**: 429-435 (1988).

PRENKI M. & MATSCHINSKY F.M. Ca^{2+}, cAMP, and phospholipid-derived messengers in coupling mechanisms of insulin secretion. *Physiol. Rev.* **67**: 1185-1248 (1987).

RORSMAN P., BOKVIST K., AMMALA C., ARKHAMMAR P., BERGGREN P-O, LARSSON O., WAHLANDER K. Activation by adrenaline of a low-conductance G protein-dependent K^+ channel in mouse pancreatic B cells. *Nature* **349**: 77-79 (1991).

SMITH P. A., ASCHCROFT F. M. & RORSMAN P., Simultaneous recordings of glucose dependent electrical activity and ATP-regulated K^+-currents in isolated mouse pancreatic ß-cells, *FEBS lett.* **261**, 187-190 (1990).

WORLEY III J. F., MCINTYRE M. S., SPENCER B. & DUKES I. D., Depletion of intracellular Ca^{2+} stores activates a maitotoxin-sensitive nonselective cationic current in b-cells, *J. Biol. Chem.* **269**, 32055-3258 (1994).

APPENDIX

A) Low-Threshold Transient Current:

$$I_{LT} = \bar{g}_{LT} \, m_\infty^3 \, h \, (V - V_{LT})$$

where $h_\infty = \dfrac{1}{1 + \exp\left[\frac{V_h - V}{S_h}\right]}$ and $\tau_h^{-1} = \lambda_h \left(\exp\left[\dfrac{V_h - V}{2 S_h}\right] + \exp\left[\dfrac{V - V_h}{2 S_h}\right] \right)$

B) Ca^{2+} Current:

$$I_{Ca} = p_{Ca} \, df_\infty \, \frac{2FV}{RT} \left\{ \frac{[Ca^{2+}]_{out} - [Ca^{2+}]_{in} \exp\left(\frac{2FV}{RT}\right)}{1 - \exp\left(\frac{2FV}{RT}\right)} \right\}$$

where $d_\infty = \dfrac{1}{1 + \exp\left[\frac{V_d - V}{S_d}\right]}$, $\tau_d^{-1} = \bar{\tau}_d \left(\exp\left[\dfrac{V_d - V}{2 S_d}\right] + \exp\left[\dfrac{V - V_d}{2 S_d}\right] \right)$, and $f_\infty = \dfrac{K_{Ca}}{K_{Ca} + [Ca^{2+}]_i}$

E) Cationic Non-Selective Inward Current:

$$I_{NS} = \bar{g}_{NS} \, \frac{K_{NS}^2}{K_{NS}^2 + [Ca^{2+}]_{lum}^2} \left(\frac{V - V_{NS}}{1 - \exp(0.1(V_{NS} - V))} - 10 \right)$$

D) Delayed-Rectifying K^+ Current:

$$I_{K\text{-}DR} = \bar{g}_{K\text{-}DR}\, n^4 (V - V_K)$$

where $n_\infty = \dfrac{1}{1 + \exp\left[\frac{V_n - V}{S_n}\right]}$ and $\tau_n^{-1} = \lambda_n \left(\exp\left[\dfrac{V_n - V}{2\,S_n}\right] + \exp\left[\dfrac{V - V_n}{2\,S_n}\right]\right)$

E) Calcium-Sensitive K^+ Current:

$$I_{K\text{-}Ca} = \bar{g}_{K\text{-}Ca}\, \frac{\left[Ca^{2+}\right]_i^3}{K_{Ca}^3 + \left[Ca^{2+}\right]_i^3} (V - V_K)$$

F) ATP-Sensitive Inward-Rectifying K^+ Current:

$$I_{K\text{-}ATP} = g_{K\text{-}ATP} (V - V_K)$$

G) Na^+ Leak Current:

$$I_{Na,L} = g_{Na,L} (V - V_{Na})$$

The basic parametric values in the model are as follows: $C_m = 1\,mF\ cm^{-2}$, $g_{LT} = 600\ mS\ cm^{-2}$, $P_{Ca} = 2.0\ nA\ cm^{-2}$, $g_{K\text{-}DR} = 600\ mS\ cm^{-2}$, $g_{K\text{-}Ca} = 5.0\ mS\ cm^{-2}$, $g_{NS} = 5.0\ mS\ cm^{-2}$, $g_{K\text{-}ATP} = 2.0\ mS\ cm^{-2}$, $g_{Na,L} = 0.3\ mS\ cm^{-2}$, $V_{LT} = 80\ mV$, $V_K = -75\ mV$, $V_{NS} = -20\ mV$, $V_{Na,L} = 80\ mV$, $V_m = -20\ mV$, $S_m = 9\ mV$, $V_h = -48\ mV$, $S_h = -7\ mV$, $V_d = -10\ mV$, $S_d = 5\ mV$, $V_n = 18\ mV$, $S_n = 14\ mV$, $l_h^{-1} = 0.08\ s^{-1}$, $l_d^{-1} = 0.4\ s$, $l_n^{-1} = 0.08\ s$, $K_{Ca} = 1.0\ mM$, $K_{NS} = 50\ mM$, $k_{Ca} = 7.0\ s^{-1}$, $k_{pump} = 30\ s^{-1}$, $k_{rel} = 0.2\ s^{-1}$, $f = 0.2$, $[Ca^{2+}]_o = 2500\ mM$, and $T = 37^\circ C$.

TOWARDS COMPUTATIONAL MODELS OF CHEMOTAXIS IN ESCHERICHIA COLI

Laurence Clark and Ray C. Paton

Department Of Computer Science
University Of Liverpool
L69 7BX
UK

1. INTRODUCTION

This report briefly reviews the progress made so far towards a computational model of the cellular signalling system used in *Escherichia coli* chemotaxis. The emphasis of the approach is to provide insights into the information processing capabilities of the system. The objective is a computational model of the system, which will hopefully be achieved using several different modelling methods. We seek to extend existing models such as Bray (1993 and 1995), Hauri *et al* (1995), Hellingwerf *et al* (1995), Barkai *et al* (1997), and Spiro *et al* (1997).

Component	Activated by...	Excites/Inhibits
Tsr (chemoreceptor)	Serine, Leucine	CheA, CheW
Tar (chemoreceptor)	Aspartate, Maltose, Co, Ni	CheA, CheW
Trg (chemoreceptor)	Ribose, Glactose	CheA, CheW
Tap (chemoreceptor)	Dipeptides	CheA, CheW
CheA	Tsr, Tar, Trg, Tap (autophosphorylation)	CheY, CheZ, CheB
CheW	Tsr, Tar, Trg, Tap	CheY, CheZ
CheY	CheA, CheW	FliM
CheZ	CheA, CheW	CheY (inhibits)
CheR	Tsr, Tar, Trg, Tap	Sites of methylation
CheB	CheA	Sites of demethylation
Flagellar Switch (FliG, FliM, FliN)	CheY	Flagellar

Table 1. Components of the chemotactic signalling system (adapted from Bourret, Borkovich and Simon 1991, and Amsler and Matsumura 1995)

E. coli uses a variety of chemoreceptors to monitor its chemical environment and to control rotation of its flagellar motors in response to attractant and repellent gradients (Parkinson 1995). The focus is on this particular bacterium because it not only handles signalling tasks common to all sensory systems, but has also been studied and analyzed to some depth. *E. coli* moves around by rotating its flagellar filament either clockwise mation is gathered by cell-surface chemoreceptors or MCPs (Methyl-accepting Chemo-taxis Proteins) which are commonly arranged in clusters, most often at one pole of the cell (Parkinson and Blair 1993). There are approximately 20 different types of receptor, although only 4 or 5 are used for chemotaxis. Each receptor attracts a different ligand (Springer *et al* 1977). They transmit information along a network of converging signalling pathways via the components detailed in Table 1.

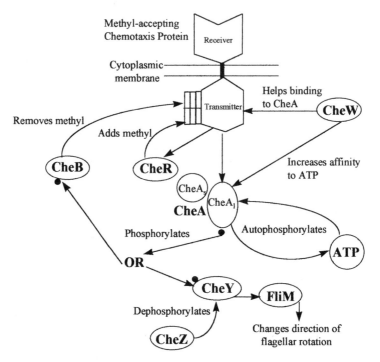

Figure 1. Diagram showing the basic interaction in the chemotactic signalling pathway (adapted from Amsler and Matsumura 1995).

THe dimer CheA interacts with the chemoreceptors and, on the binding of a ligand to a receptor, autophosphorylates and, in turn, phosphorylates either CheY (in order to influence the flagellar) or CheB (to adapt the system) (Amsler and Matsumura 1995). CheW enhances CheAs autophosphorylation by increasing its affinity to ATP. Phosphorylated CheY interacts with the flagellar motor to induce tumbling, but CheZ and CheA can dephosphorylate (deactivate) it and thus produce forward movement. CheB and CheR act like the cells memory, storing a record of the molecules that have most recently bound to a receptor. This enables the system to detect whether the level of attractant / repellent is increasing or decreasing. Attractants increase the rate of methyl groups and decrease their removal, whereas repellents have the opposite effect. There are up to 6 sites on the receptor which can be methylated by CheR and demethylated by CheB. The state of these sites represents a record of recently bound ligands.

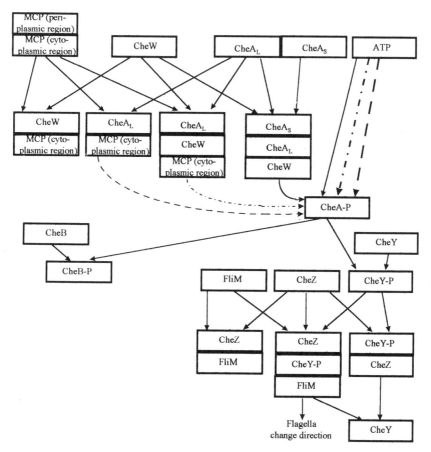

Figure 2. Diagram showing some combinations of bindings (adapted from Amsler and Matsumura 1995). Different types of line show the different ways a compound can be formed.

2. MODELLING METHODOLOGIES

In this section we assess a number of possible ways of developing computational models of the *E. coli* network. Below is a list of the factors which will have a bearing on the appropriateness of a potential model:

- *Comprehension.* Understandable to both biologists and computer scientists.

- *Explanation.* The processing should not simply be viewed as a "black box", but visible to the user in an understandable form which relates back to the biological system.

- *Abstraction.* The model must be able to cope with lots of detail, but features which are irrelevant to the overall system must be removed.

- *Synchrony.* The model should be capable of asynchronous and synchronous processing.

- *Complexity.* Many interacting signalling pathways are required in order to make the model realistic and increase its ability to adapt to different inputs.

41

- *Spatial issues.* The units relative position to others in modelled.

- *Realism.* The ability to perform the same tasks as the biological cell: signal amplification / dampening, pattern recognition, fault tolerance.

- *Analogy.* An analogous model must function in the same way as the biological system and not have inappropriate inherent features.

- *Implementation.* It should be feasible to implement the system using current technology.

- *Transportation.* Interaction between models where more than one required or different models are used for different levels of abstraction.

We now apply this criteria to the models under review.

Parallel Distributed Processing Networks

Referring to the aforementioned standard modelling criteria, PDP networks would be comprehensible to a degree to the user. Using Brays (1990) method each node has a biological counterpart, thus making the network understandable to biologists (to a certain degree). The weights and the units state may be more difficult to interpret into the real world. However as long as a decent interface is implemented, in order to aid the users examination of the networks state, then the overall system should be comprehensible. The user would be able to view the system step by step, observing the changes in the net and their effects. This system could cope with a high level of detail and complexity, dependant upon the number of nodes in the network. The level of abstraction depends upon how much detail is included (eg. the number of environmental factors accounted for in the weight system). Spatial variation could also possibly be reflected in the weights. We have already seen that an ANN can perform the same tasks as the signalling system, so that both realism is maintained, and there are no inherited features from the former. In theory, there should be no problems implementing the system and getting it to communicate with other systems. In addition, algorithms used in evolutionary nets could prove useful.

Cellular Automata

Some preliminary work has been conducted into the use of cellular automata to simulate signal propagation in a tissue of eukaryotic cells (Edwards, 1995). The results showed a correlation between the number of cells initialised at the start, and the speed which the activation pattern spreads. The end simulation seemed to be quite removed from a biological tissue, but the model could still prove useful. The space in which CAs operate is wide open to interpretation. A single cell could be used to represent either a biological cells physical location, a signal between cells, or a reaction between two cells. Intracellular signalling could be modelled if the space represented interactions between protein kinases instead. This alternative viewpoint eliminates the problems with spatial issues inherent with the original method. The overall process is broken up into discrete time steps, but synchrony is still possible. The grid is always easy to understand and the processing steps can be followed because, at any time during execution, an occupied space has a real world interpretation. Many CAs have been implemented on modern machines, with the amount of complexity dependant upon the upon neighbours, but with cell signalling theres no guarantee that one signal (ie. an occupied cell) is going to trigger a signal in that neighbourhood. Whilst there may be ways around this problem, pursuing this course would seem to be moving away from the objective.

Algebraic Machines

Holcombe (1994) uses a Petri net to model "concurrent enzyme control processing", although in his model the tokens represent enzyme resource availability. He goes on to use an alternative model called an X-machine (Holcombe 1994), which is a network of inter-connected finite-state automata machines where the labels on the arcs are functions which operate on the fundamental data-type X. This data-type is a set comprising of the inputs, outputs, and the types used during processing. An X-machine model is then developed which models a cell with two input, processing and output sites, the intention being that more detail (ie. enzyme and substrate interactions) may be included in another model showing a lower level of abstraction. Thus a hierarchy of machine-type models is required, with each level giving a different view of the system at a particular level of abstraction. The X-machine approach could also be adapted for a cell signalling system, since it has already been used for an enzyme control system. However, the functions between the nodes would be difficult to calculate and possibly irrelevant. This model would not be immediately comprehensible to the user. and well nigh impossible to follow during execution. There would still be no concept of time and space, and its impossible to say whether the model would perform like its real world counterpart.

Several component systems in the cell can be viewed as powerful parallel computer systems (Paton 1993) and may be modelled as a boolean network of many interacting two-state components (or binary automata) (Glass and Kauffman 1973, Kauffman 1991 and Weisbuch 1986). Synchrony and asynchrony are again feasible here, although the abstraction process eliminates temporal and spatial aspects. Overall, the model would be easily comprehensible to biologists, thus aiding overall understanding. If a highly detailed, sufficiently complex signalling model could be developed in this way, then this could be a useful method of implementation.

Parallel Distributed Systems

Paton et al (1995) note that the organisation of protein structure allows information to be processed in a highly parallel way, such that signal processing can be distributed over a number of independent communicating agents. Simulation packages, such as Starlogo (Resnick 1991, 1994, 1996) and SWARM (Hiebler, 1994 and Burkhart, 1994) exist for developing parallel, distributed computational models of naturally occurring systems. Starlogo is merely a cellular automata with added features for the units to communicate. SWARM is a suite of Objective C routines which can be incorporated into the users programs. This ultimately facilitates modelling on more than one level of abstraction, thus making the following theoretically possible. First, the individual organelles used in cell signalling (and their lines of communication) could be individually modelled as agents. Next, these agents could be contained within a swarm representing the cell itself. In is then possible to instantiate many cells and molecules, thus providing the right amount of complexity to emulate the biological system. A timeline is preserved within a swarm of individual, interacting agents, allowing events be no implementation problems since the system is already there. The user is able to monitor the system via an observer swarm, which can report the relevant data. In theory, there is no reason why such a model couldnt emulate the pattern-matching of a biological cell. All things considered, such general-purpose modelling systems certainly deserves further study.

3. CONCLUSION

A variety of models could be used to model a cell signalling system, and more specifically chemotaxis. All of them have their advantages and drawbacks. It seems fairly certain that a parallel distributed network model would meet the criteria laid out here, and such networks have been implemented by other authors with some success. Both cellular automata, Petri nets and boolean networks would appear to be too simplistic, and X-machines are not entirely suitable. One model worth pursuing is SWARM, which may be able to meet all of the requirements. However as was established earlier, one model will not be sufficient and combinations of different kinds of models will probably provide the way forward. Therefore future work will concentrate on applying both SWARM and PDP modelling techniques to bacterial chemotaxis.

4. REFERENCES

Amsler,C.D. and Matsumura,P. 1995. Chemotactic signal transduction in *Escherichia coli* and *Salmonella typhimurium*, in: *Two-Component signal transduction* (ed. Hoch,J. and Silhavy,T.), Blackwell Science ltd.

Barkai, N and Leibler, S (1997) Robustness in simple biochemical networks. *Nature* 387:913-917

Bourret,R.B. Borkovich,K.A. and Simon,M.I. 1991. Signal transduction pathways involving protein phosphorylation in prokaryotes. *Annu. Rev. Biochem.* 60, 401-441.

Bray,D. (1990) Intracellular signalling as a parallel distributed process. *Journal of Theoretical Biology*, 143, 215-231.

Bray D, 1993. Computer simulation of the phosphorylation cascade controlling bacterial chemotaxis. *Mol Biol Cell* 4(5), 469-482 (1993)

Bray D, 1995. Computer analysis of the binding reactions leading to a transmembrane receptor-linked multiprotein complex involved in bacterial chemotaxis. *Mol Biol Cell* 6(10), 1367-1380

Burkhart,R. (1994) The Swarm Multi-Agent Simulation System. *Position Paper for OOPSLA '94 Workshop on The Object Engine.*

Edwards,C.F. (1995) Computational models for cellular information processing systems. University of Liverpool, unpublished report.

Glass,L. and Kaufmann,S.A. (1973) The logical analysis of continuous, non-linear biochemical control networks. *J. theor. Biol.* 39, 103-129.

Hauri, D C and Ross, J (1995) A model of excitation and adaptation in bacterial chemotaxis. *Biophysical Journal* 68:708-722

Hellingwerf,K.J. Postma,P.W. Tommassen,J. and Westerhoff,H.V. (1995) Signal transduction in bacteria: phospho-neural network(s) in *Escherichia Coli? FEMS Microbiology Reviews* 16, 309-321.

Hiebler,D. (1994) The Swarm Simulation System and Individual-Based Modelling. *Proceedings of Advanced Technology for Natural Resource Management (Decision Support 2001).*

Holcombe,M. (1994) From VLSI through Machine Models to Cellular Metabolism. *Computing With Biological Metaphors*, Paton,R. (ed.), Chapman and Hall, 11-25.

Kauffman,S.A. (1991) Antichaos and adaptation. *Sci. Am.* 265, 2, 64-70.

Parkinson JS, Blair DF (1993) Does *E. coli* have a nose? *Science* 259:1701-1702

Parkinson JS, 1995. Genetic approaches for signalling pathways and proteins. in *Two-Component signal transduction* (ed. Hoch,J. and Silhavy,T.), Blackwell Science ltd.

Paton,R. (1993) Some Computational Models at the Cellular Level. *BioSystems*, 29, 63-75.

Paton,R. Staniford,G. and Kendall,G. (1995) Specifying Logical Agents in Cellular Hierarchies. *Proceedings of IPCAT (Information Processing in Cells And Tissues) 95*, Paton,R. Holcombe,M. Staniford,G. (eds.), 302-317.

Resnick, M. (1991) MultiLogo: A Study of Children and Concurrent Programming. *Interactive Learning Environments*, 1, 3, 153-170

Resnick, M. (1994) Changing the Centralized Mind. *Technology Review*, 32-40 (July).

Resnick, M. (1996) Beyond the Centralized Mindset. *Journal of the Learning Sciences*, 5, 1, 1-22.

Spiro, P A, Parkinson, J S and Othmer, H G (1997) A model of excitation and adaptation in bacterial chemotaxis. *PNAS USA* 94, 7263-7268.

Springer,M.S. Goy,M.F. and Adler,J. (1977) Sensory transduction in *Escherichia coli* - two complementary pathways of information processing that involve methylated proteins. *Proc. Natl. Sci. USA* 74, 8, 3312-3316.

Weisbuch,G. (1986) Networks of automata and biological organization. *J. theor. biol.* 121, 255-267.

THREE MODES OF CALCIUM-INDUCED CALCIUM RELEASE (CICR) IN NEURONS

David D. Friel

Department of Neurosciences
Case Western Reserve University
10900 Euclid Avenue
Cleveland, Ohio 44106-4975

INTRODUCTION

Ca^{2+} is an important signaling ion in a wide variety of cells, and many physiological stimuli produce their cellular effects by changing the intracellular free Ca^{2+} concentration ($[Ca^{2+}]$) (Clapham, 1995, Berridge, 1995). Given that cells are multicompartment systems, the action of a particular stimulus depends on the way it modifies intracompartmental $[Ca^{2+}]$, and the Ca^{2+}-sensitive effectors that are present within each compartment, poised to interpret the Ca^{2+} signal. Since $[Ca^{2+}]$ is regulated by multiple interacting Ca^{2+} transport pathways, understanding the effects of stimulation on the intracellular distribution of Ca^{2+} requires analysis of the dynamical properties of a system of coupled Ca^{2+} transporters.

In excitable cells, membrane depolarization opens voltage-sensitive Ca^{2+} channels in the plasma membrane, permitting Ca^{2+} entry across the plasma membrane and a rise in the cytosolic free Ca^{2+} concentration ($[Ca^{2+}]_i$). Although the rise in $[Ca^{2+}]_i$ is initiated by Ca^{2+} entry, the kinetics of the rise are strongly influenced by Ca^{2+} uptake and release by internal stores (for review, see Kostyuk and Verkhratsky, 1994). This chapter will focus on a particular store, the endoplasmic reticulum (ER), which expresses a particular type of Ca^{2+} release channel, the ryanodine receptor (Coronado et al., 1994, Berridge et al. 1995) named after the high-affinity ligand used in its biochemical isolation. These channels have been studied extensively in cardiac myocytes, where they open in response to elevations in $[Ca^{2+}]_i$, causing Ca^{2+} to be released into the cytosol through a process known as Ca^{2+}-induced Ca^{2+} release (CICR) (Fabiato, 1983). Ca^{2+} release channels are interesting because they render the Ca^{2+} permeability of the ER $[Ca^{2+}]_i$-sensitive, which can set the stage for positive feedback, whereby Ca^{2+} release becomes regenerative. While Ca^{2+} release channels are also found in neurons (Bezprozvanny et al., 1991; McPherson et al. 1991; for reviews, see Kuba, 1994, Verkhratsky and Shmigol, 1996) their role in neuronal Ca^{2+} signaling has been relatively unclear.

In the following, after briefly summarizing background information on cellular Ca^{2+} regulation, experimental observations will be described that implicate Ca^{2+} release channels in the modulation of $[Ca^{2+}]_i$ response kinetics during membrane depolarization in sympathetic neurons. Evidence will be presented which supports the idea that during weak depolarization, opening of Ca^{2+} release channels leads to an acceleration of $[Ca^{2+}]_i$

Information Processing in Cells and Tissues
Edited by Holcombe and Paton, Plenum Press, New York, 1998

elevations, and that paradoxically, this can occur under conditions where the ER acts as a Ca^{2+} buffer. These results can be understood in terms of a simple one-pool model of Ca^{2+} regulation that has provided a conceptual framework for the study of $[Ca^{2+}]_i$ dynamics in these cells (Friel, 1995). It is concluded that Ca^{2+} release channels are expected to have very different effects on stimulus-evoked changes in the intracellular distribution of Ca^{2+} depending on the cellular context in which they operate, illustrating the importance of interactions between different Ca^{2+} transport systems in $[Ca^{2+}]$ dynamics.

BACKGROUND

Cells are multicompartment systems in which intracompartmental Ca^{2+} concentrations are tightly regulated by a variety of Ca^{2+} transport systems. For example, the concentration of free Ca^{2+} within the cytoplasm ($[Ca^{2+}]_i$) is about four orders of magnitude lower (~100 nM) than the concentration of extracellular free Ca^{2+} (1-2 mM). This non-uniform distribution of Ca^{2+} is maintained by Ca^{2+} extrusion systems that operate in parallel with the plasma membrane's permeability to Ca^{2+}. In addition, various organelles can either accumulate Ca^{2+} from the cytosol, or release Ca^{2+} into the cytosol, depending on conditions. The endoplasmic reticulum (ER) is a membrane-delimited compartment that forms a complex meshwork within the cytosol. The concentration of free Ca^{2+} within the ER is thought to be two to three orders of magnitude above the resting $[Ca^{2+}]_i$ (~10-100 µM). This nonuniform distribution of Ca^{2+} is maintained by a particular class of Ca^{2+} ATPase (SERCA) that operates in parallel with the Ca^{2+} permeability of the ER membrane, which is revealed by the transient rise in $[Ca^{2+}]_i$ that occurs following exposure to specific SERCA pump inhibitors such as thapsigargin. Endowed with Ca^{2+} pumps and a passive Ca^{2+} permeability, such a compartment would be expected to behave as a nonsaturable Ca^{2+} buffer that accumulates Ca^{2+} from the cytosol as $[Ca^{2+}]_i$ rises during stimulation, and release Ca^{2+} into the cytosol as $[Ca^{2+}]_i$ declines after the stimulus is terminated. The overall effect would be to slow stimulus-evoked elevations in $[Ca^{2+}]_i$ and transiently elevate the intraluminal Ca^{2+} concentration, $[Ca^{2+}]_s$. While there is evidence that this can occur, the presence of Ca^{2+} release channels within the membrane of the ER causes the ER Ca^{2+} permeability to be $[Ca^{2+}]_i$-sensitive. While this is usually interpreted as a mechanism for Ca^{2+}-induced net release of Ca^{2+} from the ER, the overall impact of a $[Ca^{2+}]_i$-sensitive permeability on stimulus-evoked changes in $[Ca^{2+}]_i$ and $[Ca^{2+}]_s$ depends critically on the other transport systems operating within the cell. After describing several observations that implicate Ca^{2+} release channel activity in defining the temporal characteristics of $[Ca^{2+}]_i$ responses to membrane depolarization in sympathetic neurons, we will return to the way cellular context is expected to influence the impact of Ca^{2+} release channel gating on stimulus-evoked changes in $[Ca^{2+}]$.

CICR IN SYMPATHETIC NEURONS

Sympathetic neurons respond to membrane depolarization with a reversible rise in $[Ca^{2+}]_i$. In the presence of ryanodine, the response onset is slowed markedly, suggesting that ryanodine receptor-mediated Ca^{2+} release speeds the depolarization-induced rise in $[Ca^{2+}]_i$ (Friel and Tsien, 1992; Hua et al., 1993). Additional evidence supporting a role for CICR during $[Ca^{2+}]_i$ responses to depolarization comes from Hua et al. (1993), who showed that the relationship between the rates of depolarization-induced Ca^{2+} entry and the resulting rise in $[Ca^{2+}]_i$ is supralinear. Similar observations have been reported in cerebellar Purkinje neurons (Llano et al. 1994) and sensory neurons (Shmigol et al., 1995). These findings have been interpreted to mean that during depolarization. Ca^{2+} entry across the plasma membrane triggers net Ca^{2+} release from internal stores, causing $[Ca^{2+}]_i$ to rise under the influence of both Ca^{2+} entry and Ca^{2+} release. According to this view, in parallel with the rise in $[Ca^{2+}]_i$ is a drop in the intraluminal Ca^{2+} concentration $[Ca^{2+}]_s$.

This interpretation leads to a simple prediction. If the ER normally acts as a Ca^{2+} source during membrane depolarization, inhibition of SERCA pumps should lead to gradual depletion of Ca^{2+} from the ER with the consequence that subsequent depolarization-induced $[Ca^{2+}]_i$ responses are slowed. However, following treatment with thapsigargin, $[Ca^{2+}]_i$ response kinetics are never slowed. In fact, treatment with thapsigargin leads to an acceleration of responses to depolarization in approximately 50% of cells (Albrecht and Friel, 1997; MAA and DDF unpublished observations). This suggests that in some cells the ER accumulates net Ca^{2+} during weak depolarization, but in a manner that is modulated by Ca^{2+} release channel activity. According to this view, inhibition of Ca^{2+} accumulation by the ER would permit $[Ca^{2+}]_i$ to rise under the full impact of Ca^{2+} entry, speeding the depolarization-induced rise in $[Ca^{2+}]_i$. This would imply that normally Ca^{2+} accumulation by the ER slows the depolarization-induced rise in $[Ca^{2+}]$, but in a manner that is attenuated by Ca^{2+}-release channel opening.

CICR IN A ONE-POOL MODEL OF $[Ca^{2+}]$ REGULATION

How can these observations be explained? In the following, a simple one-pool model of Ca^{2+} regulation will be described that has been developed to account for several features of $[Ca^{2+}]_i$ dynamics in sympathetic neurons (Friel, 1995) and resembles models that have been described previously (Kuba and Takeshita, 1981; Chay, 1990; Goldbeter et al., 1990; Somogyi and Stucki, 1991; De Young and Keizer, 1992). The model includes three compartments (Fig.1): the extracellular medium with fixed Ca^{2+} concentration c_o, the cytosol (c_i) and a single internal pool (c_s), with Ca^{2+} transport across the plasma membrane governed by a 'pump' flux J_{P1} and a 'leak' flux J_{L1}; transport between the store and cytosol depends on analogous fluxes J_{P2} and J_{L2}:

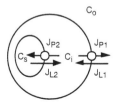

Figure 1. One-pool model of Ca^{2+} regulation. c_i and c_s designate Ca^{2+} concentrations within the cytosol and internal pool, respectively while c_o refers to the (constant) concentration of extracellular Ca^{2+}. Arrows represent the net Ca^{2+} fluxes generated by different population of Ca^{2+} transporters. Circles designate 'uphill' pump fluxes (subscript P), as distinct from passive 'downhill' leak fluxes (no circles, subscript L). Direction of arrows conforms with the net fluxes expected under physiological conditions.

It is assumed that c_i and c_s are spatially uniform within compartments, which is reasonable given that the kinetic effects that are the focus of this study outlast dissipation of spatial Ca^{2+} gradients within the cytosol by many seconds (Friel and Tsien, 1992; Hua et al., 1993).

The two dynamical variables c_i and c_s satisfy the system of differential equations:

$$dc_i(t)/dt = -\{J_{L1} + J_{P1} + \gamma(J_{L2} + J_{P2})\}$$

$$\equiv -J_1$$

$$\frac{dc_s(t)}{dt} = J_{L2} + J_{P2}$$

$$\equiv J_2$$

where the fluxes have dimensions concentration/time and γ is the ratio of pool to cytosol volume. Note that inward net Ca^{2+} fluxes that cause c_i to rise are negative to conform with electrophysiological convention. The fluxes are assumed to depend on c_i and c_s in the following way:

$$J_{L1} = \kappa_{L1}(c_i - c_o)$$

$$J_{P1} = \kappa_{P1}c_i$$

$$J_{L2} = \kappa_{L2}(c_i - c_s)$$

$$J_{P2} = \kappa_{P2}c_i.$$

With κ_{L1}, κ_{P1}, κ_{L2} and κ_{P2} constant (dimensions: time^{-1}) these equations represent a linear approximation to more general rate laws in which each flux depends on c_i and c_s but not explicitly on time, and leak fluxes reflect permeabilities and the pump fluxes are unidirectional. The steady-state solution is:

$$c_{i,ss} = c_o/(1 + \kappa_{P1}/\kappa_{L1})$$

$$c_{s,ss} = c_{i,ss}(1 + \kappa_{P2}/\kappa_{L2}).$$

Conforming with physical intuition, $c_{i,ss}$ rises with c_o and falls with (κ_{P1}/κ_{L1}) while $c_{s,ss}$ is proportional to $c_{i,ss}$ and rises with (κ_{P2}/κ_{L2}).

To include Ca^{2+} release channels in the model, the rate coefficient κ_{L2} is assumed to increasing sigmoidally with c_i (Fig.2):

$$\kappa_{L2} = \kappa_{L2}^{(0)} + \kappa_{L2}^{(1)}/(1 + (K_d/c_i)^n).$$

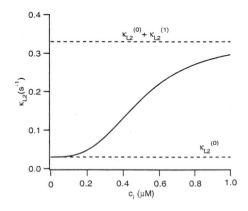

Figure 2. c_i-dependence of κ_{L2}. Curve describes $\kappa_{L2} = \kappa_{L2}^{(0)} + \kappa_{L2}^{(1)}/(1 + (K_d/c_i)^n)$ where $\kappa_{L2}^{(0)} = 0.03$ s^{-1}, $\kappa_{L2}^{(1)} = 0.3$ s^{-1}, $K_d = 0.5$ μM, and n=3. Dashed lines indicate the values of κ_{L2} at low and high c_i.

$\kappa_{L2}{}^{(0)}$ and $\kappa_{L2}{}^{(1)}$ are constants describing the background permeability of the store and the maximal c_i-sensitive permeability, respectively, K_d is the value of c_i that causes half-maximal channel activation, and n describes the steepness of activation by c_i (cf. Kuba and Takeshita, 1981). For example, this definition of κ_{L2} would describe two populations of Ca^{2+} channels, one consisting of $N^{(0)}$ channels with constant open probability $P_0{}^{(0)}$ and unitary Ca^{2+} permeability $\rho^{(0)}$ that open and close randomly under resting conditions, imparting to the ER membrane a background permeability $\kappa_{L2}{}^{(0)} = N^{(0)}P_0{}^{(0)}\rho^{(0)}$, the other consisting of $N^{(1)}$ channels with c_i-dependent open probability $P_0{}^{(1)} = (1+(K_d/c_i)^n)^{-1}$ and unitary permeability $\rho^{(1)}$, endowing the ER membrane with a c_i-sensitive permeability whose maximal value is $\kappa_{L2}{}^{(1)} = N^{(1)}\rho^{(1)}$. With this definition of κ_{L2}, the system of equations becomes nonlinear.

Flux/Concentration Relations

According to this model, c_s changes under the influence of the net flux J_2 between the store and cytosol ($J_2 = J_{L2} + J_{P2}$), and c_i changes under the influence of a flux J_1 that is the sum of J_2 and the net flux across the plasma membrane $J_{L1} + J_{P1}$. One way to analyze the behavior of this system is to examine how J_1 and J_2 depend on c_i and c_s (Fig.3):

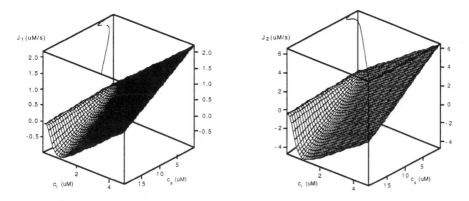

Figure 3. c_i- and c_s-dependence of J_1 and J_2. Parameter values are: $\kappa_{L1}=5\times10^{-6}$ s^{-1}, $\kappa_{P1}=0.13$ s^{-1}, $\kappa_{L2}{}^{(0)}=$ 0.03 s^{-1}, $\kappa_{L2}{}^{(1)}=0.3$ s^{-1}, $\kappa_{P2}=1$ s^{-1}, $\gamma=0.24$, n=3, $K_d=0.5$ µM. Nullclines are indicated.

For example, J_1 and J_2 could represent fluxes measured under conditions where c_i and c_s are 'clamped'. Alternatively, J_1 and J_2 could be viewed as instantaneous fluxes under conditions where c_i and c_s have specified initial values but are free to change with time.

In contrast to the linear system where $\kappa_{L2}{}^{(1)} = 0$, J_1 and J_2 do not increase monotonically with c_i and c_s. For example, raising c_i at constant c_s with $c_i \ll K_d$ causes J_1 and J_2 to increase essentially as in the linear model with $\kappa_{L2} = \kappa_{L2}{}^{(0)}$. However, as c_i nears K_d, both J_1 and J_2 turn sharply downward toward more negative values, reaching minima before turning upward again, eventually increasing as in the linear model with $\kappa_{L2} = \kappa_{L2}{}^{(0)} + \kappa_{L2}{}^{(1)}$. For a given set of parameters, the downturn in J_1 and J_2 becomes more pronounced as c_s increases. For example, with the parameters used in Fig.3, when c_s is small (store is under-loaded), J_1 and J_2 increase monotonically with c_i and become positive (outward relative to the cytosol) when c_i exceeds the steady-state value $c_{i,ss}$. However, as c_i approaches K_d, the rise in κ_{L2} slows the rise in J_1 and J_2 (see Fig.4, below). In this case, the store behaves as a c_i-regulated Ca^{2+} buffer whose buffering strength declines as c_i approaches K_d from below. On the other hand, if c_s is large (store is overloaded), both J_1 and J_2 can become negative (inward relative to the cytosol), and increase in magnitude as c_i approaches K_d. In this case, the store is a Ca^{2+}-sensitive Ca^{2+} 'source'.

Whether activation of the c_i-sensitive permeability modulates net Ca^{2+} accumulation or triggers net Ca^{2+} release depends on the sign of J_2, which in turn depends on the relationship between its components J_{P2} and J_{L2}. The 'pump' flux $J_{P2} = \kappa_{P2}c_i$ is always positive, and the 'leak' flux $J_{L2} = \kappa_{L2}(c_i - c_s)$ is always negative as long as $c_i < c_s$, which is assumed in the following. As a result, the sign of J_2 depends on the relative magnitudes of J_{P2} and J_{L2}. Therefore, assuming a fixed initial value of $(c_i - c_s)$, a sudden rise in c_i would be expected to increase κ_{L2}, causing J_{L2} to increase in magnitude, representing Ca^{2+}-induced Ca^{2+} release. However, the internal compartment will gain or lose Ca^{2+} depending on the relative magnitudes of J_{L2} and J_{P2}. This illustrates the general point that the impact of a particular Ca^{2+} transport system depends on the context in which it operates.

Figure 4 shows how J_1 and J_2 vary with c_i at constant c_s ($=c_{s,ss}$) for several values of the nonlinear parameter $\kappa_{L2}^{(1)}$.

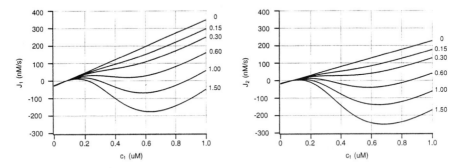

Figure 4. c_i-dependence of J_1 and J_2 at constant c_s for different values of $\kappa_{L2}^{(1)}$ over a range of c_i from 0 to $2K_d$. Values of $\kappa_{L2}^{(1)}$ are given to the right of each curve. Note that each curve represents a slice at constant c_s from surfaces like those in Fig.3 with different values of $\kappa_{L2}^{(1)}$. c_s was chosen to be the steady-state value $c_{s,ss}$ for each value of $\kappa_{L2}^{(1)}$. Note that when c_i is near K_d, the signs of J_1 and J_2 can have the three combinations (+,+), (+,-), (-,-) depending on whether $\kappa_{L2}^{(1)}$ is small, intermediate or large.

Note that when $\kappa_{L2}^{(1)}$ is small there is one (stable) steady state $(c_{i,ss})$ where $J_1 = 0$ and $\partial J_1/\partial c_i > 0$. In contrast, as $\kappa_{L2}^{(1)}$ is increased, another (unstable) steady-state emerges where $J_1 = 0$ and $\partial J_1/\partial c_i < 0$, setting the stage for positive feedback.

As c_i is raised beyond the resting value $c_{i,ss}$, J_1 and J_2 change in a manner that depends on the magnitude of $\kappa_{L2}^{(1)}$ and its relation to the other system parameters. Three ranges of $\kappa_{L2}^{(1)}$ can be distinguished in terms of the signs of J_1 and J_2:

Case 1 ($J_1 > 0$, $J_2 > 0$). When $\kappa_{L2}^{(1)}$ is small (e.g. ≤ 0.3 s^{-1}), J_1 and J_2 are positive as long as $c_i > c_{i,ss}$, and reflect outward fluxes from the cytosol toward the extracellular compartment and the internal store, respectively. The effect of the nonlinear term is to cause these outward fluxes to rise less steeply with c_i as c_i approaches K_d, and more steeply with c_i when $c_i >> K_d$, than they do in the linear model ($\kappa_{L2}^{(1)} = 0$). In this case, activation of the c_i-sensitive permeability causes the pool to be more permeable to Ca^{2+}, rendering it a less effective Ca^{2+} buffer.

Case 2 ($J_1 > 0$, $J_2 < 0$). With intermediate values of $\kappa_{L2}^{(1)}$ (e.g. 0.60 s^{-1}), as c_i approaches K_d, J_2 becomes negative but J_1 remains positive. In this case, activation of the c_i-sensitive permeability causes the store to act as a Ca^{2+} source. However, in this regime, J_2 is smaller in magnitude than the net flux across the plasma membrane, so J_1 remains positive.

Case 3 ($J_1 < 0$, $J_2 < 0$). Here, $\kappa_{L2}^{(1)}$ is large enough (e.g. ≥ 1.0) that both J_1 and J_2 become negative as c_i approaches K_d. In this case, the inward flux J_2 become so large that it overwhelms Ca^{2+} extrusion across the plasma membrane, causing J_1 to become negative. Under these conditions, CICR can be regenerative (see below).

Dynamics of c_i and c_s

How does the Ca^{2+}-sensitive permeability influence the dynamics of c_i and c_s when these variables are both free to change? Figure 5 illustrates how c_i and c_s relax toward their steady-state values if c_i is initially displaced from its steady-state value to 400 nM and $c_s = c_{s,ss}$. When $\kappa_{L2}^{(1)}$ is small (≤ 0.3 s^{-1}) $J_2 > 0$ and $J_1 > 0$, causing c_s to rise transiently and c_i to fall. In this case the store acts as a buffer whose strength is reduced by the c_i-sensitive permeability. When $\kappa_{L2}^{(1)}$ has an intermediate value (0.6 s^{-1}), $J_2 < 0$ and $J_1 > 0$, so that c_i and c_s both decline initially, with the decline in c_i slowed by net Ca^{2+} release from the internal pool. Finally, when $\kappa_{L2}^{(1)}$ is large (≥ 1s^{-1}), $J_1 < 0$ and $J_2 < 0$, and net Ca^{2+} release causes c_i to rise and c_s to fall regeneratively before relaxing toward their steady-state values.

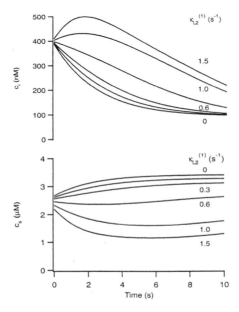

Figure 5. Early relaxations of c_i (top) and c_s (bottom) from initial values of 400 nM and $c_{s,ss}$ for different values of the nonlinear parameter $\kappa_{L2}^{(1)}$. Note that three relaxation patterns can be distinguished based on the signs of J_1 and J_2 (see Fig.4) and the initial rates of change of c_i and c_s.

Another approach to illustrating the three dynamical patterns is to compare the way c_i and c_s change in response to an externally applied step of injected Ca^{2+} of magnitude J_{ext} (Fig.6). For example, this could approximate the effects of Ca^{2+} entry through non-inactivating voltage-gated Ca^{2+} channels during a step depolarization. For reference, when $\kappa_{L2}^{(1)} = 0$ both c_i and c_s increase biexponetially in response to J_{ext} (Fig.6 trace a). In this case, the internal compartment behaves as a nonsaturable buffer, causing c_i to increase more slowly than would be the case were the store to be replaced by an equal volume of cytosol. When $\kappa_{L2}^{(1)} = 0.1$ (trace b), as c_i rises, the store becomes more permeable to Ca^{2+}, rendering it a less effective buffer. As a result, c_i approaches $c_{i,ss}$ more rapidly and c_s rises more slowly toward a lower steady-state level, compared to the linear case. Finally, when $\kappa_{L2}^{(1)} = 1.5$ (trace d), c_i rises even more rapidly, transiently overshooting its steady-state level, reflecting net Ca^{2+}-induced Ca^{2+} release. In this case, CICR is regenerative in the sense that c_i continues to rise and c_s continues to fall even after Ca^{2+} injection ends (not shown). Note

53

that when $\kappa_{L2}^{(1)}$ takes on the intermediate value 0.6 (trace c, dotted), the rise in c_i is initially accompanied by net Ca^{2+} accumulation by the pool and an elevation of c_s, which slows the rise in c_i. The rise in c_s, in concert with the c_i-dependent elevation in κ_{L2} then favors net Ca^{2+} release, which speeds the rise in c_i.

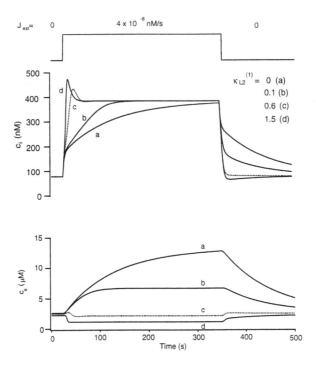

Figure 6. Effects of injected Ca^{2+} at rate J_{ext} into the cytosol on c_i and c_s for four different values of the nonlinear parameter $\kappa_{L2}^{(1)}$.

Three Modes of CICR

In each of the cases (b-d) in Fig. 6, a stimulus-evoked rise in c_i increases the store's permeability to Ca^{2+} and increases the passive net flux of Ca^{2+} from store to cytosol, which seems appropriate to describe as Ca^{2+}-induced Ca^{2+} release. However, in one case (b) c_i and c_s both rise in response to Ca^{2+} injection, while in another (d) c_i rises and c_s falls. Also, in one case CICR is weak in the sense that continued release requires continued Ca^{2+} entry, while in another, it is strong in that, once triggered, it becomes self-sustaining. From a functional point of view, these are clearly very different cases. One way to distinguish between them is to define three modes of CICR in terms of the signs of J_1 and J_2 at each instant in time:

Mode 1 ($J_1 > 0$, $J_2 > 0$): In this mode, CICR via Ca^{2+} release channels provides only a small Ca^{2+} flux from store to cytosol that cannot overcome the effects of Ca^{2+} uptake by SERCA pumps. As a result, the store acts as a $[Ca^{2+}]_i$-regulated buffer. This might occur, for example, if Ca^{2+} release channels are present at low density compared to SERCA pumps. The pattern of response to stimulated Ca^{2+} entry includes a rise in both c_i and c_s, with Ca^{2+} release channel activation accelerating the rise in c_i and slowing the rise in c_s. This mode of

CICR provides a reasonable explanation for the kinetic properties of $[Ca^{2+}]_i$ responses elicited by weak depolarization in sympathetic neurons described above.

Mode 2 ($J_1 > 0$, $J_2 < 0$): Here, CICR is powerful enough to cause net Ca^{2+} release from the store, but release is still not powerful enough to overwhelm Ca^{2+} extrusion across the plasma membrane (or Ca^{2+} removal by other intracellular buffers), preventing regenerative CICR. This might occur if Ca^{2+} release channels are present at high density compared to SERCA pumps, but the ER occupies a small fraction of total cell volume. The pattern of response to Ca^{2+} entry includes a rise in c_i and a fall in c_s.

Mode 3 ($J_1 < 0$, $J_2 < 0$): In this case, CICR is sufficiently powerful to overwhelm available Ca^{2+} removal systems. This might occur if the ER occupies a large fraction of the cell volume, or the rate of Ca^{2+} extrusion across the plasma membrane is small compared to the rate of Ca^{2+} release. In this case, CICR causes c_i to rise and c_s to fall regeneratively.

CONCLUDING REMARKS

Sympathetic neurons respond to membrane depolarization with a rise in $[Ca^{2+}]_i$ that reflects Ca^{2+} entry through voltage-gated Ca^{2+} channels in the plasma membrane and Ca^{2+} uptake and release from intracellular stores. This chapter has focused on the role of Ca^{2+} uptake and release by the ER in shaping stimulus-evoked elevations in intracellular $[Ca^{2+}]$. Particular attention was given to the impact of Ca^{2+} release channels that open in response to elevations in $[Ca^{2+}]_i$ and increase the Ca^{2+} permeability of the ER. Experimental results in sympathetic neurons suggest that while opening of Ca^{2+} release channels during weak depolarization speeds the rise in $[Ca^{2+}]_i$, the ER accumulates net Ca^{2+}. A simple explanation for these observations is that CICR operates in a low gain mode whereby Ca^{2+} release channel opening attenuates the buffering strength of the ER. It should be noted that in principle, this could also account for supralinear relations between the rate of Ca^{2+} entry and the rate at which $[Ca^{2+}]_i$ rises (cf. Hua et al., 1993; Llano et al., 1994; Shmigol et al., 1995). A simple one-pool model was used as a conceptual framework for considering the conditions under which this behavior might occur. Three modes of CICR were distinguished in terms of the way intracompartmental Ca^{2+} concentrations change in response to stimulated Ca^{2+} entry. It is interesting to speculate that differences in ryanodine receptor density or state of modulation might contribute to cell-type specific patterns of activation of Ca^{2+} sensitive enzymes within different intracellular compartments.

REFERENCES

Albrecht, M.A. and Friel, D.D., 1997, Ryanodine-induced enhancement of Ca^{2+} sequestration by intracellular stores in sympathetic neuronsm, *Biophys. J.* 72:A298.

Berridge, M.J., 1995, The elemental principles of calcium signaling, *Cell* 83:675-678.

Berridge, M.J., Cheek, T.R., Bennett, D.L., Bootman, M.D., 1995, Ryanodine receptors and intracellular calcium signaling, in: *Ryanodine Receptors.* V. Sorrentino. ed., CRC Press.

Bezprozvanny, I., Watras, J., Ehrlich, B.E., 1991, Bell-shaped calcium response curves of Ins(1,4,5)P3- and calcium-gated channels from endoplasmic reticulum of cerebellum, *Nature* 351:751-754.

Chay, 1990, Electrical bursting and intracellular Ca^{2+} oscillations in excitable cell models. *Biol. Cybern.* 63:15-23.

Clapham, D.E., 1995, Calcium Signaling. *Cell* 80:259-268.

Coronado, R., Morrissette, J., Sukhareva, M., Vaughan, D.M., 1994, Structure and function of ryanodine receptors, *Am. J. Physiol.* 266:C1485-C1504.

De Young, G.W., Keizer, J., 1992, A single-pool inositol 1,4,5-trisphosphate-receptor- based model for agonist-stimulated oscillations on Ca^{2+} concentration. *Proc. Natl. Acad. Sci. USA* 89:9895-9899.

Fabiato, A., 1983, Calcium-induced release of calcium from the cardiac sarcoplasmic reticulum, *Am. J. Physiol.* 245:C1-C14.

Friel, D.D., Tsien, R.W. 1992, A caffeine and ryanodine-sensitive Ca^{2+} store in bullfrog sympathetic neurons modulates the effects of Ca^{2+} entry on $[Ca^{2+}]_i$, *J. Physiol.* 450:217-246.

Friel, D.D., 1995, $[Ca^{2+}]_i$ oscillations in sympathetic neurons: an experimental test of a theoretical model, *Biophys. J.* 68:1752-1766.

Goldbeter, A., Dupont, G., Berridge, M.J., 1990, Minimal model for signal-induced Ca^{2+} oscillations and for their frequency encoding through protein phosphorylation, *Proc. Natl. Acad. Sci. USA* 87:1461-1465.

Hua, S.Y., Nohmi, M., Kuba, K., 1993, Characteristics of Ca^{2+} release induced by Ca^{2+} influx in cultured bullfrog sympathetic neurones, *J. Phyisol.* 464: 245-272.

Kostyuk, P., Verkhratsky A., 1994, Calcium stores in neurons and glia, *Neuroscience* 63:381-404.

Kuba, K., 1994, Ca^{2+}-induced Ca^{2+} release in neurons, *Jap. J. Physiol.* 44:613-650.

Kuba, K., Takeshita, S., 1981, Simulation of intracellular Ca^{2+} oscillation in a sympathetic neuron, *J. Theor. Biol.* 93:1009-1031.

Llano, I., DiPolo, R., Marty, A., 1994, Calcium-induced calcium release in cerebellar Purkinje cells, *Neuron* 12:663-673.

McPherson, P.S., Kim, Y.K., Valdivia, H., Knudson, C.M., Takekura, H., Franzini-Armstrong, C., Coronado, R., Campbell, K.P., 1991, The brain ryanodine receptor; a caffeine sensitive calcium release channel, *Neuron* 7:17-25.

Shmigol, A., Verkhratshy, A., Isenberg, G., 1995, Calcium-induced calcium release in rat sensory neurons, *J. Physiol.* 489:627-636.

Somogyi, R., Stucki, J., 1991, Hormone-induced calcium oscillations in liver cells can be explained by a simple one pool model, *J. Biol. Chem.* 266:11068-11077.

Verkhratsky, A., Shmigol, A., 1996, Calcium-induced calcium release in neurones. *Cell Calcium* 19:1-14.

Θ-NEURON, A ONE DIMENSIONAL SPIKING MODEL THAT REPRODUCES *IN VITRO* AND *IN VIVO* SPIKING CHARAC-TERISTICS OF CORTICAL NEURONS.

Boris S. Gutkin* and G. Bard Ermentrout

Department of Mathematics, University of Pittsburgh
Pittsburgh Pa USA

INTRODUCTION

Models in neurobiology have played a major role in our understanding of neural information processing. Modeling strategies typically take two divergent routes: detailed compartmental models of single neurons and theoretical models of large scale networks. Compartmental models strive to examine in detail the consequence of biophysical characteristics of the neural membrane on the activity of the cell and elucidate the biological mechanism underlying the cells behavior (for example see [Jaeger et al., 1997; Segev et al., 1989]). Due to the computational requirements simulations of extended networks of such *in computo* neurons are not feasible. Large scale models on the other hand concern themselves with information processing in large assemblies of neural units which are often ad hoc abstractions of neurons or local neural circuits (e.g. binary neurons [Van Vreeswejk Sompolinsky, 1996] or rate-coding units [Ermentrout Cowan, 1980]). In fact the single most popular computational unit in spiking neural networks is the integrate-and-fire neuron (*IFN*) which is computationally cheap, but carries with it a number of pathological characteristics. Furthermore the *IFN* is a phenomenological model and has become popular in the literature largely out of convenience. We suggest that much of future progress in theoretical neuroscience shall depend on connecting the two modeling approaches by developing methods for formal reduction of detailed models to simpler, more tractable ones. Here we exhibit a technique for reduction and the resulting one-dimensional neural model.

In the resulting model, the θ-neuron, instead of tracking the membrane voltage and various conductance activation variables, we describe the neurons activity by a phase variable θ. Thus phase can be thought of as the location along a spike trajectory in the phase plane. That is the resting potential of the cell is at some value $\theta = \theta_{rest}$, and when the phase θ traverses from 0 to 2π, the corresponding membrane voltage goes through the a ction potential. The dynamics of the phase under noisy input are governed by the following Langevin differential equation:

$$d\theta/dt = q(1 - \cos\theta) + (1 + \cos\theta)(\beta + W_t), \theta \in [0, 2\pi], \theta(0) = \theta(2\pi) \qquad (1)$$

where W_t is a noisy conductance modeled by white noise with intensity σ, or can be a sum of Poisson timed excitatory and inhibitory inputs; β reflects the strength of the constant biasing current. While the dynamics of the model are defined by the evolution of the phase θ, if we take the observable quantity to be $(1+cos(\theta))$ spikes become readily apparent.

TYPE I VS. TYPE II MEMBRANES

The major assumption for this work is a particular dynamical mechanism for the spike generating membrane: type I dynamics. In this section we shall review the characteristics of the two membrane types using the Morris-Lecar neural model as a generic example. The general idea is to classify the cells by their repetitive firing characteristics in response to an injected current. A more complete discussion of this classification can be found in [Rinzel, 1989].

A type I membrane is characterized by a continuous frequency to input response (f-i) curve that shows oscillations arising with infinitely long periods, or equivalently with low frequencies, see Fig. 1a. The f-i curve shows that the type I cell is capable of a wide range of firing frequencies. This suggests that in the excitable regime, when the time independent injected bias current is below the threshold, a number of dynamical behaviors is possible depending on the temporal and amplitude characteristics of the time dependent stimulus inputs. In fact, in Fig. 1b we see that the delay to the spike for a type I neural model is strongly sensitive to the magnitude of the supra- threshold stimulus. The type I membrane will integrate the time-dependent unitary inputs (e.g. Poisson timed epsp's and ipsp's) and convert the variability in the timing of these inputs to variability in output spike timing. The phase plane for type I membrane helps us understand why this happens. The voltage nullcline intersects the activation variable nullcline forming an attractive steady state and a saddle node which acts as a threshold for the spike generation, see fig 2. Any stimulus pushing the voltage past the saddle node will result in a spike of constant shape, but with a varying delay. This is because in the vicinity of the saddle node threshold the membrane integration time is long (thus the membrane depolarizes slowly), yet the membrane is most sensitive to small size inputs. Farther away from the saddle node, the velocity of depolarization increases, reducing the integration time. However the membrane becomes insensitive to inputs because the internal dynamics dominate. The velocity of motion slows once again in the vicinity of the steady state, and the model slowly repolarizes with the membrane once again becoming sensitive to inputs. Thus given a small supra threshold input (e.g. a single epsp) the neuron will slowly depolarize then producing a fast spike, with a larger input (e.g. several epsp's in quick su ccession), the neuron depolarized faster and the spike onset is advanced.

The above argument holds for a type I membrane in the excitable regime. As the bias current is increased past a critical value, the model switches into an intrinsically oscillating regime, where the motion around the phase plane is more uniform, and the

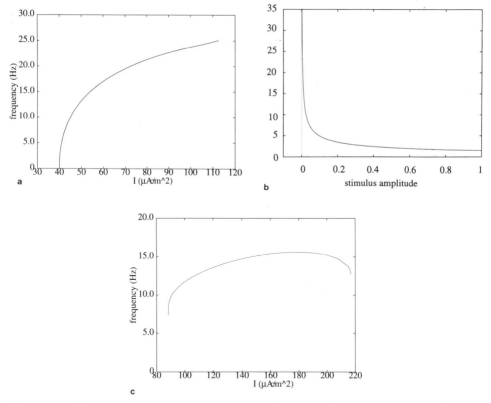

Figure 1: a. Firing frequency to input current (f-i) plot for type I membrane shows oscillations appearing with arbitrarily low frequencies. b. The delay to spike in the type I model depends on the amplitude of the supra-threshold pulse stimulus c. Firing frequency to input current plot for the type II shows oscillations arising with non-zero frequency. The type I membrane model is the $\theta - neuron$, while type II dynamics are illustrated by a Morris-Lecar model with parameters as in [Rinzel, 1989]

sensitivity to the timing of inputs is reduced. We should note that a number of well studied cortical cell models exhibit type I dynamics (e.g. Connors model, Traub model, Bower model).

Type II neurons on the other hand are characterized by discontinuous f-icurve with the oscillations arising with a non-zero frequency and exhibiting a rather limited range of firing frequencies. There is also no threshold for the appearance of spikes, which arise instantaneously but amplitude dependent on the size of the pulse stimulus. The oscillations appear through a sub-critical Hopf bifurcation appearing with a non-zero frequency.

REDUCTION TO Θ-MODEL

While complete details of the mathematical reduction by which the θ-neuron is derived are beyond the scope of this report, we shall present an outline of the process. For a more detailed description see [Ermentrout, 1996].

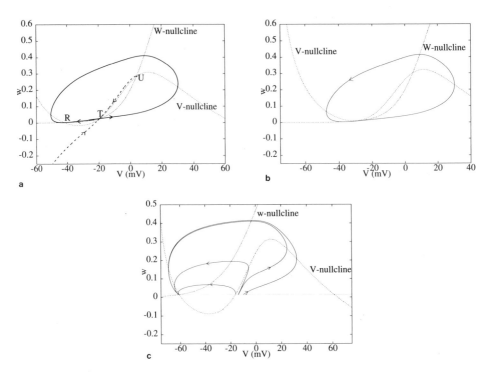

Figure 2: Phase plane for Type I and Type II neural membranes. Here we use the Morris-Lecar model as an example, with w being the activation variable and parameters set as in [Rinzel 1989] a. Phase plane for a Type I membrane in the excitable regime. The stimulus induced processions around the phaseplane are of constant size. However the rise time of spikes depends on stimulus amplitude. b. Phaseplane for Type I membrane in oscillatory regime, once again the spike are of constant amplitude. The voltage nullcline has been lifted by the added constant bias current. c. Phaseplane for Type II membrane in the excitable regime. The spikes are of variable amplitude and there isno delay.

The reduction relies on the fact that for Type I membranes the route to oscillations is through a saddle node bifurcation with the amplitude of the limit cycle, remaining largely invariant. Heuristically, this is why we can describe the behavior of the neuron near the onset of oscillations by a phase variable. In mathematical terms, the saddle node corresponds to a passage through a zero eigenvalue for the linearized dynamical system in the neighborhood of the bifurcation. For the reduction let us consider a generic conductance model:

$$dV/dt = F_0(V) + \epsilon^2 N(V) \tag{2}$$

Here $F_0(V)$ is the non-linear function that includes the membrane properties of the conductance model, and $N(V)$ is the input. We shall take $N(V) = I$ and ϵ small. We assume that when $\epsilon = 0$, there exists an invariant circle with a single fixed point that persists on both sides of the criticality. Let the saddle node disappear at the critical values of the voltage V^* (corresponding to $I = I_{critical}$). We expand the $F_0(V)$ around that value. We note that the Jacobian of $F_0(V)$ at V^* has a zero eigenvalue and call the eigenvector corresponding to this eigenvalue $\vec{e}$. Letting $V = V^* + \epsilon * z * \vec{e}$, the dynamics of (3) near the bifurcation are governed by the standard form for a saddle node bifurcation:

$$dz/dt = \epsilon(\beta + qz^2) + h.o.t. \tag{3}$$

Where β is proportional to the bias current and q summarizes the details of the active currents. Both of these can be computed directly from conductance models and are in general functions of time and voltage. We now make a change of coordinates $\tau = \epsilon t$ and $z = \tan(\theta/2)$ arriving at

$$d\theta/d\tau = q(1 - \cos\theta) + \beta(1 + \cos\theta), \theta \in [0, 2\pi], \theta(0) = \theta(2\pi) \tag{4}$$

IN VITRO AND *IN VIVO* FIRING PATTERNS OF PYRAMIDAL NEURONS

Here we review *in vitro* properties of cortical neurons that our model reproduces. In particular we are concerned with input/output properties observed in intracellular experiments. One of the more noticeable characteristics of cortical non-bursting neurons (both the pyramidal cells, as well as interneurons) is the appearance of long delays from the onset of near-threshold stimulus to the action potential [McCormick et al. 1985]. Such a delay is subserved by a slowly activating outward currents, such as the A-current [Gutnick Crill1995]. The delay to spike is reproduced by our model and is important for generating the in vivo statistics of firing. This delay has been generally presented as a instantaneous firing frequency to current (f-i) curve with high gain (see [Troyer Miller 1996]). That is the stimulus driven repetitive activity is seen to arise with arbitrarily low frequencies with the f-i curve showing a near square-root dependence of the frequency on the stimulus amplitude. A number of detailed compartmental models of the cortical neuron, one that include the tell-tail slow conductances, exhibit such f-i curves [Bush Sejnowski, 1996]. When examined as non-linear dynamical systems these models are classified as type I by the classification introduced by Rinzel et al., [1989] and whose properties we briefly reviewed above.

The results of the *in vitro* experiments are highly reproducible, with neurons exhibiting virtualy no variability in the spike train from trial to trial. This is in marked contrast with *in vivo* observations [Holt et al., 1994]. In fact cortical spike trains exhibit a puzzling level of variability and histograms with long tails and coefficients of variation (cv) near unity have been reported in a number of experimentalconditions [Noda et al., 1970, Burns et al., 1976, Softky et al., 1993, Dean et al., 1981, McCormick et al., 1985]. A lively debate has ensued in search of a mechanism that could account for the high ISI variability, and reconcile such with the information coding properties of cortical neurons. Most of the previous work focused on requirements for the statistics of input currents and consequences for the information processing role of single cells, e.g. temporal integration vs. Coincidence detection. Unlike in previous work, which rely on complicated compartmental models [Softky et al., 1994, Bell et al., 1995] , or networks of *IFN* neurons with a balance of excitation and inhibition [Usher et al., 1995] , we shall show that our simple model can account for the high ISI variability under noisy input without the balance of excitation and inhibition.

DETERMINISTIC DYNAMICS OF Θ-NEURON REPRODUCE *IN VITRO* DATA

The deterministic behavior of the model has been described in detail in [Ermentrout,

1996]. The dynamics of the model are determined by the constant bias current, which acts as a control parameter to move the system through its bifurcation. Since the formal reduction described above is valid near the criticality we consider values of the bias parameter on a small interval near zero. For negative values of bias the model has an unstable node and a stable node on an invariant circle (here the circle is the action potential trajectory which for type I models has a constant shape), see Fig. 3). The unstable node acts as a threshold and the model remains at the rest state unless it is perturbed beyond the threshold. In case of a subthreshold input the model returns quickly to the rest state. On the other hand a supra-threshold impulse stimulus induces a procession away from the unstable node and around the circle back to the stable node, producing a "spike". We call this regime the excitable regime. The non-linear phase dynamics reflect the combine absolute and relative refractory periods of the model neuron as well as non-linear summation of epsp's and intrinsic all-or-none spikes. The θ-neuron also shows the delay-to-spike sensitivity to stimulus amplitude characteristic of type I conductance models. Thus, just like in the *in vitro* cortical preparations the θ-neuron produces long delays from the onset of the supra-threshold stimulus, or the occurrence of a supra-threshold pulse stimulus (e.g. an epsp) to the correspond ing spike.

Spontaneous oscillations arise in the model as the bias current is increased past zero; this is the oscillating regime. The mathematical mechanism underlying the onset of the oscillations is identical to the type I membrane conductance models. With the bias nearing zero the two critical points move closer, at $\beta = 0$ these coalesce to form a saddle node with an infinite period homoclinic orbit around the circle. As the bias is moved past zero the saddle node disappears leaving limit circle with a finite period of the oscillation. The amplitude of the positive bias determines the period of the oscillation. Thus oscillations arise with arbitrarily low frequencies and the θ-neuron f-i curve reproduces qualitatively the cortical f-i curves without any parameter fitting. Quantitative agreement for the experimental and model f-i curves can be easily obtained.

DYNAMICS OF THE STOCHASTIC MODEL ACCOUNT FOR *IN VIVO* SPIKING STATISTICS

Modelling the random inputs

We include a noise term in the input term of the θ-neuron to model the influence of a large number of excitatory and inhibitory inputs of random strength. We suggest that an additive Gaussian white noise in the right hand side of the phase equation fails to reflect either the non-linear summation of epsp's or the periodic nature of the phase θ. Also the input term should reflect the observation that the neuron is most sensitive to its inputs at rest but insensitive during the spike, when the voltage dynamics are dominated by spike generating currents, and during the refractory period the follows immediately after the spike. Thus we want the inputs in the θ-neuron to have the most effect on the phase θ when the cell is close to the resting potential and little, or no effect when the cell is traversing through the spike. In fact we see from the reduction method that the θ-neuron equivalent of the additive outside input N(V) in the full model (2)

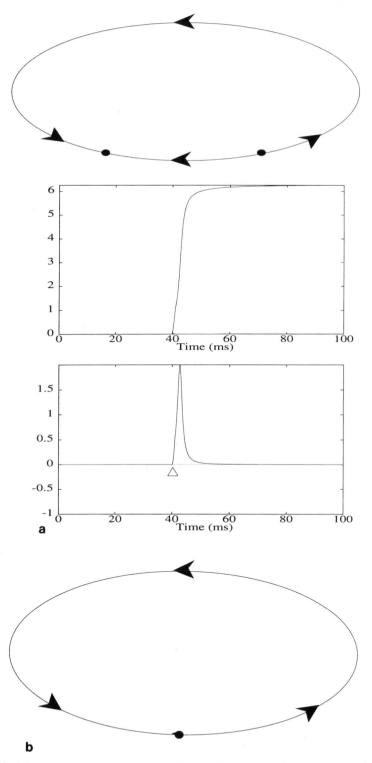

Figure 3: Saddle node dynamics on an invariant circle a. excitable regime, stimulus marked by triangle in bottom graph b. bifurcation with saddle node point. c. oscillatory regime The phase plane is in the top graph, behavior of θ on $[0,2\pi]$ in the middle and the correspondng "spikes" in $(1+cos\theta)$ in the bottom

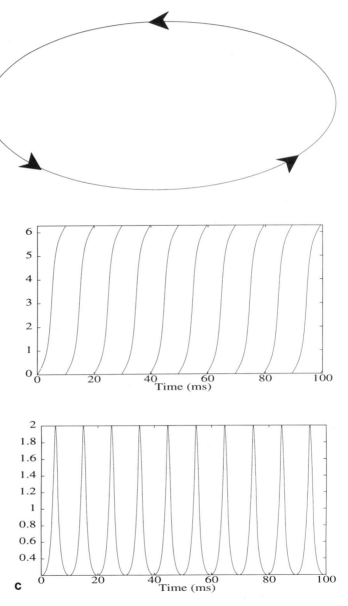

Figure 3: continued.

is given by the $\beta(1 + cos(\theta))$ term. Then clearly by extending this term to include the random white noise inputs,

$$\beta + \sigma * W_\tau)(1 + cos(\theta) \tag{5}$$

we arrive at the appropriate model for the random inputs. The o. d. e. for the phase (4) then becomes the Langevin d.e.

$$d\theta/dt = q(1 - \cos\theta) + (1 + \cos\theta)(\beta + W_\tau), \theta \in [0, 2\pi], \theta(0) = \theta(2\pi) \tag{6}$$

The noise in (5) has the variance that is dependent on the phase θ with a peak at the rest value $\theta = 0$ and a zero at $\theta = \pi$, the top of the spike. Note that the specific model for the noise inputs is motivated purely by the reduction method and not by an ad hoc fitting to the desired high c.v. behavior described below.

Similar arguments would lead to the following model for Poisson timed excitatory and inhibitory inputs (N_{ex} and N_{in}, each parameterized by an arrival rate) with constant amplitudes g_{ex} and g_{in} :

$$d\theta/dt = (\beta * (1 - cos\theta) + (1 + cos\theta)(\beta + (g_e N_{ex} - g_i N_{in}), \theta \in [0, 2\pi], \theta(0) = \theta(2\pi) \quad (7)$$

Noise modulated oscillations

At super-critical bias values, $\beta > 0$, the stochastic θ-neuron is in the oscillatory regime. It exhibits dynamics that are qualitatively similar to the deterministic behavior, with the noise modulating the mean frequency of oscillation toward higher values. In this regime the firing of the model is regular with low cv's and short tailed ISI histograms. A more formal analysis of the stationary probability distribution as well as the noise induced frequency shift will be presented elsewhere.

Noise induced firing in the excitible regime: high ISI variability

At the critical value of the bias, $\beta = 0$ firing is induced purely by the noisy inputs. This behavior persists into the excitable regime, $\beta < 0$ where the corresponding deterministic system remains quiescent. ISI histograms of the noise driven system in this regime show a characteristic peak. That is, for a given β the noise induces a characteristic mean ISI for the spike train process. The location of the peak in the ISI histogram is controlled both by the values of the constant bias and the variance of the noise process. That is, as bias moves to more positive value or σ increases the peak in the ISI histogram moves to the left (see fig. 4). The a bias changes both the location of the peak and the scale of the ISI histogram, with more negative bias values producing histograms with longer tails. Thus the coefficient of variation decreases as bias moves toward positive and the model moves out of the excitable regime and into the noise modulated oscillations.

On the other hand with $\beta < 0$ and fixed, increasing the variance of the noise inputs has no effect on the scale of the histogram while moving the location of the mode in ISI histogram to the left. As can be seen in Fig. 4, for β=-1, the high noise histogram has qualitatively equal mass in the tail as the low noise one. Consequently the cv remains invariant under changing noise intensity, suggesting that it is a basic property of the membrane dynamics. In fact, the cv is largely constant for wide range of firing rates in the noise induced regime, see Fig.5, but decreases linearly with change in bias toward positive (fig 5). This is in agreement with our observations for the type I Morris-Lecar and for the θ-neuron with Poisson timed discrete inputs. The bias and the noise thus play different roles, with the bias strongly influencing both the irregularity of the spike train and the mean firing rate and the noise level modulating only the firing rate.

We note here that our simulations are in line with analytical results for a general saddle node stochastic oscillator. [Siggeti, 1988].

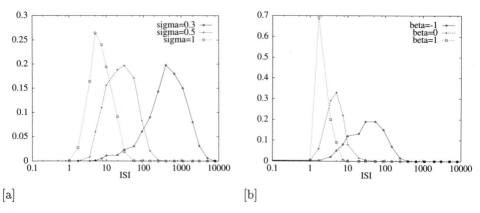

[a]   [b]

Figure 4: Normalized histograms for $\theta - neuron$, ISI's plotted on log scale. a. β=-0.3 and noise amplitude varied. b. noise amplitude=1 and β varied.

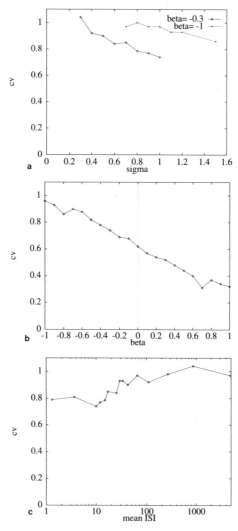

Figure 5: CV results for θ-model. a.cv remains high across wide range of noise parameters. b. cv decreases linearly with β. c. cv remains high across a wide range of firing rate, when such is controlled by noise, here $\beta = -1$, σ varied

DISCUSSION

The strength of the above results is that with a generic model, based on one major assumption about the membrane dynamics of the cell, we can account for the *in vitro* response characteristics of non-bursting cortical neurons as well as *in vivo* statistics of cortical spike trains. In fact we believe this is the first time a one dimensional intrinsically spiking model has been presented. In contrast to the integrat e-and-fire neurons, which are ad hoc approximations of more complex models, we provide a formal mathematical method to derive the θ-neuron from conductance based models, and thus can give specific physiological meaning to the parameters. Our reduction method makes it clear why multiplicative, not additive noise is appropriate for inputs in this situation. Interspike interval adaptation is not included in our model, but can be by coupling a slow time scale equation, making the model 2-dimensional. A similar approach can include a mechanism for bursting. The model also suggests the observable quantity which allows spikes to be observed in the model dynamics. The θ-neuron has been previously used to study phase dynamics of networks of type I oscillators, showing a novel mechanism for appearance of synchrony in cortical net works [Ermentrout, 1996].

REFERENCES

A. Bell, Z. Mainen, M. Tsodyks, and T. Sejnowski, "Balancing" of conductances may explain irregular cortical spiking Institute for Neural Computation, UCSD, San Diego (1995) Technical report INC-9502

B.D. Burns, The spontaneous activity of neurones in the cats visual cortex. Proc. R. Soc. London (Biol.) 194:211 (1976)

P. Bush, and T. Sejnowski, Inhibition synchronises sparsely connected cortical neurons within and between columns in realistic network models. J. Comp. Neurosci.3:91 (1996)

A. Dean, The variability of discharge of simple cells in the cat striate cortex. Exp. Brain Res. 44:437

D. Jaeger, E. De Shutter, J.M. Bower, The role of synaptic and voltage-gated currents in the control of purkinje cell spiking: a modelling study. J. Neurosci 17:91 (1997)

G.B. Ermentrout, J.D. and Cowan, Large Scale Spatially Organized Activity in Neural Nets. SIAM J. Appl. Math., 38 (1980)

G.B. Ermentrout, Reduction of conduction based models to neural networks. Neural Computation 6:679 (1994)

G.B Ermentrout, Type I Membranes, Phase resetting Curves, and Synchrony. Neural Computation 8:979 (1996)

M.J. Gutnick, W. Crill, The cortical neuron as an electrophysiological unit. in: *The Cortical Neuron* , M. Gutnick, and I. Mody, eds., Oxford Unitversity Press, Oxford, UK (1995)

G.R. Holt, W.R. Softky, K. Koch, R.J. Douglas, A comparison of discharge variability *in vitro* and *in vivo* in cat visual cortex neurons. pre-print (1994)

D.A. McCormick, B.W. Connors, J.W. Lighthall, D.A. Prince, Comparative electrophysiology of pyramidal and sparsely spiny stellate neurones of the neocortex. J. Neurophysiol. 54:782 (1985)

H. Noda, R. Ada., Firing variability in cats association cortex during sleep and wakefullness Brain Res. 18:513 (1978)

J. Rinzel and G.B. Ermentrout, Analysis of neural excitability and oscillations in *Methods in Neuronal Modelling*, K. Koch and I. Segev eds., MIT Press, Cambridge, Mass. (1989)

I. Segev, J.W. Fleshman, R.E. Burke, Compartmental models of complex neurons. in *Methods in Neuronal Modelling*, K. Koch and I. Segev eds., MIT Press, Cambridge, Mass. (1989)

M.N. Shadlen and W.T. Newsome Noise, neural codes and cortical organization Current Opinions in Neurobiology 4 :569 (1994)

D.E. Sigetti, *Universal Results for the Effects of Noise on Dynamical Systems* Doctoral Thesis, University of Texas at Austin, (1988)

W. Softky and K. Koch, The highly irregular firing of cortical cells is inconsistent with temporal integration of random EPSPs. J. Neurosci. 13:334 (1993)

T.W. Troyer and K.D. Miller, Physiological gain leads to high ISI variability in a simple model of a cortical regular spiking cell. Neural Computation in press

M. Usher, M. Stemmler, C. Koch, and Z. Olami, Network amplification of local fluctuations causes high spike rate variability, fractal firing patterns and oscillatory local field potentials Neural Computation 6-5:795 (1994)

C. van Vreeswejk and H. Sompolinsky, Irregular spiking in cortex through inhibition/excitation balance. poster presented at Computational Neural Systems Conference '96 Cambridge, Mass. (1996)

A SIMULATION OF GROWTH CONE FILOPODIA DYNAMICS
BASED ON TURING MORPHOGENESIS PATTERNS

Tim A. Hely, Arjen van Ooyen, David J. Willshaw

Centre for Cognitive Science
University of Edinburgh
2 Buccleuch Place
Edinburgh

INTRODUCTION

The neuronal growth cone is a dynamic, "shape changing" structure which guides the developing neurite to a distant target (Figure 1). The growth cone membrane is constantly creating **filopodia**, long, thin structures which grow and shrink into the extra–cellular space. The exact causes of filopodial excursions are unknown. However, experimental work by Davenport (1992) linked filopodial outgrowth to the local concentration of calcium, which Hentschel (1994) proposed as a morphogen regulating neuronal dendrite growth. We suggest that calcium acts as a morphogen to directly regulate the pattern of filopodial outgrowth and subsequent retraction. Turing (1952) developed a mathematical basis for **"morphogenesis"**, the process underlying the development of the shape of an organism. This provides an appropriate framework for modelling the neuronal growth cone.

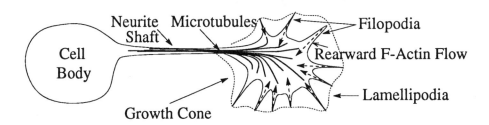

Figure 1. The growth cone.

TURING PATTERNS

Turing patterns can occur when two (or more) chemicals diffuse and interact. The two chemicals are normally an activator A and inhibitor B. A activates itself and B. B inhibits itself and A. When diffusion is included this can be written:

$$\frac{\delta A}{\delta t} = k_1 A - k_2 B + D_A \nabla^2 A \qquad \frac{\delta B}{\delta t} = k_3 A - k_4 B + D_B \nabla^2 B \qquad (1)$$

k_1, k_2, k_3, k_4 are constants, and D_A and D_B are the diffusion constants. For pattern formation to occur, the activator A must diffuse slower than the inhibitor B. The events leading up to pattern formation are as follows (see Figure 2).

(1) The system is at steady state. (2) A random perturbation leads to an increase in activator A. (3) Inhibitor B increases. If there is no diffusion this halts the process. (4) The inhibitor diffuses faster than the activator. (5) The activator peak grows (6) Surround inhibition leads to a characteristic pattern width.

Turing patterns have been used to explain animal coat patterns such as tiger and zebra markings, seashell stripes, and drosophila wing prepatterns. The advantage of using Turing equations for this simulation is that they are able to generate a large number of different behaviours - stripes, blobs, oscillations, moving waves - with only a small change needed in any one of the parameters.

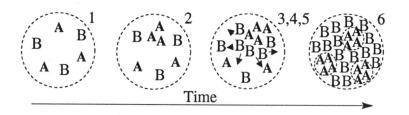

Figure 2. Reaction–diffusion events leading to pattern formation.

THE BIOLOGY OF THE GROWTH CONE

The growth cone is a structure which develops at the tip of the neurite to guide the growing axon to its target. When a growth cone reaches its correct target, it halts, and remodels itself to form a synapse. Figure 1 shows the major regions of the growth cone, the central, transitional and peripheral zones.

The central zone is dominated by microtubules, a polymerised protein made up of tubulin monomers which is important in giving structural support to the axon. Microtubules in the growth cone undergo rapid periods of growing and shrinking known as dynamic instability. The peripheral domain has few microtubules present and a high concentration of the **F–Actin** polymer. Actin normally polymerizes at the leading edge of the growth cone, and then flows rearward towards the central zone at a rate of 3–6 μm per minute. Growth cone advance is inversely proportional to retrograde F–Actin flow. Greater adhesion to the substrate changes the default retrograde movement into a forward movement of the growth cone (Lin, 1995). Filopodia growth and shrinkage is then due to localised variations in the binding of the F–Actin network to the substrate.

CALCIUM AS A MORPHOGEN

Calcium regulates many proteins which alter the F–Actin network. These include α-actinin an actin bundling protein, and gelsolin, an actin severing protein. Localised changes in growth cone calcium concentration alter F–Actin dynamics and consequently filopodial behaviour. Growth occurs with a bell shaped dependency on calcium, peaking at a certain calcium concentration (Mattson, 1987) (Figure 3).

An increase in calcium can be due to either calcium influx across the membrane or calcium induced calcium release (CICR) from internal stores (Davenport 1992;1996). Calcium can be viewed as an activator as both mechanisms result in rapid filopodial extension from the membrane surface. A candidate molecule for an inhibitor is cAMP. The diffusive constant of cAMP $\approx 4.6 \times 10^{-7} cm^2 s^{-1}$ is faster than that of calcium $\approx 10^{-8} cm^2 s^{-1}$ (Safford, 1977). Thus calcium could act as a short range activator and cAMP as a long range inhibitor in the growth cone. Previously Goodwin (1985) proposed cAMP and calcium as a possible reaction-diffusion pair in the marine algae Acetabularia.

THE SIMULATION

The model chosen to simulate calcium and cAMP concentrations uses simplified equations based on a standard pattern generator developed by Gierer and Meinhardt (Edelstein, 1988, p. 531), with similar properties to (1).

$$\frac{\delta A}{\delta t} = \frac{k_1 A^2}{B} - k_2 A + D_A \Delta^2 A \qquad \frac{\delta B}{\delta t} = k_3 A^2 - k_4 B + D_B \Delta^2 B \qquad (2)$$

In the equations, A=calcium, B=cAMP. Calcium activates itself and cAMP with rate constants k_1 and k_3. A rise in calcium levels due to CICR leads to a rise in cAMP levels, which inhibit further calcium increase. Both equations have inhibitory decay terms, with constants k_2 and k_4. (For calcium this can also be viewed as a term describing membrane influx and pumping). The equations are implemented on a finite element grid with e.g. 100×100 pixels, representing a square of side 10 to 100 μm (Figure 4). At the start of the simulation the growth cone is circular, with diameter half the grid size. Initially, all pixels within the growth cone have steady state concentration values for calcium and cAMP with $\pm 1\%$ random noise added. Changes in the values of calcium and cAMP are calculated and updated at each time step.

Pixels outside the growth cone represent the substrate and are initially empty. Pixels on the growth cone membrane have a moveable position co-ordinate which allows them to expand radially into the empty pixels. Calcium and cAMP can then diffuse into the new space. Maximum outgrowth of 6 μm/minute occurs at a concentration of 200 μM calcium. At very low and very high calcium concentrations, the flow is negative, representing lack of F–Actin coupling to the substrate. In the present simulation, outgrowth and shrinkage only occurs from a fixed number of membrane pixels, while the remainder are static.

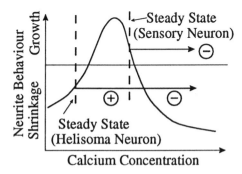

Figure 3. Calcium-dependent outgrowth.

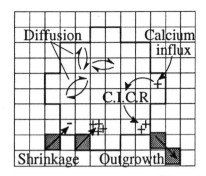

Figure 4. The model simulation.

RESULTS

Figure 5 shows the results of a simulation with $k_1=k_2=k_4=1$, $k_3=0.01$, $D_A = 10^{-8}cm^2s^{-1}$, $D_B = 10^{-7}cm^2s^{-1}$. Initially the growth cone has a calcium concentration of 100nM and filopodia sprout in all directions. After 40 seconds, a hot spot begins to develop, leading to extended growth in this area, and after 90s filopodia have retracted at the opposite pole.

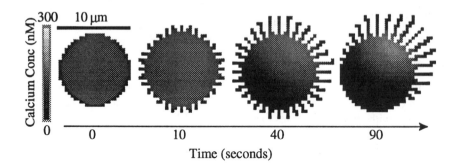

Figure 5. Time sequence of calcium concentration leading to stable pattern formation.

The growth cone pattern can be altered by increasing the size of the cell (Figure 6) or by altering the decay parameters k_2, k_4, and keeping the size fixed. In the present model, the width of the calcium hotspots determines both where the filopodia sprout, and their maximum extension. Explicitly modelling the calcium influx from the external medium would alter the maximum filopodia extension. As filopodia have a large surface membrane to volume ratio, a small calcium influx significantly affects the calcium concentration. Including this influx in the model may lead to narrow spacing of filopodia in the growth cone with long extensions.

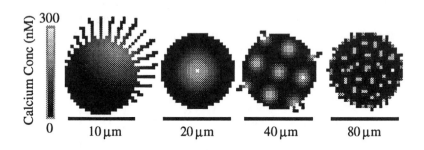

Figure 6. Stable patterns obtained at increasing length scales.

In both Figures 5 and 6, the patterns are stable once the hotspots are established, and existing filopodia remain extended. Continuous extension and retraction of the filopodia would require an underlying calcium concentration with temporal and spatial variations. Unstable oscillations generated by the system are shown in Figure 5 ($k_1=k_2=150, k_3=1$, $k_4=100, D_A=2.5 \times 10^{-8}cm^2s^{-1}$). These oscillations are too rapid for stable outgrowth to occur but similar variations over a timescale of minutes would lead to filopodial extension and retraction.

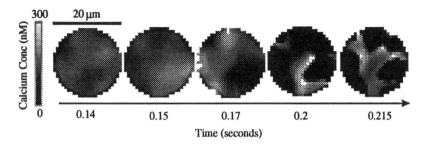

Figure 7. Time sequence of rapid, instable patterns.

CONCLUSION

This is the first model to simulate the causes of the "random and spontaneous" excursions of neurite growth cone filopodia. The simulation demonstrates that any mechanism which results in a slowly varying temporal and spatial calcium pattern would alter the local neurite geometry, resulting in filopodial creation, outgrowth and shrinkage. In the current model, the number of membrane points allowed to develop into filopodia are limited. Although it is possible that biological factors such as local calcium stores, and radially oriented micro-tubules may create discrete sites for filopodia creation, it is more likely that filopodia could develop continuously along the membrane. We aim to extend the present model to include this behaviour. This model provides a starting point for understanding filopodia dynamics from a new computational perspective.

ACKNOWLEDGEMENTS

Tim Hely thanks the Wellcome Trust for financial support.

REFERENCES

Davenport, R.W. and Kater, S.B., 1992, Local increases in intracellular calcium elicit local filopodial responses in Helisoma neuronal growth cones, *Neuron* 9:405.

Davenport, R.W., Dou, P., Mills, R.L., and Kater, S.B., 1996, Distinct calcium signalling within neuronal growth cones and filopodia, *J. of Neurobiology* 31:1.

Edelstein-Keshet, L., 1988, *Mathematical Models in Biology*, Birkhauser Math. Series.

Goodwin, B.C., and Trainor, L.E.H., 1985, Tip and whorl morphogenesis in Acetabularia by calcium-regulated strain fields, *J. Theor. Biol.* 117:79.

Hentschel, H., 1994, Instabilities in cellular dendritic morphogenesis, *Ph. Rev. Lett.* 73:3592.

Lin, C.H., and Forscher, P., 1995, Growth cone advance is inversely proportional to retrograde F-actin flow, *Neuron* 14:763.

Mattson, M.P., and Kater, S.B., 1987, Calcium regulation of neurite elongation and growth cone motility, *The J. of Neuroscience* 7:4034.

Safford, R.E., and Bassingthwaite, J.B., 1977, Calcium diffusion in transient and steady states in muscle. *Biophys. J.* 20:113.

Turing, A., 1952, The chemical basis for morphogenesis, *Phil. Tr. Roy. Soc. Lon.* 237:37.

ORGAN FUNCTION AND CELL BEHAVIOUR: SIMULATING DISTURBANCES IN VENTRICULAR PROPAGATION

A V Holden[1], G P Kremmydas[1,2], and A Bezerianos[2]

[1]Computational Biology Group, CNLS,
University of Leeds, Leeds LS2 9JT, England
[2]Department of Medical Physics, Medical School,
University of Patras, Patras 261 10, Greece

INTRODUCTION

Applications of nonlinear dynamics to mathematical cardiology and cardiac electrophysiology use a range of different models – circle and interval maps, differential and partial differential systems, and spatially discrete models such as cellular automata and coupled map lattices. These different mathematical structures are used to model phenomena in a structural hierarchy, from intracellular phenomena, through single isolated cells, small pieces of tissue, and in perfused and in vivo heart. They also model behaviour in a functional hierarchy, where the rhythmic beating is produced by propagating electrical waves in anisotropic, heterogeneous tissue built up from coupled contractile cells, whose mechanical activity is triggered by intracellular calcium fluxes evoked by the electrical waves.

Our approach to integrative physiology (Holden *et al.* 1995, Panfilov & Holden 1997) is computational and constructive: understanding the behaviour of a physiological system involves understanding the behaviour of current models of that system, which, given the complexity of biological systems, must be computational models.

In this paper we illustrate the construction of a computational model for the mammalian ventricles. The model is in the form of a coupled map lattice whose geometry is derived from anatomical measurements of ventricular fibre position and orientation. This geometric model allows us to simulate the normal and abnormal propagation of waves of electrical activity in the heart. The precise time course and properties of the excitation waves requires more detailed, membrane excitation equations that are typically high order, stiff nonlinear differential systems derived from extensive voltage clamp investigations.

Information Processing in Cells and Tissues
Edited by Holcombe and Paton, Plenum Press, New York, 1998

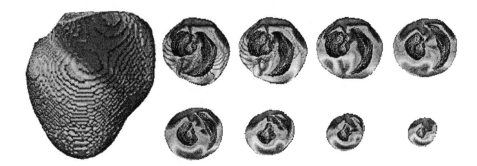

Figure 1. Ventricular geometry is represented by the anterior view of the ventricles, together with a series of slices through the ventricular chambers. The muscle fibre orientation is colour coded as the cosine of the local muscle fibre projection onto one Cartesian coordinate; the spiral arrangement of the muscle fibres, and the changes in fibre orientation going from the endo to epicardial walls is apparent.

Membrane excitation equations are of the form:

$$C\partial u/\partial t = f(u, v, w)$$
$$\partial v/\partial t = g(u, v, w)$$
$$\partial w/\partial t = h(u, v, w)$$

where $u = u(t)$ is the transmembrane voltage, C is specific membrane capacitance, f is transmembrane current density, vector $v = v(t)$ describes the fast gating variables, and vector $w = w(t)$ comprises slow gating variables and intra- and extra-cellular ionic concentrations, and g and h describe their kinetics. The variables u and v have comparable characteristic times.

Boyett *et al* (1997) review the origin and properties of these equations; a key point is that they are continually being upgraded to incorporate new experimental information. As an example, one form of long QT syndrome (LQT3) has been identified with specific mutations in the gene encoding the cardiac voltage-gated Na^+ channel; the effects on ventricular action potentials and progaagtion can be computed by incorporating the resultant changed Na^+ kinetics into the membrane excitation equations (Biktashev & Holden 1996a). Here we combine geometry and anisotropy in propagation velocity with local excitation equations by coupling nodes in the CML model to membrane excitation equations appropriate for that region of the ventricle. This allows us to combine an anatomical overview of propagating activity in a whole ventricle model with a detailed description of regional differences in cellular behaviour.

HEART GEOMETRY AND ANISOTROPY

We now begin by considering a static data set that describes the shape, size and fibre orientation of a dead (non-contracting) dog ventricle.

Heart geometry

We derive a set of model lattice points from experimental ventricular data supplied

Figure 2. The normal pattern of activation of the ventricles is endocardial, via the Purkině fibre conducting system. The exploded views show the full cardiac cycle from depolarisation to repolarisation. The first two panels show the spread of activation from the endocardial to the epicardial surface while the following three panels correspond to the ventricular action potential which lasts a few hundred ms. The last two panels show the gradual repolarisation of the ventricles.

by P J Hunter of the University of Auckland; see (Nielsen *et al.*, 1991). We define a straightforward discrete integer approximation $\mathcal{H} \subseteq \mathbf{Z}^3$, such that each lattice point $A_{i,j,k} \in \mathcal{H}$ approximates a cuboid of muscle of side 1 millimetre, containing some 10^3 cells. Let the set $\mathcal{H}$ also index a set of network inputs that supply external stimuli. The set H contains in the order of 10^5 points, and is illustrated in Fig. 1.

To each point $A_{i,j,k} \in \mathcal{H}$, we assign an internal neighbourhood $\mathcal{N}_{i,j,k}$ of orthogonal nearest neighbours by

$$
\begin{aligned}
\mathcal{N}_{i,j,k} = \ & \{A_{i,j,k}, A_{i-1,j,k}, A_{i+1,j,k}, A_{i,j-1,k}, A_{i,j+1,k}, A_{i,j,k-1}, A_{i,j,k+1}, A_{i+1,j+1,k}, \\
& A_{i-1,j-1,k}, A_{i+1,j-1,k}, A_{i-1,j+1,k}, A_{i,j+1,k+1}, A_{i,j-1,k-1}, A_{i,j+1,k-1}, A_{i,j-1,k+1}, \\
& A_{i+1,j,k+1}, A_{i-1,j,k-1}, A_{i+1,j,k-1}, A_{i-1,j,k+1}, A_{i+1,j+1,k+1}, A_{i-1,j-1,k-1}, \\
& A_{i+1,j+1,k-1}, A_{i-1,j-1,k+1}, A_{i+1,j-1,k-1}, A_{i-1,j+1,k+1}, \\
& A_{i-1,j+1,k-1}, A_{i+1,j-1,k+1}\} \cap \mathcal{H}.
\end{aligned}
$$

Intuitively, a point representing a cuboid of myocardium deep within the heart wall will have twenty seven such neighbours, whereas a point representing a cuboid near the epi- or endo-cardial surfaces may have fewer. We shall assume, however, that the set $\mathcal{H}$ is such that all points have at least three neighbours. We may enumerate these neighbourhoods using maps $p : \mathcal{H} \to \mathbf{N}$ and $\alpha : \mathcal{H} \times \mathbf{N} \to \mathcal{H}$ defined, for all $A_{i,j,k} \in \mathcal{H}$, such that $p(i, j, k) = |\mathcal{N}_{i,j,k}|$ and

$$
\mathcal{N}_{i,j,k} = \{\alpha(i, j, k, 1) = (i, j, k), \dots, \alpha(i, j, k, p(i, j, k))\}.
$$

We define the external neighbourhood of each point $A_{i,j,k} \in \mathcal{H}$ simply by

$$
\mathcal{E}_{i,j,k} = A_{i,j,k}
$$

which is trivially enumerated by maps $q : \mathcal{H} \to \mathbf{N}$ and $\beta : \mathcal{H} \times \mathbf{N} \to \mathcal{H}$, defined such that $q(i, j, k) = 1$ and $\beta(i, j, k, 1) = (i, j, k)$.

Fibre orientation

For isotropic tissue a simple diffusive CML approximation for the Laplacian with a constant diffusion coefficent would be used, as in (Biktashev and Holden, 1995, 1996a,b).

However, propagation velocity in cardiac tissue is anisotropic; this anisotropy is due to lower coupling resistance between the ends of approximately cylindrical cardiac cells, compared to the transverse coupling resistance. Thus the diffusion coefficient for a partial differential equation model, or the coupling coefficient for a CML model, depends on the local fibre orienation. Hunter's experimental data contains direction cosines for cardiac muscle fibre orientation. Fibre orientation determines the strength of coupling between neighbours, and hence anisotropy in the conduction velocity of the muscle. The smaller the angle fibres make between neighbouring points, the higher the coupling strength between them. Let $\varepsilon_{min}, \varepsilon_{max} \in \mathbf{R}$ be the minimum and maximum coupling strengths for neighbouring points perpendicular to fibres and parallel with fibres, respectively.

Consider, for example, point $A_{i,j,k} \in \mathcal{H}$, and the neighbouring point $A_{i+1,j,k}$ in the direction of the x axis. Let the muscle fibres make an angle of φ_x with the x axis at point $A_{i,j,k}$. We define the *coupling coefficient* $\varepsilon_{i+1,j,k} \in \mathbf{R}$ from $A_{i+1,j,k}$ to $A_{i,j,k}$ by

$$\varepsilon_{i+1,j,k} = \varepsilon_{min} + (\varepsilon_{max} - \varepsilon_{min})(\cos \varphi_x)^2.$$

We see that, for $\cos \varphi_x = 0$, the coupling strength from $A_{i+1,j,k}$ to $A_{i,j,k}$ is the minimum value, ε_{min}, and for $\cos \varphi = \pm 1$, the coupling strength is the maximum value, ε_{max}. Generalising this to all orthogonal neighbours of each point $(i,j,k) \in \mathcal{H}$ using all three fibre direction cosines φ_x, φ_y and φ_z determines the following coupling coefficients:

$$\varepsilon_x = \varepsilon_{i-1,j,k} = \varepsilon_{i+1,j,k} = \varepsilon_{min} + (\varepsilon_{max} - \varepsilon_{min}) \cos{(\varphi_x)}^2$$
$$\varepsilon_y = \varepsilon_{i,j-1,k} = \varepsilon_{i,j+1,k} = \varepsilon_{min} + (\varepsilon_{max} - \varepsilon_{min}) \cos{(\varphi_y)}^2$$
$$\varepsilon_z = \varepsilon_{i,j,k-1} = \varepsilon_{i,j,k+1} = \varepsilon_{min} + (\varepsilon_{max} - \varepsilon_{min}) \cos{(\varphi_z)}^2$$

$$\varepsilon_{i+1,j+1,k} = \varepsilon_{i-1,j-1,k} = \varepsilon_{min} + (\varepsilon_{max} - \varepsilon_{min})(\cos(\varphi_x) + \cos(\varphi_y))^2 / 2$$
$$\varepsilon_{i+1,j-1,k} = \varepsilon_{i-1,j+1,k} = \varepsilon_{min} + (\varepsilon_{max} - \varepsilon_{min})(\cos(\varphi_x) - \cos(\varphi_y))^2 / 2$$
$$\varepsilon_{i,j+1,k+1} = \varepsilon_{i,j-1,k-1} = \varepsilon_{min} + (\varepsilon_{max} - \varepsilon_{min})(\cos(\varphi_y) + \cos(\varphi_z))^2 / 2$$
$$\varepsilon_{i,j+1,k-1} = \varepsilon_{i,j-1,k+1} = \varepsilon_{min} + (\varepsilon_{max} - \varepsilon_{min})(\cos(\varphi_y) - \cos(\varphi_z))^2 / 2$$
$$\varepsilon_{i+1,j,k+1} = \varepsilon_{i-1,j,k-1} = \varepsilon_{min} + (\varepsilon_{max} - \varepsilon_{min})(\cos(\varphi_x) + \cos(\varphi_z))^2 / 2$$
$$\varepsilon_{i+1,j,k-1} = \varepsilon_{i-1,j,k+1} = \varepsilon_{min} + (\varepsilon_{max} - \varepsilon_{min})(\cos(\varphi_x) - \cos(\varphi_z))^2 / 2$$

$$\varepsilon_{i+1,j+1,k+1} = \varepsilon_{i-1,j-1,k-1} = \varepsilon_{min} + (\varepsilon_{max} - \varepsilon_{min})(\cos(\varphi_x) + \cos(\varphi_y) + \cos(\varphi_z))^2 / 3$$
$$\varepsilon_{i+1,j+1,k-1} = \varepsilon_{i-1,j-1,k+1} = \varepsilon_{min} + (\varepsilon_{max} - \varepsilon_{min})(\cos(\varphi_x) + \cos(\varphi_y) - \cos(\varphi_z))^2 / 3$$
$$\varepsilon_{i+1,j-1,k-1} = \varepsilon_{i-1,j+1,k+1} = \varepsilon_{min} + (\varepsilon_{max} - \varepsilon_{min})(\cos(\varphi_x) - \cos(\varphi_y) - \cos(\varphi_z))^2 / 3$$
$$\varepsilon_{i-1,j+1,k-1} = \varepsilon_{i+1,j-1,k+1} = \varepsilon_{min} + (\varepsilon_{max} - \varepsilon_{min})(\cos(\varphi_x) + \cos(\varphi_y) + \cos(\varphi_z))^2 / 3$$

We also define the total coupling at each point $A_{i,j,k} \in \mathcal{H}$ by $1 - \varepsilon_{i,j,k}$ with

$$\varepsilon_{i,j,k} = \sum_{l=2}^{p(i,j,k)} \varepsilon_{\alpha(i,j,k,l)}.$$

Note that as $\cos{(\varphi_x)}^2 + \cos{(\varphi_y)}^2 + \cos{(\varphi_z)}^2 = 1$, then for a mid wall point $A_{i,j,k}$ (i.e. a point with 27 internal neighbours), $\varepsilon_{i,j,k} = 18\varepsilon_{min} + 8\varepsilon_{max}$ (which we assume to be less than 1); for a boundary point, the value of $\varepsilon_{i,j,k}$ will be smaller.

Ventricular Propagation

The coupling coefficients were chosen to give a solitary plane wave propagation velocity along the fibre axis of 0.4 ms^{-1}, corresponding to a homogenous partial differential equation model of mammalian ventricular tissue (Biktashev & Holden, 1996b)

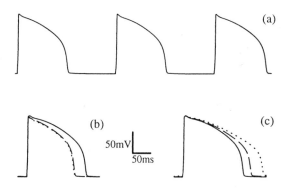

Figure 3. Oxsoft ODE timeseries for the simplified guinea pig ventricle model (Biktashev & Holden 1996): (a) Periodic behaviour at a stimulation rate of 3.7 s^{-1} (b) Reduction of action potential duration due to repetitive (5.0 s^{-1}) stimulation. The trace shows the first three action potentials superimposed. (c) Prolongation of action potential duration to simulate LQT3 syndrome (Biktashev & Holden 1996).

diffusion coefficient of 0.4125 cm^{2}s^{-1}, or intercellular (gap junction) conductance of 2.0 μS for 80 μm long cylindrical cells

Figure 2 illustrates frames from an exploded visualisation movie of ventricular activation produced by a simultaneous activation of the endocardial ventricular surfaces, using the three-variable version of Chialvo map (Chialvo 1995) at each node of the CML model of the ventricle (Holden, Poole and Tucker 1996).

THE VENTRICULAR ACTION POTENTIAL

We construct a model for mammalian ventricular tissue by incorporating ordinary differential equations for ventricular cell excitability into the CML model for ventricular geometry and anisotropy. There are a number of models available for ventricular excitation that summarize the results of voltage clamp experiments — these include the Beeler-Reuter (1989) model, the Oxsoft guinea-pig ventricular cell model specified in (Noble 1991) and the phase 2 Luo-Rudy (1994) model. None of these models are definitive, they all represent steps in an on-going process of modelling the behaviour of ventricular cells by a description of membrane currents and pumps, and intracellular ion binding and concentration changes (Noble, 1995).

In this paper we use equations of the Oxsoft guinea pig ventricle model. These equations provide a convenient starting point, as they have been extensively used to reproduce experimental results e.g. see Noble *et al.* (1991), Kiyosue *et al* (1993), LeGuennec & Noble (1994), Rice *et al* (1995). The Oxsoft equations for cell excitability are incorporated into the coupled map lattice model equation with appropriate voltage diffusion coefficient (hence, spatial scaling) selected to give an appropriate conduction velocity.

Experimental studies have illustrated electrophysiological differences between the epicardium and the endocardium of the ventricular wall (reviewed by Antzelevitch *et al* 1995). The epicardial action potential shows a typical "spike-and-dome" configuration an observation that has been confirmed in canine, rabbit and human ventricular tissue. A pronounced transient outward current I_{to} in epicardial ventricular tissue has been suggested as the underlying mechanism for the "spike-and-dome" configuration but also

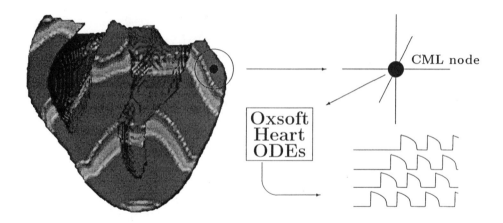

Figure 4. The anatomically accurate CML model can be coupled to current biophysical excitation equation models of ventricular activity by taking the activity at a CML node and using it to drive (as a brief, suprathreshold current pulse) the ODE model for that part of the ventricle. This coupling can be at any selected points in the heart for the CML and ODE models to run concurrently only a modest number of nodes can be coupled to ODEs.

differences in other currents ($I_{K1}, I_K, I_{Ca}, I_{Cl}, I_{K-ATP}$) have been reported. Modification of the Oxsoft Heart current kinetics allows for the reconstruction of the "spike-and-dome" configuration and such an example is shown in figure 6 where epicardial action potentials are modeled by altering I_{to}, I_K and I_{Ca} kinetics.

In a similar way the model of LQT3 ventricular cell illustrated in figure 3(c) is constructed from the Oxsoft model for guinea-pig ventricular cells, with a fraction of persistently open h-gates of the fast Na$^+$ current, equivalent to a modification of opening and closing rate coefficients (α and β) for this gate.

HYBRID MODELLING

Coupling to ODEs.

In this approach cardiac electrical activity is modeled in two different temporal and spatial scales by different dynamical systems. At the macroscopic whole heart level cardiac dynamics is described by a return map that reproduces the restitution properties of cardiac excitability. Each node on the lattice describing cardiac anatomy is diffusively coupled to its neighbouring points. Thus, propagation phenomena are described in the whole ventricle and cardiac electrical activity can be visualized as propagation of excitation waves.

At the microscopic level (either a single cell or a small cuboid of tissue) we use the *Oxsoft Heart* ODEs to describe excitability and the ionic processes underlying it.

ventricle.

The coupling of the ODE system to a single CML node is implemented by assuming that the electrical activity at a given CML node acts as an external forcing on the detailed *Oxsoft Heart* equations. Thus, an "action potential" on the CML node triggers

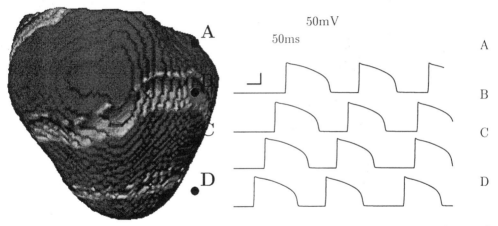

Figure 5. Simulation of quiescent whole ventricle model to periodic stimulation of apex of

an action potential in the ODE system. By this approach the dynamics of the CML can be coupled to the *Oxsoft Heart* model at any selected points on the lattice and therefore the CML description of cardiac dynamics is translated into the behaviour of an ODE detailed model.

Time Scaling.

A technical problem related to the coupling of ODE models to the CML whole-heart model is that of time scaling. The CML coupling coefficients have been chosen to match experimental conduction velocity measurements.

The *Oxsoft Heart* ODEs do not have a space scale, but have a time scale directly obtained from experiment. This gives the derived time scales of action potential duration shown in Figure 3. Such observables include the action potential duration and the absolute refractory period as well as the timecourse of ionic currents during action potential. Such information is not available in simple return maps models.

For the ODE model to be coupled to the CML activity an appropriate time scaling has to be introduced. It is obvious that appropriate time scaling can be selected so that one can match different action potential characteristics: duration, refractory period or re-entry period. In the simulations shown here the CML time is scaled so that the restitution characteristics in the two models match.

PROPAGATION DURING ARRHYTHMIA

If all the ventricular cells have the same properties (i.e. same excitation equation) then the response at any two recording sites will be the same, delayed only in time. The differences in potential with distance will drive the local circuit currents that produce propagation in the heart. Regional differences in action potential waveform and duration, as seen in endo - and epicardial action potential under the same conditions, are illustrated in Figure 6. Under the normal pattern of endo- to epicardial activation (Figure 2) and even during apical stimulation (Figures 5 and 6), these differences in

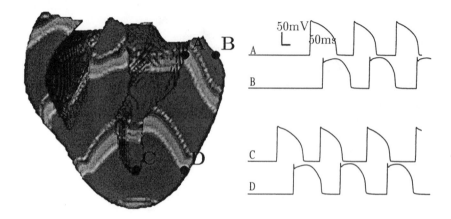

Figure 6. Endocardial and epicardial activity through a slice of the ventricle driven by apical stimulation.

Figure 7. Visualisation of re-entrant scroll wave (monomorphic ventricular tachycardia) on surface and through section of heart

action potential duration are not arrhythmogenic, as any resultant current flow is in the refractory period of the cells with the shorter action potential duration. However, prolongation of the action potential as in LQT syndrome illustrated in Figure 3c can allow these differences to be arrhythmogenic, when heart rate (and hence action potential duration) is changing differently over different parts of the heart.

The result could be the formation of an ectopic action potential, or of a re-entrant wave (illustrated in Figure 7); both of which could trigger fibrillation (illustrated in Figure 8). In these figures the arrhythmias have been produced purely by the CML model, and result from appropriate initial conditions to establish re-entrant propagation.

To generate arrhythmias produced by local differences in cell properties and action potential characteristics either requires a bidirectional coupling between cell ODE

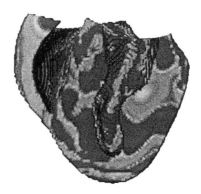

Figure 8. Visualisation of irregular activity of fibrillation produced by breakdown of scroll wave of Figure 7.

model and CML node, in which changing aspects of the ODE model behaviour (action potential duration, refractory period) are mapped in changed parameters in the CML model; or means embedding the cell excitability models in an anisotropic, 3-D model partial differential equation of the entire heart. Since the latter is computationally too demanding, the first alternative is being developed.

ACKNOWLEDGEMENTS

This work was supported by grant from EPSRC (GR/K/49775) and the Wellcome Trust (044365).

REFERENCES

Antzelevitch, C., Sicouri, S., Lukas, A., Nesterenko, V.V., Liu, D.W., Di Diego, J.M., 1995, Regional differences in the electrophysiology of ventricular cells: physiological and clinical implications. In D.P. Zipes & J. Jalife (Eds.) *Cardiac Electrophysiology: from cell to bedside* (2nd Ed.), W.B. Saunders Company, New York.

Biktashev, V.N. & Holden, A.V., 1995, Control of re-entrant activity in a model of mammalian atrial tissue, *Proc. Roy. Soc. Lond. B, Series B* **260**, 211-217.

Biktashev, V.N., & Holden, A.V., 1996a, Computation of re-entrant propagation in a two-dimensional model of ventricular tissue simulating long QT syndrome *Eur. Heart J.* **17**, 597 [Abstract P3212].

Biktashev, V.N. & Holden, A.V., 1996b, Re-entrant activity and it's control in a model of mammalian ventricular tissue, *Proc. Roy. Soc. Lond. B, Series B* **263**, 1373-1382.

Boyett, M.R., Clough, A., Dekanksi, J., & Holden, A.V., 1997, Modelling cardiac excitation and excitability In *Computational Biology of the Heart* ed A.V.Panfilov & A.V.Holden. John Wiley: Chichester , p 1-47

Chauvet, G., 1996, *Theoretical Systems in Biology: Hierarchical and Functional Integration 1, 2 and 3* (Pergamon, Oxford).

Chialvo, D.R., 1995, Generic excitable dynamics on a two-dimensional map, *Chaos, Solitons & Fractals* **5**, 461-479.

El Naschie, M.S. & Holden, A.V. (eds), 1995, *Nonlinear Wave Phenomena in Excitable Physiological Systems*, special issue of *Chaos, Solitons and Fractals* **5**, 317-726.

Holden, A.V., Markus, M. & Othmer, H.G. (eds), 1990, *Nonlinear Wave Processes in Excitable Media* (Plenum, New York).

Holden, A.V., Poole, M.J. & Tucker, J.V. , 1995, Reconstructing the Heart, *Chaos, Solitons & Fractals* **5**, 691-704.

Kiouse, T., Arita, M., Spindler, A.J. & Noble, D., 1993] Ionic mechanisms of action potential prolongation at low temperature in guinea-pig ventricular myocytes, *J. Physiol. Lond.* **468**, 85-106.

LeGuennec, J.Y. & Noble, D., 1994] Effects of rapid changes of external Na$^+$ at different moments during the action potential in guinea-pig myocytes, *J. Physiol. Lond.* **468**, 85-106.

Luo, C.H. & Rudy, Y., 1994, A dynamic model of the cardiac ventricular action potential I & II *Circ. Res.* **74**, 1071-1096, 1097-1113.

Nielsen, P.M.F., Le Grice, L.J., Smaill, B.H. & Hunter, P.J., 1991] Mathematical model of geometry and fibrous structure of the heart, *American Journal of Physiology* **260**, H1365-H1378.

Noble, D., 1995, The development of mathematical models of the heart, *Chaos, Solitons & Fractals* **5**, 321-333.

Noble, D., Noble, S.J.,Bett, G.C.L., Earm, Y.E., Ho, W.K., & So, I.K., 1991, The role of sodium-calcium exchange during the cardiac action potential, *Ann. N.Y. Acad. Sci* **639**, 334-353.

Panfilov, A.V. & Holden, A.V. eds., 1997, *The Computational Biology of the Heart.* (John Wiley, Chichester).

Panfilov, A.V. & Keener, J.P., 1995, Re-entry in three-dimensional Fitzhugh-Nagumo medium with rotational anisotropy, *Physica-D* **84** 545-552.

Rice, J.J., Winslow, R.W., & Kohl, P., 1995, Gap junction distribution in the heart: functional relevance,. In D.P. Zipes & J. Jalife (Eds.) *Cardiac Electrophysiology: from cell to bedside* (2nd Ed.), W.B. Saunders Company, New York.

EXTRACELLULAR SIGNALING IN AN OSCILLATORY YEAST CULTURE

Marc Keulers and Hiroshi Kuriyama

National Institute of Bioscience & Human-Technology
Biochemical Engineering Laboratory
Tsukuba
Ibaraki 305 Japan

SUMMARY

When the yeast *Saccharomyces cerevisiae* was grown under aerobic condition a metabolic oscillation appeared. This oscillation was observed due to synchronization of oscillatory metabolism in a population. The synchronized metabolic oscillation was dependent on the aeration rate, high aeration stopped the oscillation suggesting that synchronization was caused by a volatile compound in the culture. Ethanol, acetate, acetaldehyde and oxygen were found not to be the synchronizer of the oscillation. Stepwise increase in carbon dioxide concentration of the gas flow rate ceased synchronization, but the oscillation continued in each individual cell. Stepwise increase of the aeration rate keeping carbon dioxide at oscillatory condition did not cease the oscillation. Based on these facts it is postulated that carbon dioxide, through the influence of its dissociation to bicarbonate or through its direct effect on major metabolic pathways, could be the synchronization affecter of the metabolic oscillation of *S. cerevisiae*.

KEY WORDS

Oscillation, Synchronization, *Saccharomyces cerevisiae*, Intercellular communication, Carbon dioxide, Metabolism.

INTRODUCTION

Several types of spontaneous and forced autonomous oscillations have been observed for *S. cerevisiae*. First, *S. cerevisiae* shows oscillations which were shown to be due to the tendency of budding yeast to self-synchronize under aerobic conditions in a chemostat culture, referred to as: cell-cycle synchronized oscillation (Kuenzi and Fiechter, 1969; Porro *et al.*, 1988). Second, under anaerobic conditions yeast suspensions showed glycolytic oscillations after a pulse of glucose and KCN, the latter which was added to uncouple respiration, referred to as: glycolytic oscillation (Chance *et al.*, 1964; Aon *et al.*, 1992). Third, a metabolic oscillation was found in our laboratory under aerobic conditions, referred to as: metabolic oscillation (Satroutdinov *et al.*, 1992; Keulers *et al.*, 1996b).

The metabolic oscillation was first reported by Satroutdinov *et al.* (1992). This report described synchronized oscillation during which many parameters changed cyclically, such as, ethanol, acetate, glycogen, dissolved oxygen and intracellular pH. During continuous operation of the fermentor no shift in period or phase was observable for more than four weeks, implying the sustainability of the oscillation. The oscillation was not caused by fluctuations in glucose feed, the concentration was constant at a level of 0.1 mM, or by any other external trigger and was thought to be autonomous. However, a certain level of ethanol was necessary for the existence of the oscillation. The metabolic oscillation was different in period, approximately 40 minutes, from cell-cycle synchronized and glycolytic oscillation. The period of the oscillation found in cell-cycle synchronized oscillation has been reported to be between two and 70 hours, depending on dilution rate. The period of glycolytic oscillation was found from less than one minute (Aon *et al.*, 1992) to 30 minutes (Das *et al.*, 1990).

Oscillations in a yeast chemostat culture could occur due to a certain feedback interaction (Harrison and Topiwala, 1974). These interactions could be either between cells and extracellular parameters or between linked intracellular reactions. The autonomous cell-cycle synchronized oscillation was linked to secretion of ethanol in the budding-cell phase and subsequently utilization of the ethanol in the single cell phase (Porro *et al.*, 1988). The glycolytic oscillation was first thought to be synchronized by some glycolytic intermediate (Aldridge and Pye, 1976) but recently Richard *et al.* (1996) showed that extracellular acetaldehyde was most likely the synchronization mediator through its direct influence on the NAD/NADH redox balance of the cells. The metabolic oscillation was postulated to be synchronized by carbon dioxide (Keulers *et al.*, 1996a), in this communication we will further examine the extracellular signaling mechanism of the metabolic oscillation by means of new results.

MATERIALS AND METHODS

The microorganism used in this study was *S. cerevisiae*, diploid strain IFO 0233, which was a wild type strain used for distillery and bakers' yeast in Germany. The strain was kept on an agar slope at 4 °C. Inocula were prepared by transferring a colony to a test tube containing 20 cm^3 YPG, (3 g dm^{-3} Yeast extract, 5 g dm^{-3} Pepton, and 20 g dm^{-3} Glucose). The culture was kept in a rotary shaker incubator at 30 °C and 170 rpm for 16 hours. Next the experiments started with a batch culture for about 40 hours. The medium composition of the batch culture was the following: glucose monohydrate, 22 g kg^{-1}; $(NH_4)_2SO_4$, 5 g kg^{-1}; KH_2PO_4, 2 g kg^{-1}; $CaCl_2.2H_2O$, 0.1 g kg^{-1}; $MgSO_4.7H_2O$, 0.5 g kg^{-1}; $FeSO_4.7H_2O$, 0.02 g kg^{-1}; $ZnSO_4.7H_2O$, 0.01 g kg^{-1}; $CuSO_4.5H_2O$, 0.005 g kg^{-1}; $MnCl_2.4H_2O$, 0.001 g kg^{-1}; 70% H_2SO_4, 1 cm^3 kg^{-1}; Antifoam agent (Adecanol LG-294; Asahidenka, Japan), 0.6 cm^3 kg^{-1}; Yeast Extract (Difco), 1 g kg^{-1}. Next the fermenter was operated in continuous mode using the same medium composition as with the batch culture. During the batch and the continuous culture the pH of the medium was kept at 4 (glucose medium) or 3.4 (ethanol medium), using 2.5N NaOH. If the pH was above set point no actions were taken. The fermenter (BioFlo, New Brunswick, NJ) was operated (oscillatory conditions) at a temperature set point of 30 °C, a stirrer rate of 800 rpm, a working volume of 1.2 dm^3, a gas flow rate of 180 cm^3 min^{-1}, and a dilution rate of 0.085 h^{-1} (unless otherwise specified). Ethanol-based medium had the following composition: Ethanol, 15.7 g kg^{-1}; Polypepton (Nihon Seiyaku), 1 g kg^{-1}; the rest of the composition was the same as glucose-based medium exclusive glucose monohydrate. The same instrumentation was used as described in Keulers *et al.* (1994). The same analytical procedures as in Satroutdinov *et al.* (1992) were used.

RESULTS

Metabolic oscillation of *S. cerevisiae* was found in our laboratory under aerobic continuous culture with a medium containing glucose or ethanol as carbon source (Satroutdinov *et al.*, 1992; Keulers *et al.*, 1996b). The period of this oscillation was about 40 minutes and various parameters, such as, oxygen uptake rate, carbon dioxide evolution rate, ethanol, acetate and intracellular pH, changed cyclically. The shape of oscillation is periodic and it can be sustained for more than four weeks (see Figure 1). The metabolic oscillation occurred under strict environmental conditions, pH 4 or lower and a low aeration rate, that is lower than 600 ml min⁻¹. In case of growth on ethanol a change in the extracellular ethanol concentration was observed. Here we describe our experiments which were partly published in Keulers *et al.*, (1996a).

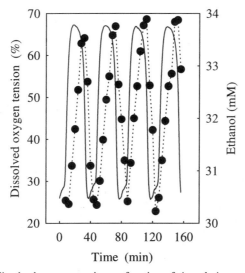

Figure 1. Ethanol and dissolved oxygen tension as function of time during oscillatory condition. ●, and dotted line, measured ethanol (mM); solid line, dissolved oxygen tension in % of air saturation (100% = 7.5 ppm).

Aeration rate

The influence of aeration rate on the oscillation was investigated by applying several aeration rate steps from 150 to 1000 ml min⁻¹ for both glucose and ethanol medium. During every aeration step the sustainability of the oscillation was tested and several parameters were determined. The oscillation was sustained for aeration rates ranging from 150 to 600 ml min⁻¹, outside this area the oscillation ceased. The 150 till 600 ml min⁻¹ range had fairly constant parameters, except dissolved oxygen tension in the culture and carbon dioxide content in the exhaust gas. Biomass concentration was around 8 g dm⁻³ for glucose medium and 6.5 g dm⁻³ for ethanol medium and constant. Ethanol was around 27 mM for glucose medium and 175 mM for ethanol medium. No oscillation was observed at

1000 ml min⁻¹, ethanol concentration was almost zero for glucose medium. This value differed remarkably compared with the one in the 150 till 600 ml min⁻¹ region.

Carbon pulses

Ethanol, acetate, and acetaldehyde were tested as possible synchronization affecters of the metabolic oscillation by various experiments under both growth on glucose and ethanol. The results of these experiments for growth on glucose can be found in Keulers *et al.*, (1996a) except for acetate and acetaldehyde which have not been published. Dissolved oxygen tension is taken as the standard oscillation and changes in its pattern describe the changes in oscillatory patterns.

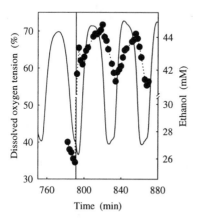

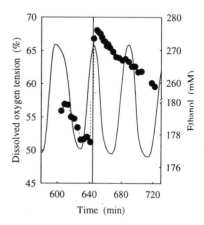

Figure 2. The response of ethanol concentration and dissolved oxygen tension as function of time to a pulse of 25 mM ethanol at high respiration during glucose medium oscillation. The solid vertical line indicates the time of ethanol addition; ●, and dotted line, ethanol concentration (mM); solid line, dissolved oxygen tension in % of air saturation (100% = 7.5 ppm).

Figure 3. The response of ethanol concentration and dissolved oxygen tension as function of time to a pulse of 100 mM ethanol at low respiration during ethanol medium oscillation. The solid vertical line indicates the time of ethanol addition; ●, and dotted line, ethanol concentration (mM); solid line, dissolved oxygen tension in % of air saturation (100% = 7.5 ppm).

Ethanol was added to the medium, 25 mM for glucose medium oscillation and 100 mM for ethanol medium oscillation, at both high and low dissolved oxygen tension (e.g., Figure 2 and 3) for both media. For glucose medium no changes in phase or period of the oscillation were observed except for small changes in the oscillation amplitude. Ethanol medium oscillation showed a small change in phase and a substantial increase in mean level but almost no change in amplitude. High concentrations of added ethanol (500 mM) stopped the metabolic oscillation. Acetate was added in a similar way as ethanol to the medium and the results were that no effect could be noticed at low concentrations of added acetate (1 mM for both media), whereas high concentrations of added acetate (20 mM for both media) stopped the metabolic oscillation. Acetaldehyde was added to glucose medium oscillation at various oscillation phases. Results of acetaldehyde addition up to 4.5 mM are summarized in Figure 4. The added acetaldehyde axis represents the amount of acetaldehyde added to the medium, the phase of addition axis represents the oscillation phase at which acetaldehyde was added to the medium, and the phase shift axis represents the difference in oscillation phase caused by addition of acetaldehyde. The closed circles

(•) form a sequence of added acetaldehyde of approximately 2.5 mM. From the figure one can see that added levels of acetaldehyde up to 2.0 mM had almost no effect on the oscillation phase. Levels higher than 2.0 mM added acetaldehyde between oscillation phase 180° and 360° resulted in a positive phase shift where the phase shift increased with increasing oscillation phase. Increased amounts of acetaldehyde added near oscillation phase 360° showed increased phase shifts. Added acetaldehyde levels higher than 2.0 mM between oscillation phase 0° and 180° resulted in a negative phase shift where the negative phase shift increased with decreasing oscillation phase of acetaldehyde addition. Increased amounts of acetaldehyde added at oscillation phase 180° showed only slightly increased phase shifts. The same sequel of acetaldehyde additions as under glucose medium oscillation were done under ethanol medium oscillation. The result of these additions was that no pattern as depicted in Figure 4 was found with acetaldehyde addition during ethanol medium oscillation. Acetaldehyde shots of approximately 22.5 mM ceased oscillation both for glucose and ethanol medium for a couple of hours (results not shown).

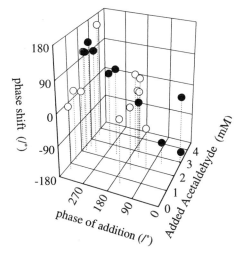

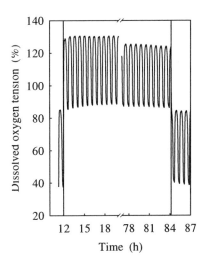

Figure 4. 3-D representation of the effect of increasing concentration of added acetaldehyde at different oscillation phases on the oscillation phase expressed in phase shift during glucose medium oscillation. ●, added acetaldehyde of approximately 2.5 mM at different oscillation phases.

Figure 5. The change in dissolved oxygen tension as function of time. Areas 1 and 3 have a gas flow of 180 ml min^{-1} air, area 2 has a gas flow of 180 ml min^{-1} 30% O_2 and 70% N_2 mixture; solid line, dissolved oxygen tension in % of air saturation (100% = 7.5 ppm).

Oxygen and carbon dioxide

The influence of a change in oxygen concentration in the inlet gas flow was investigated through a change of inlet gas flow from normal air to oxygen enriched gas (O_2:30% + N_2: 70%) keeping the flow rate constant (180 ml min^{-1}). Changes in dissolved oxygen tension from air to oxygen enriched gas and back were very smooth (Figure 5). No change was observable in the oscillation in terms of period or phase. This also held immediately after the changes from normal air to oxygen enriched air and back. The amplitude of the oscillation changed slightly, dissolved oxygen tension oscillated between

80% and 130%. The oscillatory change of ethanol concentration in the culture was also measured and showed no difference to the oscillatory change found under oscillatory conditions (data not shown). Carbon dioxide concentration of the gas flow rate was changed stepwise from normal air to carbon dioxide enriched gas (180 ml min^{-1} air and 5.6 ml min^{-1} CO_2). Increase in carbon dioxide content in the gas flow caused ceasing of the oscillation after approximately 10 periods (Figure 6). This was seen both in dissolved oxygen tension and carbon dioxide concentration in the outlet gas. During 17 hours no stable oscillatory pattern was recognized, except small changes in dissolved oxygen tension with an amplitude of approximately 10% of the stable oscillation's amplitude. After changing the gas flow back to normal air the oscillation reappeared becoming stable after 10-11 periods (Figure 6). The oscillatory change of ethanol concentration in the culture was also measured and showed oscillatory behaviour as mentioned above during normal conditions and a non oscillating pattern during carbon dioxide enriched gas conditions (data not shown). During ethanol medium oscillation the air flow was increased to 700 ml min^{-1} but at the same time CO_2 was added (2.2%) such that its concentration in the air would remain the same as before the increase in airflow rate. The result is shown in Figure 7 where dissolved oxygen tension and CO_2 concentration in the outlet gas are depicted and it can be seen that the oscillation did not cease but continued with a smaller amplitude. This smaller amplitude was expected due to the higher gas flow rate.

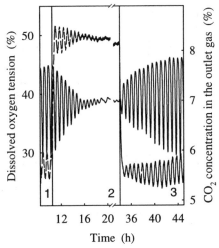

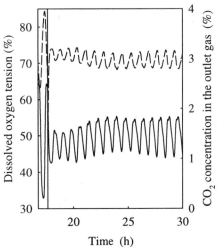

Figure 6. Change of dissolved oxygen tension and carbon dioxide in the outlet gas as function of time during enhanced carbon dioxide concentration in the inlet gas during glucose medium oscillation. Areas 1 and 3 have a gas flow of 180 ml min^{-1} air, area 2 has a gas flow of 180 ml min^{-1} air and 5.6 ml min^{-1} pure CO_2 gas; dashed line, carbon dioxide concentration in the outlet gas; solid line, dissolved oxygen tension in % of air saturation (100% = 7.5 ppm).

Figure 7. Change of dissolved oxygen tension and carbon dioxide in the outlet gas as function of time during enhanced carbon dioxide concentration in the inlet gas during ethanol medium oscillation. Area before the vertical solid line has a gas flow of 180 ml min^{-1} air, area after the solid line has a gas flow of 700 ml min^{-1} air and 2.2 ml min^{-1} pure CO_2 gas; dashed line, carbon dioxide concentration in the outlet gas; solid line, dissolved oxygen tension in % of air saturation (100% = 7.5 ppm).

DISCUSSION

The metabolic oscillation was found under both growth on glucose and ethanol, both period and shape did not differ much. From this observation it was concluded that the oscillation was not regulated by rate changes in the glycolysis caused by switching glycogen metabolism (Satroutdinov *et al.*, 1992) as the glycogen concentration was found not to oscillate under growth on ethanol (Keulers *et al*, 1996b). Intermediates associated with the glycolysis were thus excluded from being the source of synchronization. The cause of the oscillation was restricted to the uptake of ethanol, the pathway of conversion into acetyl-CoA, the TCA-cycle and the glyoxylate bypass, which was active under growth on ethanol (de Jong-Gubbels *et al.*, 1995).

The effect of aeration on the synchronization of the metabolic oscillation was investigated. High aeration rate (1000 ml min^{-1}) stopped the oscillation. Two conditions changed by the increase in aeration rate: dissolved oxygen tension increased and volatiles present in the liquid were stripped faster by the higher aeration rate. From the result obtained with oxygen enriched gas mixture with the same gas flow rate it is concluded that oxygen had no influence on the synchronization (Figure 5). The oscillation showed no change in period or phase during a shift from air to oxygen enriched gas or back. Ethanol concentration showed no change in its oscillatory behaviour either. Even under high oxygen tension of 80-130% no changes were noticeable. If there had been an oxygen influence, distinct changes in the oscillation's phase or ceasing of the oscillation should have occurred. It can be concluded that oxygen was not the synchronization affecter of the oscillation. The effect of increasing the gas flow rate at oscillatory condition was investigated (Keulers *et al.*, 1996a). Increasing the flow rate to 700 ml min^{-1} should result in a ceasing of the oscillation if a volatile compound in the culture caused the synchronization. Increasing the gas flow caused the oscillation to cease for both glucose

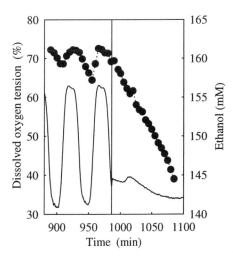 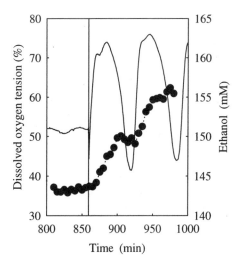

Figure 8. The response of ethanol concentration and dissolved oxygen tension as function of time to a shift up in aeration rate from 180 to 700 ml min^{-1} during ethanol medium oscillation. The solid vertical line indicates the time of aeration change; ●, and dotted line, ethanol concentration (mM); solid line, dissolved oxygen tension in % of air saturation (100% = 7.5 ppm).

Figure 9. The response of ethanol concentration and dissolved oxygen tension as function of time to a shift down in aeration rate from 700 to 180 ml min^{-1} during ethanol medium oscillation. The solid vertical line indicates the time of aeration change; ●, and dotted line, ethanol concentration (mM); solid line, dissolved oxygen tension in % of air saturation (100% = 7.5 ppm).

(Keulers *et al.*, 1996a) and ethanol medium (Figure 8 and 9). In case of growth on ethanol the ceasing of the oscillation was almost immediately as was the restart of the oscillation after switching back from high to normal flow rate. In case of growth on glucose the ceasing of the oscillation was smooth and gradually. The mean value of dissolved oxygen tension remained approximately at the same level before and after the ceasing of the oscillation for glucose medium oscillation. This change was considered to be caused by ceasing of synchronization, but oscillation itself continued in each cell, suggesting that the ceasing of the synchronization was caused by stripping of a volatile compound. These observations suggest that a volatile could be the synchronization affecter.

Ethanol was considered to be the synchronization affecter of cell-cycle oscillation (Martegani *et al.*, 1990) and was therefore the first synchronization candidate of the metabolic oscillation. However, an ethanol pulse during glucose medium oscillatory condition did not show any changes in period and phase of the oscillation (Figure 2). The results with ethanol medium were more complex as there was a slight shift in phase and an increase in mean level indicating an increase in the uptake of ethanol due to the higher levels of ethanol present in the medium. Both results suggested that ethanol concentration could not be the synchronization affecter considering the relative small amplitude (1~1.9 mM) changes in ethanol concentration during oscillation. If ethanol was the synchronization affecter, an 1.9 mM increase in ethanol during glucose medium oscillation should stop ethanol production and force cells to take up the ethanol, an 180 degree phase shift. Nevertheless, an 25 mM increase in ethanol through a pulse addition did not stop ethanol production and cells continued to oscillate under a rather high ethanol concentration of about 43 mM during glucose medium oscillation. Acetate results showed no influence on phase, period or amplitude at low added concentrations but ceased oscillation at high concentration probably due to acidic uncoupling effects. These results suggested that acetate was not involved in the synchronization of the metabolic oscillation.

In our previous communication on synchronization of metabolic oscillation (Keulers *et al.*, 1996a) the effect of acetaldehyde on synchronization was not investigated. However, acetaldehyde was considered the synchronization mediator of glycolytic oscillation (Richard *et al.*, 1996) and was therefore a possible synchronization candidate for the metabolic oscillation as well. Although acetaldehyde exhibits various properties required (Winfree, 1980) for a synchronization affecter, it permeates biological membranes (Stanley and Pamment, 1993) and the phase shift it induces in oscillating cells depends on the phase of addition and the amount added under glucose medium oscillation (Figure 4), it will be shown that acetaldehyde was most likely not the synchronization effector of the metabolic oscillation.

The effect of acetaldehyde added during various oscillation phases at various amounts during glucose medium oscillation showed phase shifts from added amounts of 2.27 mM onwards, but did not show phase shifts at lower levels (Figure 4). The added amounts of acetaldehyde for both glucose and ethanol medium oscillations which gave rise to phase shifts were higher than the amounts present during oscillation (Figure 10). If acetaldehyde concentration was the synchronization affecter, a small addition of 0.5 mM acetaldehyde given during glucose or ethanol medium oscillation at a certain oscillation phase should stop acetaldehyde excretion and force the cells into acetaldehyde uptake, a positive phase shift of 180°. Oscillatory pattern of acetaldehyde during glucose and ethanol medium oscillation was roughly the same except for a difference in mean value (Figure 10) suggesting that phase shifts resulting from acetaldehyde addition at the same oscillation phase under ethanol and glucose medium phase should be the same. Acetaldehyde (>2.0 mM) added during glucose medium oscillation showed roughly negative phase shifts when added at oscillation phases 0° to 150° and positive phase changes at other phases. The effect of acetaldehyde addition during ethanol medium oscillation showed no coherent phase changes as during glucose medium oscillation. Moreover, oscillation was also found under acetaldehyde medium (data not shown) in which case there was a constant positive flux of acetaldehyde into the cell. The changes in acetaldehyde concentration during acetaldehyde medium oscillation gave rise to changes in the acetaldehyde flux of about 5% (data not shown), which level was too low to induce synchronization as also suggested for ethanol during glycolytic oscillation by Richard *et al.* (1996). These results suggested that acetaldehyde concentration was not the synchronization mediator of the oscillation.

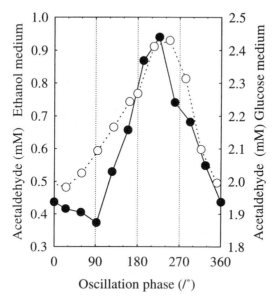

Figure 10. Acetaldehyde concentration during glucose and ethanol medium oscillation as function of oscillation phase. O, dotted line, acetaldehyde concentration (mM) during ethanol medium oscillation; ●, solid line, acetaldehyde concentration (mM) during glucose medium oscillation.

Carbon dioxide was the last volatile reported in the results that could give rise to synchronized metabolic oscillation. Unfortunately, experiments as done with acetaldehyde, that is pulse addition at different phases, were not feasible at our laboratory, instead we used the step-wise increase or decrease of added carbon dioxide as described in the results section. In Keulers *et al.*, 1996a, we postulated a hypothesis for the synchronization of the metabolic oscillation based on the effect of carbon dioxide. Carbon dioxide would regulate the intracellular pH which is known to be a regulator of metabolic rates, a lower intracellular pH leads to lower respiration (Madshus, 1988). Thus changes in carbon dioxide levels in the culture were considered to regulate indirect respiratory activity and to force synchronization of the oscillation among cells in the culture. In our next paper (Keulers *et al.*, 1996b) we found intracellular pH not to be the regulator of the oscillation because intracellular pH oscillation phase was 180° different from intracellular pH oscillation phase found under glucose medium.

The synchronization affecter was influenced by a shift up in the aeration rate for both ethanol and glucose medium oscillation, such that it was concluded that a volatile was involved in the synchronization of the cells. None of the volatile chemicals mentioned in Results showed an effect relevant to a synchronization affecter except for carbon dioxide. An upward shift in the carbon dioxide concentration in the inlet gas caused the synchronization to cease, but the oscillation continued in each individual cell (Figure 6). The fact that the oscillation continued was concluded from the observation that the mean value of the dissolved oxygen tension before the ceasing of the synchronization remained at the same level as after ceasing of the synchronization. This means individual cells were not oscillating in phase with each other anymore (synchronized) but had random phase. On the other hand, the results shown in Figure 6 suggested that when carbon dioxide concentration in the medium is high, synchronization is (gradually) lost due to a decrease in sensitivity of the synchronization affecter. Compare with male fireflies congregated in trees that flash in synchrony only after the sun has gone down and the ambient light has decreased (Strogatz and Stewart, 1993). This comparison was however not supported by the observation that added carbon dioxide to a high aeration rate during ethanol medium oscillation (Figure 7) could support the oscillation. It is so far not clear how carbon dioxide exactly influences both synchronization and oscillation but it is thought that carbon dioxide plays a major role in the metabolic oscillation.

Our main finding is that a volatile is the synchronization affecter of the short period sustained oscillation of S. cerevisiae. Of all synchronization candidates considered, carbon dioxide is suggested as the most likely volatile candidate for synchronization. This is based on the hypothesis that carbon dioxide will influence cell metabolism through intracellular bicarbonate or has a direct influence on carbon dioxide producing pathways involved in major metabolisms.

REFERENCES

Aldridge, J., and Pye, E.K., 1976, Cell density dependence of oscillatory metabolism, *Nature* 259:670-671.

Aon, M.A., Cortassa, S., Westerhoff, H.V., and van Dam, K., 1992, Synchrony and mutual stimulation of yeast cells during fast glycolytic oscillations, *J. Gen. Microbiol.* 138:2219-2227

Chance, B., Hess, B., and Betz, A., 1964, DPNH oscillations in a cell-free extract of *S. carlbergensis*, *Biochem. Biophys. Res. Comm.* 16:182-187.

Das, J., Timm, H., Busse, H.-G., and Degn, H., 1990, Oscillatory CO_2 evolution in glycolysing yeast extracts, *Yeast* 6:255-261.

de Jong-Gubbels, P., VanRolleghem, P., Heijnen, S., van Dijken, J.P., and Pronk, J.T., 1995, Regulation of carbon metabolism in chemostat cultures of *S. cerevisiae* grown on mixtures of glucose and ethanol, *Yeast* 11:407-418.

Harrison, D.E.F. and Topiwala, H.H., 1974, Transient and oscillatory states of continuous cultures, *Adv. Biochem. Engng.* 3:167:219.

Keulers, M. and Kuriyama, H., 1995, Effect of gas flow rate and oxygen concentration on the damping (filtering) action of fermenter head space, *Biotechn. Letters* 17:675-680.

Keulers, M., Asaka, T., and Kuriyama, H., 1994, A versatile data acquisition system for physiological modelling of laboratory fermentation processes, *Biotechn. Techn.* 8:879-884.

Keulers, M., Satroutdinov, A.D., Suzuki, T., and Kuriyama, H., 1996a, Synchronisation affecter of autonomous short period sustained oscillation of *S. cerevisiae, Yeast* 12:673-682.

Keulers, M., Suzuki, T, Satroutdinov, A.D., and Kuriyama, H., 1996b, Autonomous metabolic oscillation in continuous culture of *S. cerevisiae* grown on ethanol, *FEMS Microb. Letters* 142:253-258.

Madshus, I.H., 1988, Regulation of intracellular pH in eukaryotic cells, *Biochem. J.* 250:1-8.

Martegani, E. Porro, D., Ranzi, B.M., and Alberghina, L., 1990, Involvement of a cell size control mechanism in the induction and maintenance of oscillations in continuous cultures of budding yeast cells, *Biotechn. Bioeng.* 36:453-459.

Porro, D., Martegani, E., Ranzi, B.M., and Alberghina , L., 1988, Oscillations in continuous cultures of budding yeast; a segregated parameter analysis, *Biotechn. Bioeng.* 32:411-417.

Richard, P., Bakker, B.M., Teusink, B., van Dam, K., and Westerhoff, H.V., 1996, Acetaldehyde mediates the synchronization of sustained glycolytic oscillations in populations of yeast cells, *Eur. J. Biochem.* 235:238-241.

Satroutdinov, A.D., Kuriyama, H., and Kobayashi, H., 1992, Oscillatory metabolism of *S. cerevisiae* in continuous culture, *FEMS Mircob. Letters* 98:261-268.

Stanley, G.A. and Pamment, N.B., 1993, Transport and intracellular accumulation of acetaldehyde in *S. cerevisiae*, *Biotechn. Bioeng.* 42: 24-29.

Strogatz, S.H. and Stewart, I., 1993,Coupled oscillations and biological synchronization, *Scientific American* December issue: 68-75.

Winfree, A.T., 1990, *The Geometry of Biological Time*. Springer-Verlag, New York.

DOPAMINE-MEDIATED DEPHOSPHORYLATION OF N/P-TYPE CALCIUM CHANNELS IN STRIATAL NEURONS: A QUANTITATIVE MODEL

Rolf Kötter,[1,2] Dirk Schirok,[2] and Karl Zilles[2]

[1]Institut für Morphologische Endokrinologie u. Histochemie
[2]C. u. O. Vogt-Institut für Hirnforschung
Heinrich-Heine-Universität Düsseldorf
Düsseldorf, Germany

INTRODUCTION

Research into receptor-stimulated intracellular signalling pathways in the striatum has discovered a multitude of complex regulatory mechanisms (for reviews see Walaas and Greengard, 1991; Kötter, 1994), which require higher order analyses to identify their individual roles and to predict their combined actions (Cuthbertson et al., 1996). We combine kinetic analyses and computer modeling in order to evaluate further the intracellular pathways in striatal principal neurons (Kötter, 1994; Kötter et al., 1996) that lead from dopamine receptor activation to the modulation of calcium currents (Bargas et al., 1994).

In a recent sophisticated study Surmeier and colleagues reported the effects of D1 dopaminergic agonists on depolarization-evoked high voltage activated (HVA) calcium currents in striatal principal neurons and the modification of peak currents by substances that affect specific steps in the intermediate protein kinase/phosphatase cascades (Surmeier et al., 1995): Inward calcium currents were measured in whole-cell patch clamp configuration and the temporal variation of peak currents noted in response to bath application of agonists or antagonists affecting the cAMP-mediated signalling pathway. Experimental observations led these authors to propose a model involving the cellular signalling pathways shown in Fig. 1. The aim of this study is to extend and to improve this informal model in several respects: 1) underlying assumptions are made explicit, 2) missing regulatory pathways are added, 3) the informal model is cast into a quantitative form, 4) kinetic constraints for its correct function are specified.

Information Processing in Cells and Tissues
Edited by Holcombe and Paton, Plenum Press, New York, 1998

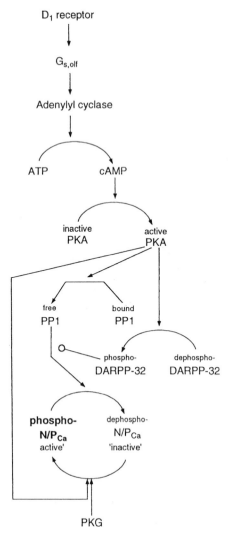

Fig. 1. Informal model of intracellular pathways mediating D1 dopamine-induced dephosphorylation of N/P-type calcium channels in striatal neurons (adapted from Surmeier et al., 1995).

METHODS

In order to evaluate the effects of agonists and inhibitors applied in the experiments by Surmeier and colleagues we extended their informal model of D1 dopamine receptor-dependent regulation of N/P-type calcium channels (Fig. 1) and added several modifiers. The next step was to translate the individual reactions into differential equations taking into account the biochemical mechanisms and the experimentally observed kinetic behaviour of the overall system (see Table 1). Then, we extracted kinetic parameter values from the time constants of modulator actions by fitting the time course of current change following onset or end of modulator administration (see Table 2). These time constants represent the overall change of peak calcium currents, which is often caused by several interacting mechanisms.

Table 1. Equations describing second-messenger cascade mediating D1 dopaminergic regulation of N/P-type calcium channels.*

$$d\ Ract\ /\ dt = kAPB * Rinact * APB - k\_APB * Ract$$

$$d\ Rinact\ /\ dt = 1 - Ract$$

$$cAMP = cAMPbasal + sfR * Ract * (1 - 0.5 * PDE)$$

$$d\ PKAact\ /\ dt = kcAMP * PKAinact * cAMP - k\_PKAact * PKAact - kH89 * PKAact * H89 + k\_H89 * PKAactin$$

$$d\ PKAinact\ /\ dt = k\_PKAact * PKAact - kcAMP * PKAinact * cAMP$$

$$d\ PKAactin\ /\ dt = kH89 * PKAact * H89 - k\_H89 * PKAactin$$

$$d\ PKGact\ /\ dt = 1 * k\_H89 * PKGactin - kH89 * PKGact * H89$$

$$d\ PKGactin\ /\ dt = kH89 * PKGact * H89 - k\_H89 * PKGactin$$

$$fPP1 = sfPKA * PKAact * (1 - pDARPP)$$

$$d\ NPact\ /\ dt = sfPKA * kNPPKA * NPinact * PKAact + kNPPKG * NPinact * PKGact- kNPPP1 * NPact * fPP1 - kNPPP2B * NPact * PP2B$$
$$- kCd * NPact * Cd + k\_Cd * NPactinCd - kTx * NPact * Tx + k\_Tx * NPactinTx$$

$$d\ NPinact\ /\ dt = kNPPP1 * NPact * fPP1 + kNPPP2B * NPact * PP2B - sfPKA * kNPPKA * NPinact * PKAact - kNPPKG * NPinact * PKGact$$
$$- kCd * NPinact * Cd + k\_Cd * NPinactinCd - kTx * NPinact * Tx + k\_Tx * NPinactinTx$$

$$d\ NPactinTx\ /\ dt = kTx * NPact * Tx - k\_Tx * NPactinTx$$

$$d\ NPactinCd\ /\ dt = kCd * NPact * Cd - k\_Cd * NPactinCd$$

$$d\ NPinactinTx\ /\ dt = kTx * NPinact * Tx - k\_Tx * NPinactinTx$$

$$d\ NPinactinCd\ /\ dt = kCd * NPinact * Cd - k\_Cd * NPinactinCd$$

$$Ica = sfNP * Npact$$

* Ract = fraction of active D1 dopaminergic receptors; Rinact = fraction of inactive D1 dopaminergic receptors; PKAact = phosphorylated protein kinase A; PKAinact = dephosph. PKA; PKAactin = inhibited phosph. PKA; PKGact = active protein kinase G; PKGactin = inhibited PKG; fPP1 = free protein phosphatase 1; NPact = phosphorylated N/P-type calcium channels; NPinact = dephosph. N/P channels; NPactinTx, NPactinCd, NPinactinTx, NPinactinCd = fractions of N/P channels inhibited by toxins or cadmium, respectively; Ica = calcium current. Some hypothetical channel states involving combinations of inactivating factors (such as NPactinTxCd or the conversion of NPactinCd into NPininCd) have been omitted since their effects on currents were found to be negligible. For explanation of parameters see Table 3.

Therefore, we started from final reaction steps during blockade of interfering components and, working backwards, separated the contributions of different reactions by temporal constraints. For example, the sequence of two obligatory reaction steps A → B and B → C cannot be slower than the experimentally observed global reaction leading from A to C. Thereby, we obtained a complete set of initial parameters. The Gear and the backward Euler integration algorithms implemented in the programme XPP were used in numerical integration of the ordinary differential equations (Ermentrout, 1994). Minimal time steps were 0.1 and 0.2 s, the tolerances 10^{-4} and 10^{-7}, respectively. We evaluated both the steady state and the kinetic behaviour of the model system under basal (unstimulated) conditions as well as several stimulated conditions. The resulting values of calcium channel phosphorylation were compared to the corresponding experimental data induced by administration and washout of various modulators. Repeated evaluation and optimization was carried out to find parameters that satisfactorily matched the graphs provided by Surmeier et al. considering both the steady state behaviour and the dynamic responses of the system.

RESULTS

Correct function of the informal model depends on several important assumptions:
- Dopamine-dependent HVA calcium currents in striatal principal neurons are mediated by N/P-type calcium channels.
- Phosphorylation of the respective channel proteins increases peak calcium currents, whereas dephosphorylation decreases calcium currents.
- Under basal conditions N/P type calcium channels are almost completely phosphorylated by protein kinase G (PKG).
- The net effect of protein kinase A (PKA) on calcium channels is a dephosphorylation through PKA-dependent release of protein phosphatase 1 (PP1), which exceeds the direct phosphorylation.

The predominant involvement of N/P-type calcium channels follows from the observation that a combination of the N-channel blocker ω-conotoxin GVIA (1-2 µM) and the P-channel blocker ω-agatoxin IVA (100-200 nM) completely eliminates D1 dopaminergic receptor-mediated modulation in most cells (Surmeier et al., 1995). In a subset of cells additional L-type calcium currents are demonstrated. The regulation of L-type currents by D1 agonists, however, can be distinguished by its slow time course (onset after 15-20 s) and is not considered further in this paper. It is known from several tissues and species that a positive relationship exists between calcium currents and phosphorylation of calcium channel proteins (Trautwein and Hescheler, 1990; Gross et al., 1990; Hartzell et al., 1991; Mogul et al., 1993). A strong phosphorylation of N/P-type calcium channels at rest is postulated by Surmeier et al. in order to accomodate the surprising observation that activation of cAMP-mediated pathways reduces peak calcium currents. Since cAMP-activated PKA may have contrasting effects on channel proteins (see Fig. 1) a high basal phosphorylation and its reversal by PKA-induced release of protein phosphatase 1 (PP1) is a candidate mechanism for the explanation of this phenomenon.

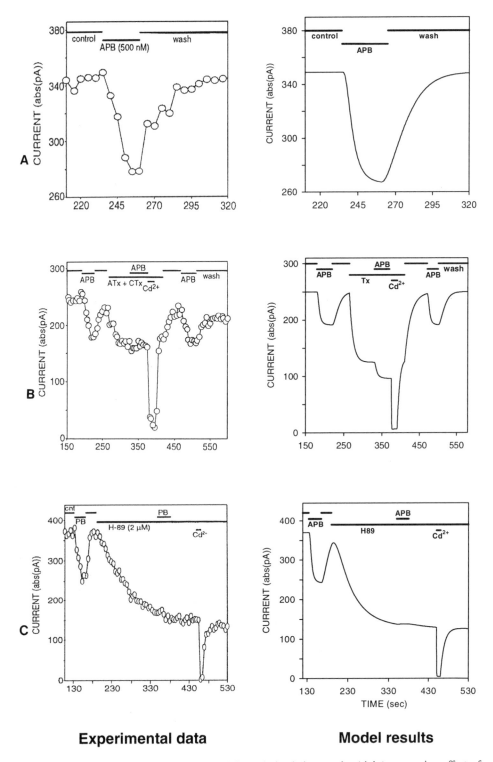

Experimental data   **Model results**

Fig. 2 Comparison of experimental data (left) and simulation results (right) concerning effect of various manipulations (top of plots) on peak calcium currents in striatal neurons. For explanation of abbreviations see Table 2.

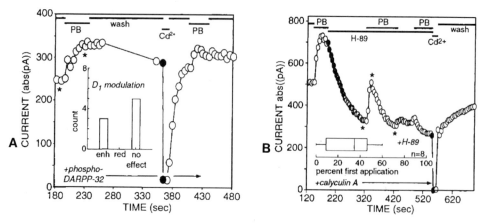

Fig. 3. Further plots from Surmeier et al. (1995) illustrating influence of D1 agonists on L-type calcium channels.

Based on these assumptions we have implemented the model proposed by Surmeier et al. using the set of differential equations shown in Table 1. In addition to those shown in Fig. 1, these equations accomodate all regulatory steps that are required to reconstruct the crucial experiments performed by these authors in order to analyze the contribution of individual intracellular pathways to the overall effect of D1 agonists on N/P-type calcium currents. This includes, for example, a phosphodiesterase (PDE) activity, the kinase inhibitor H89, and several blockers of calcium currents.

The process of evaluation and optimization has focused our attention on three critical issues:

1) Bath application of the protein kinase inhibitor H89 results in a slow exponential decrease of peak calcium currents (Fig. 2C). The slow time course could be due to a slow action of H89, leading to reduced phosphorylation of channel proteins. Alternatively, it may reflect the rate of unopposed dephosphorylation of calcium channels by the remaining phosphatases, or a mixture of both. Variation of model parameters showed that a slow on-rate of H89 would be insufficient to suppress further reduction of calcium currents by application of the D1 dopaminergic agonist APB (Fig. 2C at 350s). In addition, other experiments (Fig. 3B) indicate that some activation of PKA is still available 120 seconds after the start of H89 application, but then disappears within 80 seconds with steadily applied D1 receptor agonist. This apparent discrepancy can be resolved assuming that PKA is only amenable to inhibition by H89 after PKA has been activated (since release of the catalytic unit of PKA is stimulation-dependent). From this follows a comparatively fast rate constant for the action of H89 on PKA of about 0.15 /s. The slow decrease of peak currents in Fig. 2C should then be mainly due to a basal phosphatase activity and results in an apparent dephosporylation rate of about 0.0135 /s.

2) Application of H89 evokes a decrease of calcium currents despite the presence of calyculin A, a blocker of PP1 (see Fig. 3B). If the reduction of peak calcium currents is a measure of the phosphorylation state of channel proteins then the involvement of a further phosphatase activity in the regulation of N/P-type calcium currents has to be postulated. The characteristics of this phosphatase activity, which is not blocked by calyculin A, match the properties of protein phosphatase 2 B (PP2B) (Klee and Cohen, 1988), which is present in

high concentrations in striatal principal neurons (Goto et al., 1987). Since PP2B is not affected by modulation of dopamine receptor-dependent pathways it can be considered as a constant in this model with an estimated rate of 0.005 /s for N/P-type calcium channels. Altogether, we now have to consider at least four different enzymes that affect the phosphorylation state of N/P-type calcium channels: the protein kinases PKA and PKG, and the protein phosphatases PP1 and PP2B. The dephosphorylation of N/P-type calcium channels as shown in Fig. 2A with an apparent rate constant of 0.0135 /s thus reflects a combination of the background activity of PP2B and of PP1 activity. The true dephosphorylation rate constant must be even higher than 0.0135 /s since a small activity of PKG and PKA remains present due to the mass reaction of inhibitor H89.

3) Returning to Fig. 2C the agonist added at 350 s should cause considerable activation of PKA despite a fast onset of H89 action on activated PKA. However, no effect on the time course of peak calcium currents is visible. This phenomenon could be explained if the opposite effects of channel phosphorylation and dephosphorylation neutralized each other. This occurs at the point where the portions of substrates (i.e. phosphorylated channels in the case of PP1 and dephosphorylated channels in the case of PKA) correlate to the respective catalytic activities of PP1 or PKA. If the channels are largely phosphorylated PKA finds only a small amount of substrate to phosphorylate. When the portion of dephosphorylated channel proteins increases, the PKA-mediated phosphorylation become more effective and, finally, outweighs the dephosphorylation mediated by PP1. Applying this to Fig. 2A the curve during continuous application of APB reflects not only PP1-mediated dephosphorylation of N/P-type calcium channels, but also an increasing phosphorylation by PKA and PKG. After removal of the D1 receptor agonist, the N/P-type current slowly returns to baseline within 60 seconds. The apparent time constant of 13 s depends mainly on the deactivation of PKA and PP1 and the background activity of PKG. We establish the phosphorylation rate constant of PKA at 0.25 /s from Fig. 2C in order to compensate the PP1 activity leaving an apparent phosphorylation rate constant of 0.077. The relative contributions of PP1 and PKG are constrained by the assumption that N/P type channels are largely phosphorylated at rest. Thus we estimate a phosphorylation rate constant of PKG of 0.25 /s, which seems plausible because of the similarities between PKG and PKA (Glass, 1990). This leaves a dephosphorylation rate constant around 0.25 /s for PP1 in order to achieve the time course of calcium channel dephosphorylation after D1 receptor agonist application described by Surmeier et al.

Table 3 gives the final set of parameters obtained by this optimization process. The goodness of match can be judged from Fig. 2.

DISCUSSION

Knowledge about the components of intracellular signalling pathways has steadily increased over recent years. Simultaneously, it has become more and more difficult to conceptualize and evaluate this multitude of components and pathways. Surmeier et al. make a remarkable attempt not only to systematically investigate intracellular pathways involved in D1 dopamine-dependent regulation of cellular responses but also to link conceptually biochemical and electrophysiological mechanisms into a coherent model. At present there are not enough quantitative data to prove that the pathways actually work in the proposed way. A thorough analysis, however, can formalize the model such that the role of its components is more clearly recognized. Furthermore, a formal model based on many qualitative observations can make use of internal constraints to produce not only qualitative but quantitative statements.

The list of assumptions that affect the validity of the model appear plausible. Some, however, require further specification: Surmeier et al. demonstrate that D1 dopaminergic stimulation of striatal neurons decreases peak calcium currents. This result contradicts prevailing concepts, which hold that D1-mediated stimulation of PKA leads to phosphorylation of calcium channels and facilitation of calcium fluxes. Although this contradiction is conceptually resolved by inclusion in analogy to muscle cells of a PKA-activated pathway that releases bound PP1 the responsibility of this mechanism remains to be shown experimentally. Surmeier et al. pre-incubated cells with phosphorylated DARPP-32 (Fig. 3A) and never observed D1 receptor-mediated reduction of calcium currents under this condition. This is interpreted as a stronger PP1 release than inhibition of PP1 by DARPP-32 at least during the observation period. A slow onset of inhibition by DARPP-32 cannot be excluded, however, since no temporal information is provided. Consequently, the model is limited to steady-state effects of phosphorylated DARPP-32, which can be omitted under resting conditions because of the known low basal phosphorylation state of DARPP-32 (Lewis et al., 1990). If significant portions of DARPP-32 were phosphorylated the suppression of relevant PP1 activity would require higher dephosphorylation rates of calcium channels by PP1. The other prerequisite for a dephosphorylating effect of PKA activation is an almost full phosphorylation of N/P-type calcium channels at rest. Surmeier et al. show that the PKG inhibitor Rp-8-pCPT-cGMPS can occlude the effects of H89 and conclude that PKG is involved in the maintenance of calcium currents. Again, there are not sufficient data to specify the individual contribution of PKG to the phosphorylation of N/P-type calcium channels.

We have extended the proposed model of signal transduction pathways from receptor to target calcium channels by inclusion of a dopamine-insensitive phosphatase activity (most likely PP2B) that is required in order to explain the decrease of peak calcium currents during blockade of the phosphatases PP1 and PP2A. Although we cannot exclude that further components may be involved in the experimentally observed phenomena, the addition of at least one phosphatase activity is essential for explanation of the data and correct function of the model. Less obvious is the interaction of various kinases and phosphatases even in the unstimulated state. This quantitative model shows that basal enzyme activities provide a sensitive background regulation such that they not only control the unstimulated phosphorylation state of phosphoproteins and calcium channels but also act as substantial counterforces during fast activation processes. Therefore, terms such as "active" or "inactive" do not adequately characterize the regulatory states of several of these components. It appears more useful to specify the ratio of opposing forces in order to judge their net effect.

In Table 3 we have provided a tentative list of quantitative parameters for D1 dopaminergic receptor-dependent pathways. Activities are expressed as relative values within a range from 0 to 1, where 0 denotes no activity and 1 represents maximal activity. Relative values have the advantage of clearly indicating the degree of activation achieved by different regulatory mechanisms. Multiplication of these values with respective maximal

Table 2. Survey of effects of various modulators on N/P-type calcium currents as observed by Surmeier et al. (1995).

Modulator*	Effectors involved in modulation^ (D1 AC PDE PKA PKG PP1 PP2B N/P)	Max.change of current (%)	Apparent time constants° — τ_{on} (s)	τ_{off} (s)
APB		-20; -11; -12	11(2A);6(4A);4(4A)	13(2A);8(4A);15(4A)
PB		+52	10	28(4B)
Sp-cAMPS		-33	8	6(3C)
8-cpt-cAMP		-10;-36	5(4A); 13(5A)	9(4A); 14(5A)
H89		-59	74(4c); 19(4D)	45(4D)
cpt-cAMP + OA		+23	45(5A)	
PB(+pDARPP32)		+35	17(5C)	
Cd(+pDARPP32)		-93	0.2(5C)	10(5C)
Sp-cAMPS(+OA)		+20	43(7A)	
PB(+Calyculin A)		+43	8(7C)	
H89(+Calyculin A)		-53	49(7C)	
Cadmium		-90	3(2D)	
ATx + CTx		-30	18(2D)	19(2D)

* Modulators in brackets are preincubated and, thus, affected equilibrium conditions. APB=(±)SKF82958; PB=(±)SKF81297; Sp-cAMPS=adenosine-3'-5'-cyclic monophosphorothioate; 8-cpt-cAMP=8-(4-chlorophenylthio)-AMP; OA=Okadaic acid; ATx=omega-agatoxin IVA; CTx=omega-conotoxin GIVA.

^ ||| indicates dopamine dependent reactants in the signalling pathway, whereas = marks dopamine-insensitive reactants. Cross-hatched areas represent reactants that were inhibited by the respective modulators. D1=D1 dopamine receptor; AC=adenylate cyclase; PDE=phosphodiesterase; PKA=protein kinase A; PKG=protein kinase G; PP1= protein phosphatase 1; PP2B=protein phosphatase 2B; N/P=N/P-type calcium channels.

° Time constants are derived from the respective plots (in brackets) of Surmeier et al. (1995).

Table 3. Parameter values used in the model equations.

Parameter	Value	Description
APB	0 or 1	relative activity of 6-chloro APB
kAPB	0.5/s	estimated on rate for APB
k_APB	0.5/s	estimated off rate for APB
sfR	2	scaling factor providing full range of relative D1 receptor activity from 0 to 1
PDE	1	relative activity of phosphodiesterase
cAMPbasal	0.04	relative basal cAMP concentration
kcAMP	1/s	apparent on rate for PKA activation
k_PKAact	0.1/s	apparent off rate for PKA deactivation
H89	0 or 1	relative activity of H89
kH89	0.15/s	estimated on rate for H89 binding to PKA or PKG
k_H89	0.001/s	estimated off rate for H89 dissociating from PKA or PKG
sfPKA	2	scaling factor providing full range of relative PKA activity from 0 to 1
pDARPP	0 or 1	relative activity of phospho-DARPP32
PP2B	1	protein phosphatase 2B
Cd	0 or 1	relative activity of cadmium
kCd	2.5/s	apparent on rate for cadmium block of calcium channels
k_Cd	0.1/s	apparent off rate for dissociation of cadmium from N/P-type calcium channels
kNPPKA	0.25/s	apparent on rate for phosphorylation of N/P-type calcium channels by PKA
kNPPKG	0.25/s	apparent on rate for phosphorylation of N/P-type calcium channels by PKG
kNPPP1	0.25/s	apparent on rate for dephosphorylation of N/P-type calcium channels by PP1
kNPPP2B	0.005/s	apparent on rate for dephosphorylation of N/P-type calcium channels by PP2B
Tx	0 or 1	relative activity of toxins (AgTx and CgTx)
kTx	0.056/s	apparent on rate for toxin block of N/P-type calcium channels
k_Tx	0.056/s	apparent off rate for removal of N/P-type channel toxins
sfNP	250 - 450 pA	factor scaling relative size of N/P-type calcium current to absolute values*

* fitted to match peak calcium currents under basal conditions, which vary between experiments.

activities is a convenient way to arrive at absolute activity values. These kinetic values are based on experimental time course as noted after start or end of application of substances and modified according to system constraints. Generally, rate constants are not corrected for the time required for the substance to diffuse and penetrate the target cells. Only in the case of cadmium Surmeier et al. specified the pure time constant, which is by far the smallest (400 ms $\Rightarrow$ kCd = 2.5 /s). The comparatively fast on rate constant of cadmium is not surprising since it blocks calcium channels directly by entering the pores, whereas intracellularly acting substances such as cAMP analogues have to penetrate the cell membrane before they have can act. Nevertheless, it is possible that the true intracellular time constants are smaller than those found here since we cannot distinguish diffusion and membrane permeation from true rates of onset. The relative values of time constants provide valuable information. Thus, time constants near the start of the signalling cascade are smaller than those further downstream. They are largest where channels proteins are affected but no larger than 200 s. This indicates that early components in signal transduction pathways are more rapidly activated than late components. The latter may take up to a minute to reach a new equilibrium.

The differential equations presented in Table 1 reflect a compromise between the intentions of providing a detailed description of biochemical processes and focusing on relevant regulatory mechanisms. It would be possible to introduce more molecular detail concerning e.g. diffusion, G protein activation or release of catalytic subunits from cAMP-activated PKA. The challenge, however, is to create a simple but meaningful model without sacrificing important regulatory mechanims. We believe that the present model has some of these qualities since all equations correspond to separable biochemical reactions but only those mechanisms are included that are relevant to the explanatory model proposed by Surmeier et al. and their experimental study. When analyzing the model behaviour it became clear that a phosphatase activity had to be included in order to explain the reduction of calcium currents during inhibition of calyculin-sensitive components. An inclusion of G proteins, however, would increase the number of parameters without a corresponding gain in explanatory power. Our next step will be to enhance to significance of the model by inclusion of the regulation of L-type calcium currents by the same signalling pathway.

It has been proposed that multiple interactions in striatal principal neurons occur between calcium-dependent and dopamine-dependent pathways (Smith and Bolam, 1990; Kötter, 1994; Cooper et al., 1995; Kötter et al., 1996). Calcium-dependent mechanisms have not been investigated in the present model and had been eliminated in the experimental study by use of 0.1 μM of BAPTA as an efficient intracellular calcium buffer. The regulation of calcium currents by D1-dopamine-mediated pathways shows that D1 agonists will have at least indirect effects on calcium-dependent processes. Further interactions are likely to occur as a result of calcium influx through glutamate-activated NMDA receptors. The convergence of cortical glutamatergic and nigral dopaminergic afferents on individual dendritic spines of striatal principal neurons (Freund et al., 1984) makes these spines excellent object for the study of such interactions. Because of their small size diffusion processes are not a major complication but observation of intracellular processes is difficult. Using calcium and cAMP-sensitive dyes in combination with confocal microscopy (Augustine, 1994) it should be possible to unravel more details of these interactions. For example, calcium-dependent pathways may interact with dopamine-sensitive pathways at several levels from cAMP to DARPP-32 and calcium channel proteins with synergistic or antagonistic effects (Kötter, 1994). Considering the present model D1-dopaminergic stimulation and intracellular calcium would be antagonistic at the levels of cAMP and DARPP-32 but synergistic at N/P-type calcium channels.

ACKNOWLEDGMENT

Supported by a Helmholtz Scholarship to R.K.

REFERENCES

Augustine, G. J., 1994, Combining patch-clamp and optical methods in brain slices, *J. Neurosci. Meth.* 54:163.

Bargas, J., Howe, A., Eberwine, J., Cao, Y., and Surmeier, D. J., 1994, Cellular and molecular characterization of Ca^{2+} currents in acutely-isolated, adult rat neostriatal neurons, *J. Neurosci.* 14:6667.

Cooper, D. M. F., Mons, N., and Karpen, J. W., 1995, Adenylyl cyclases and the interaction between calcium and cAMP signalling, *Nature* 374:421.

Cuthbertson, R., Holcombe, M., and Paton, R., 1996, *Computation in cellular and molecular biological systems*, World Scientific, Singapore.

Ermentrout, G. B., 1994, The mathematics of biological oscillators, *Meth. Enzymol.* 240:198.

Freund, T. F., Powell, J. F., and Smith, A. D., 1984, Tyrosine hydroxylase-immunoreactive boutons in synaptic contact with identified striatonigral neurons, with particular reference to dendritic spines, *Neuroscience* 13:1189.

Glass, D.B., 1990, Substrate specificity of the cyclic GMP dependent protein kinase, in: *Peptides and Protein Phosphorylation,* B. E. Kemp, Ed., CRC, Boca Raton.

Goto, S., Matsukado, Y., Miyamoto, E., and Yamada, M., 1987, Morphological characterization of the rat striatal neurons expressing calcineurin immunoreactivity, *Neuroscience* 22:189.

Gross, R. A., Uhler, M. D., and MacDonald, R. L., 1990, The cyclic AMP-dependent protein kinase catalytic subunit selectively enhances calcium currents in rat nodose neurones, *J. Physiol.* 429:483.

Hartzell, H. C., Mery, P. F., Fischmeister, R., and Szabo, G., 1991, Sympathetic regulation of cardiac calcium current is due exclusively to cAMP-dependent phosphorylation, *Nature* 351:573.

Klee, C.B., and Cohen, P., 1988, The calmodulin-regulated protein phosphatase, in: *Calmodulin,* P. Cohen, and C. B. Klee, Eds., Cohen, P., *Molecular aspects of cellular regulation*, Vol. 5 Elsevier, Amsterdam.

Kötter, R., 1994, Postsynaptic integration of glutamatergic and dopaminergic signals in the striatum, *Prog. Neurobiol.* 44:163.

Kötter, R., Schirok, D., and Zilles, K., 1996, Concerted regulation of cyclic adenosine monophosphate by calmodulin/calcium complex and dopamine: a kinetic modelling approach, in: *Computation in Cellular and Molecular Biological Systems,* R. Cuthbertson, M. Holcombe, and R. Paton, Eds., World Scientific, Singapore.

Lewis, R. M., Levari, I., Ihrig, B., and Zigmond, M. J., 1990, In vivo stimulation of D1 receptors increases the phosphorylation of proteins in the striatum, *J. Neurochem.* 55:1071.

Mogul, D. J., Adams, M. E., and Fox, A. P., 1993, Differential activation of adenosine receptors decreases N-type but potentiates P-type Ca2+ current in hippocampal CA3 neurons, *Neuron* 10:327.

Smith, A. D., and Bolam, J. P., 1990, The neural network of the basal ganglia as revealed by the study of synaptic connections of identified neurones, *Trends Neurosci.* 13:259.

Surmeier, D. J., Bargas, J., Hemmings jr., H. C., Nairn, A. C., and Greengard, P., 1995, Modulation of calcium currents by a D1 dopaminergic protein kinase/phosphatase cascade in rat neostriatal neurons, *Neuron* 14:385.

Trautwein, W., and Hescheler, J., 1990, Regulation of cardiac L-type calcium current by phosphorylation and G proteins, *Annu. Rev. Physiol.* 52:257.

Walaas, S. I., and Greengard, P., 1991, Protein phosphorylation and neuronal function, *Pharmacol. Rev.* 43:299.

THE WORLD WIDE WEB CYTOKINE DATABASE ---
NEW TECHNIQUES OF DIAGRAMATIC
INFORMATION MANAGEMENT ON THE WEB

Xiao Mang Shou, Siobhán North
[X.Shou, S.North]@dcs.shef.ac.uk

Department of Computer Science
University of Sheffield
Sheffield S1 4DP, UK

Over the last few years, the WWW has become a popular repository of multimedia information. Biological and medical applications are among the most rapidly developing areas on the Web today. This project is to explore methods of constructing a cytokine database on the Web which will provide on-line access to a full scale compendium of cytokines and their receptors across the Internet. This compendium will include different representations of data types such as text information, signal pathway diagrams and 3D images.

At present, the major content of the database describes most of the known cytokines. Each cytokine entry includes information under the headings of crossreactivity, physicochemical properties, gene structure, amino acid sequences, signal transduction, receptors, references, etc. Most of the information currently available on this subject is text description with its obvious limitations[1]. By constructing the cytokine database on the Web, true cross-reference links can be made between different cytokines, cells, receptors and references. This provides flexible and simple access from one kind of data to another. The cross-reference links within the cytokine database will also make the management and maintenance of the database easier. The field of cytokine research is so complicated that it is important to be easily able to identify interrelated data both for data retrieval and update, so that when one element has to be changed as result of a new discovery or modification, all related data can be drawn to the attention of whoever is changing the database with a view to changing them too. Furthermore, by using the Web as the platform, the WWW cytokine database will be able to present some of the cytokine attributes like the amino acid sequence and signal transduction in more intuitive ways as multimedia information such as images and diagrams rather than static text.

The primary principles of the design of the cytokine database must cater for frequent update and easy maintenance. This is because the research in the cytokine area is very active today, so the information stored in the database is expected to change frequently.

Once the databases is set up on the Web, it will be maintained by cytokine scientists rather than computer specialists, so the easy maintenance of the database for non computer specialists is very important.

There are many biological databases on the Web today, for instance, the Protein Data Bank[2], the Swiss-Prot databank[3], the GenBank database[4], etc. Actually, some of the cytokine data, like the amino acid sequence, can be found in these databases as well. So the links between the cytokine database and other databases are necessary for users to retrieve broader information. Most of the existing biological databases on the Web support static text information only. A few of the databases do provide 3D images, but those pictures are also static without any editing and viewing functions such as rotate and structure change. The distinguish features of the cytokine database from other existing biological databases can be summarized as:

1. Dynamic information generation

   Most of the Web pages are not statically prepared. They are generated dynamically in response to a user's requests.

2. Dynamic information collection

   Since the data stored in the cytokine database are expected to change frequently, it will be difficult for the database administrator/webmaster to manage the database update alone. Users of the database will be able to contribute new information or modify the existing data to the database interactively on the Web. Security and version control must be conducted to ensure the database accuracy and consistency. The data that can be collected from users are diagrams as well as text because cytokine signal transfer pathways are one of the most rapidly change fields of the database.

3. Dynamic information display

   3D images of cytokine protein sequences can be viewed in different formats, users can rotate and alter the display mode of the images directly on the Web.

To display 3D images of protein sequence, there is a powerful molecular visualization tool called RasMol[5] which supports many editing and viewing functions, but it cannot run on the Web. The WWW cytokine database aims to provide a real interactive tool for 3D image rotating, altering and viewing directly on the Web. One of the possible solution to this problem is to develop a similar tool to RasMol using Java applet so that users can edit and view 3D structures of cytokine proteins at any angles interactively on the Web. The following shows how the same 3D structure can be displayed in two different views:

Figure 1 Different views of a protein sequence

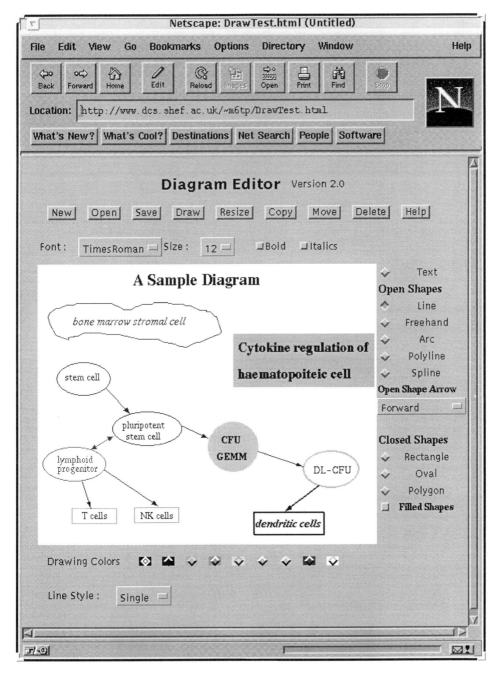

Figure 2   The interface of WIDE

Among all the cytokine information, the storage and manipulation of the signal pathways diagrams is the most complicated part. Though the WWW is popular for its capability to present multimedia information, there was no existing effective method available to handle diagrams on the Web. These diagrams could have been stored as scanned images but they would occupy too much storage space and slow down the network traffic. Furthermore, they are based on the current state of scientific understanding of these pathways and this is a field where scientific understanding is advancing rapidly. They are bound to change frequently as new pathways and methods are discovered. Therefore, it would not have been wise to consider these diagrams as static. With the help of Java development, a WWW interactive diagram editor (WIDE)[6] has been implemented to provide a tool on the Web to allow users to contribute new cytokine pathways and update existing diagrams from the cytokine database interactively. In order to avoid unauthorized update to existing diagrams, strict security control and version control must be imposed. Although the current implementation of WIDE is still being developed, it has succeeded in its attempt to allow users edit and transfer non-text data on the Web dynamically. By using WIDE for diagram submission and modification, it is possible to develop discussion groups for users to exchange different views on cytokine signal transfer pathways, this will be a significant helpful resource for cytokine researchers. The interface of WIDE can be viewed in Figure 2[7].

Other than the facts data of cytokines and their receptors, some other links such as cytokine related researches and vendors can also be usefully added to the cytokine database Web page. Some important functions such as the protein sequence comparison will also be developed. The WWW cytokine database aims to build a wide ranging home for cytokine related information and research where anyone who is interested in cytokine will benefit from the information it provides.

Reference:

1. Callard R, Gearing A; *The Cytokine Facts Book;* Academic Press; 1994.

2. PDB WWW Home Page;
   http://www.pdb.bnl.gov/

3. ExPASy – SWISS-PROT top page;
   http://expasy.hcuge.ch/sprot/sprot-top.html

4. GenBank Database;
   http://pscinfo.psc.edu/general/software/packages/genbank/genbank.html

5. RasMol V2.5 Molecular Visualization Program;
   http://www.bio.cam.ac.uk/doc/rasmol.html

6. Shou XM, North, SD; *WIDE: A WWW Interactive Diagram Editor;* submitted to IEEE Internet Computing; 1997

7. Pavlidou T; *A diagram Editor in Java;*
   MSc dissertation, Department of Computer Science, University of Sheffield; 1997

INVOLVEMENT OF RECEPTOR-KINASE IN THE BIO-CHEMISTRY UPSTREAM FROM cAMP SYNTHESIS IN CELLS

Juergen Nauroschat and Uwe an der Heiden

Department of Mathematics
University of Witten-Herdecke
Stockumer Str. 10
D-58448 Witten, Germany

INTRODUCTION

The cyclic nucleotide 3',5'-cycloadenosine monophosphate (cAMP) is a ubiquitous element of eukaryotic cells (Robison et al., 1971). As a cytosolic "second" messenger agent cAMP drives pleiotypic cascades inherent in the networks of cellular signal processing. The range of phenomena dependent on cAMP includes neuromodulation, glycogenolysis and gluconeogenesis, and cardiac contraction. Any of these examples serves to illustrate that defects in cAMP processing and/or cAMP mobilization can impose severe limitations to both the quality and duration of human life.

Mobilization of cellular second messengers different from cAMP often occurs from several distinct sources. Ionized calcium, for instance, is released from the endoplasmic or sarcoplasmic reticulum, or enters the cytosol by regulated passage of plasma membrane channels (Prentki et al., 1984; Borle and Uchikawa, 1978). Diacylglycerols with second messenger function are generated by dephosphorylation of phosphatidic acid or by phospholipase-C dependent hydrolysis of other phospholipids (Liscovitch and Cantley, 1994). In contrast, only a single source has been established for second messenger cAMP, namely *de novo* biosynthesis from adenosinetrisphosphate (ATP). This reaction is effectively catalyzed by the activated form of the integral membrane enzyme adenylate cyclase.

BIOCHEMISTRY UPSTREAM FROM cAMP SYNTHESIS IN CELLS

Activation of adenylate cyclase can be induced by a suitable extracellular signal molecule ("agonist") such as a stimulatory hormone or neurotransmitter. In spite of having several extracellular domains the adenylate cyclase, by itself, cannot detect the signal molecule. Detection of the signal requires a distinct cognate receptor-protein, typically spanning the plasma membrane seven times ("serpentine receptor").

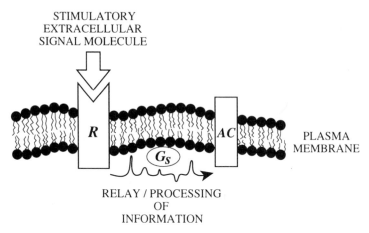

STIMULATORY
EXTRACELLULAR
SIGNAL MOLECULE

R

AC

Gs

PLASMA
MEMBRANE

RELAY / PROCESSING
OF
INFORMATION

Figure 1. Receptor (*R*), stimulatory G-protein (*G$_S$*), and adenylate cyclase (*AC*) colocalized at the lipid bilayer enclosing the cytoplasm.

Relay and, importantly, processing of stimulatory information between cognate receptor and adenylate cyclase is mediated by stimulatory heterotrimeric GTPases, see Figure 1. These so called "stimulatory G-proteins" are located at the cytosolic side of the plasma membrane and consist of three polypeptide subunits, namely a monomeric "alpha"- subunit and a dimeric "beta/gamma"-subunit complex (Hepler and Gilman, 1992). A stimulatory G-protein can be occupied by the nucleotides guanosinediphosphate (GDP) or guanosine-trisphosphate (GTP). Occupation by GDP or GTP is mutually exclusive. The nature of the bound nucleotide determines wether the G-protein attains competence for activating the adenylate cyclase or not: The GDP-bound G-protein lacks competence for activating the adenylate cyclase, and is therefore termed "inactive". On the other hand, the GTP-bound G-protein attains competence for activating the adenylate cyclase, and is therefore termed "active". A complex formed of GDP and G-protein is highly stable so that the G-protein is arrested in the inactive state. Transformation, at the G-protein level, of *GDP to GTP* ("switch-on") yet can be driven by an extracellular stimulus: Physical interaction between the extracellular signal molecule and its cognate receptor conformationally changes the latter to become a high affinity ligand for the GDP-bound G-protein. After coupling to the GDP/G-protein complex, the receptor "catalyzes" the release of GDP. The nucleotide-free G-protein preferably binds to GTP, as the latter is much more abundant in the cell than the competitive GDP. Transformation, at the G-protein level, of *GTP to GDP* ("switch-off") proceeds via hydrolysis. The latter reaction is based on the GTPase activity inherent in the G-protein.

The GTPase activity of the stimulatory G-protein defines an important mechanism for downregulation of adenylate cyclase activity. Hyperfunction of adenylate cyclase, related to insufficient G-protein GTPase activity, is observed in various diseases. An impressive example of such disease is hypersecretion of the intestine after bacterial infection with *vibrio cholerae* (Holmgren, 1981). GTPase activity, however, is not the sole determinant for downregulation of G-protein mediated adenylate cyclase activity. In fact, *hydrolytic breakdown* of active G-protein can be supplemented by *inhibited synthesis* of active G-protein, with the latter phenomenon being also dependent on the presence of active G-protein, see Figure 2:

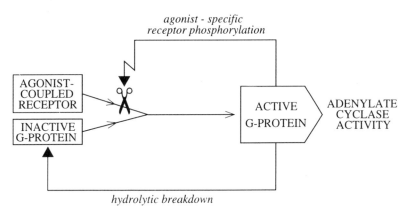

Figure 2. Downregulation of adenylate cyclase activity by agonist-specific receptor-phosphorylation and hydrolytic breakdown of active G-protein.

The prenylated beta/gamma subunit complex belonging to an active G-protein can serve to anchor a certain cytosolic serine/threonine receptor-kinase to the plasma membrane (Mueller et al., 1993). This kinase specifically phosphorylates the receptor that has coupled the extracellular agonist. Phosphorylation induces inability of the receptor to contact the inactive G-protein. Inhibited signaling between the agonist-coupled receptor and the inactive G-protein clearly implies inhibited synthesis of active G-protein. The transmembrane signaling system downregulated by the receptor-kinase is called "homologously desensitized". Sensitivity can be restored by dephosphorylation of receptors at endocytic endosomes (Yu et al., 1993).

As is the case with the G-protein GTPase activity, various recent observations of cell performance (dys)regulated by receptor-kinase can be listed. These include the following examples: First, the receptor-kinase proves to be vitally important for embryogenesis, in particular with respect to cardiac development and activity (Jaber et al., 1996). Second, the kinase downregulates histamine H_2-receptor dependent signaling, a common target of pharmaceutical gastric ulcer therapy, in gastric carcinoma (Nakata et al., 1996). Third, cardiac hyporesponsiveness to stimulatory agonist is observed in transgenic mice overexpressing receptor-kinase (Koch et al., 1995). In the clinical treatment of cardiac failure some beneficial effect is achieved with use of beta-adrenergic receptor antagonists. Interestingly, such agents are able to attenuate hyperaction of receptor-kinase. These and the many other findings indicate that receptor-kinase dependent dysregulation of the adenylate cyclase may be an important constituent of clinical pathology. In the current literature, therefore, development of strategies aiming at specific *in vivo* modulation of receptor-kinase is suggested as an attractive new target of future therapeutic research.

MATHEMATICAL MODEL AND RESULTS OF ANALYSIS

We are interested in the theoretical study of the described adenylate cyclase system. A deterministic modeling approach has been undertaken by us in three consecutive steps (Nauroschat and an der Heiden, 1997): The first step is devoted to the structural framework of the analysis. A complicated network of molecular reactions is suggested to represent the mechanistic core of the membrane machinery described in the preceding section. Among others, the set of constitutive reactants contains several (conformationally differing) types of the receptor, G-protein, and adenylate cyclase, respectively. The second step is devoted to the mathematical framework of the analysis. The law of mass action is used to define increment and decline of reactants with respect to continuous time. Rate-laws are summarized in a high-dimensional, nonlinear, system of ordinary differential equations. The third step is devoted to mathematical recasting of the system, including elimination of several time-dependent variables. Quasi-steady state conditions and deduction of conservation equations support the recasting procedure. The outcome of the three-step modeling process is the following system:

$$\dot{r}(t) = -(h(t) + c_1)r + c_2 x + c_3 \tag{1}$$

$$\dot{x}(t) = h(t)r - c_4 x - F(x)(1 - a) - Q_1(a)Q_2(x) \tag{2}$$

$$\dot{a}(t) = c_5 F(x)(1 - a) - c_6 a \tag{3}$$

$$\dot{m}(t) = G(a) - c_7 m \, . \tag{4}$$

Time is given by t. Values of $c_1,...,c_7$ are temporally invariant and positive. The nonnegative function $h(t)$ represents the extracellular signal, encoded by the receptor-specific agonist at the cell surface. The variable r defines the fraction of agonist-free cell surface receptor. Thus r measures the quality of the membrane machinery with respect to *competence for signal detection*. The variable x defines the fraction of agonist-coupled, nonphosphorylated, cell surface receptor. Thus x measures the quality of the membrane machinery with respect to *competence for signal relay* to the inactive G-protein. The variable a defines the fraction of active G-protein. Thus a measures the quality of the membrane machinery with respect to *competence for adenylate cyclase activation*. The low-pass filter relationship given by equation (4) describes how G-protein based adenylate cyclase activity controls the level of cAMP (m). Nonlinearities involving the variables x or a are referred to as $F(x)$, $G(a)$, and $Q_1(a)Q_2(x)$. The first derivatives, corresponding to x or a, of these functions are nonnegative at $x \geq 0$ or $a \geq 0$. Both F, G, and Q_2 have single roots at the origin. In contrast, $a = 0$ is a multiple root of Q_1. Equation (4) is solved by

$$m(t) = m(0)\exp(-c_7 t) + \int_0^t \exp(c_7(t' - t))G(a(t')) \, dt' \tag{4'}$$

Equations (1-3) can be studied regardless of (4'). Results of mathematical analysis of (1-3) include the following: The trajectories describing physiologically relevant solutions in phase-space are contained in a well defined subdomain of the cube $[0,1]^3$. The subdomain is prismatic. A specific vertex of the prism reflects the unique equilibrium point of the unforced (i.e. $h(t) \equiv 0$) system. This equilibrium is locally asymptotically stable. Balancing of the system to equilibrium is of nonoscillatory character. The system exhibits also a unique, locally asymptotically stable, equilibrium point in case of stationary positive forcing (i.e. $h(t) \equiv h > 0$). This result holds irrespective of the amplitude h of forcing. Contrary to the case of $h(t) \equiv 0$ the equilibrium corresponding to any value of $h(t) \equiv h > 0$ is represented by an *internal* point of the prism. Coordinates of this internal point are not explicitly known, but can be shown to be contained in specific intervals depending on the value of h. The first equilibrium coordinate (measuring the stationary level of agonist-free cell surface receptor) strictly declines with $h > 0$. The second equilibrium coordinate (measuring the stationary level of agonist-coupled, non-phosphorylated, cell surface receptor) strictly increases with $h > 0$. The third equilibrium coordinate (measuring the stationary level of active G-protein) also strictly increases with $h > 0$. The last three statements about dependence on the value of $h(t) \equiv h > 0$ are of qualitative nature. A quantitative statement can be made for the case of $h(t) \equiv h = 0$ as the corresponding equilibrium point is explicitly known. In fact, knowledge of the equilibrium point and use of the implicit function theorem quantifies the partial derivatives of the equilibrium coordinates with respect to h, at $h = 0$. Moreover, these derivatives can be used to define a smooth approximative parametrization, dependent on small $h > 0$, of the manifold of equilibrium points in the prism.

The mathematical results outlined so far cover two aspects of analysis. The first aspect is spatial structure of the manifold of physiologically relevant equilibrium points. The second aspect is dynamical behavior of autonomously operating systems near equilibrium. Dynamical behavior of nonautonomously operating systems (1-4) clearly is a third interesting aspect of analysis. Examples of related results shall be discussed in the following:

Assume system (1-4) to be equilibrated to the resting state at time $t = 0$, i.e.

$$(r(0),x(0),a(0),m(0)) = (c_3/c_1,0,0,0) \tag{5}$$

We consider responses of variables r and x to a brief signal of fixed amplitude $h > 0$. The onset of responses is determined by

$$\dot{r}(0) = -\dot{x}(0) = \phi_1(h) < 0 \tag{6}$$

$$\ddot{r}(0) = \phi_2(h) > 0 \tag{7}$$

$$\ddot{x}(0) = \phi_3(h) < 0 \tag{8}$$

with $\phi_1(h) = -c_3 h/c_1$, $\phi_2(h) = c_3 h(h+c_4)/c_1$, $\phi_3(h) = -c_3 h(h+c_4+F'(0))/c_1$. Result (6) predicts a nonsmooth onset of decline in the level of agonist-free cell surface receptor ("sharp" r-response). Conversely, (6) predicts a nonsmooth onset of increment in the level of agonist-coupled, nonphosphorylated, cell surface receptor ("sharp" x-response). The initial rates of the quantitative changes in r and x cancel each other, and vary in proportion with the signal amplitude. Results (7,8) predict the early time-courses of r and x to be of convex and concave shape, respectively. This means that absolute values of the rates of quantitative changes in r and x are maximum at the onset of stimulation, and continuously declining with time.

Figure 3 illustrates behavior of r and x as observed in two numerical simulations done with (1-5). The two simulations differ only with respect to their stimulatory input: Both simulations consider an input of fixed positive amplitude, but the values of amplitude are distinct. Duration of simulations has been chosen very short. Time courses of r and x thus appear linear, highlighting the statements deduced from (6).

Let us consider again the system (1-4) equilibrated to rest. The initial responses of variables a and m to a brief signal of fixed amplitude $h > 0$ are determined by

$$\dot{a}(0) = \dot{m}(0) = 0 \tag{9}$$

$$\ddot{a}(0) = \phi_4(h) > 0 , \ \ddot{m}(0) = 0 \tag{10}$$

$$\dddot{m}(0) = \phi_5(h) > 0 \tag{11}$$

with $\phi_4(h) = c_3 F'(0)h/c_1$ and $\phi_5(h) = c_3 F'(0)G'(0)h/c_1$. Results (9-11) predict a smooth on-

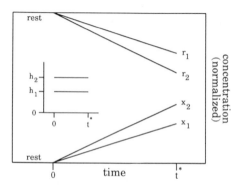

Figure 3. Response to short-term stimulation by a signal of fixed positive amplitude, after previous equilibration of the system to the resting state (c_3/c_1,0,0,0). Time-courses of agonist-free cell surface receptor r and agonist coupled, nonphosphorylated, cell surface receptor x. Shown are the early responses $r = r_1, r_2$ and $x = x_1, x_2$ corresponding to two distinct signal amplitudes $h = h_1, h_2$ ($0 < h_1 < h_2$), respectively. *Inset:* Schemes of stimulation applied to the system.

set of increment in the levels of both the active G-protein a and cAMP m ("delay" responses). Delay is more pronounced for m than is for a. Figure 4 illustrates behavior of a and m as observed in the two simulation experiments underlying Figure 3.

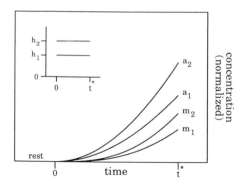

Figure 4. Response to short-term stimulation by a signal of fixed positive amplitude, after previous equilibration of the system to the resting state (c_3/c_1,0,0,0). Time-courses of active G-protein a and cAMP m. Shown are the early responses $a = a_1, a_2$ and $m = m_1, m_2$ corresponding to two distinct signal amplitudes $h = h_1, h_2$ ($0 < h_1 < h_2$) equal to those underlying Figure 3. *Inset:* Schemes of stimulation applied to the system, see Figure 3.

Note that the "sharp" r,x-responses / progressively "delayed" a,m-responses to short-term stimulation clearly reflect the chronological order of molecular events described in the preceding section.

A typical response of the resting (r,x,a,m)-system to a sustained signal of fixed positive amplitude is shown in Figure 5. The (non)smooth started increases of x,a,m-values gradually turn into decreases again, followed by equilibration of x,a,m-values to submaximum levels. Such solution behavior can be interpreted to reflect membrane hyperrespon-siveness at the early phase of stimulation, and development of membrane hyporespon-siveness under prolonged stimulation.

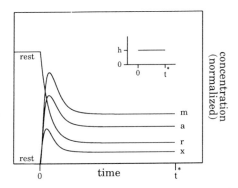

Figure 5. Typical response to long-term stimulation by a signal of fixed positive amplitude $h > 0$, after previous equilibration of the system to the resting state $(c_3/c_1,0,0,0)$. *Inset:* Scheme of stimulation applied to the system. The value of t^* has been chosen considerably larger than was done in the computation underlying Figs. 3,4.

Contribution of receptor-kinase activity to development of hyporesponsiveness is dis-cussed next: In experimental studies the development of membrane hyporesponsiveness can be inhibited by inhibition of receptor-kinase. The polyanion heparin, for instance, can be used as a potent kinase inhibitor. Note that, because of its charge, heparin presumably cannot pass the plasma membrane and therefore has to be made accessible to the cytoplasm by special technique. To obtain reliable results the experimentalist has to assure that the ap-plied technique *per se* does not affect hyporesponsiveness. An alternative method to elimi-nate receptor-kinase activity is creation of loss-of-function mutants by treating the cell with oligodeoxynucleotides antisense to the kinase. In the differential system (1-4), the func-tional expression $Q_1(a)Q_2(x)$ represents kinase activity. Definition of Q_1Q_2 as the zero-function thus models complete kinase inhibition. The effect of $Q_1Q_2 \equiv 0$ on (1-5) is illus-trated by Figure 6. The latter summarizes the results of two numerical simulations: The first simulation is for the purpose of control (i.e. Q_1Q_2 not the zero-function). Conditions are similar to those underlying Figure 5. The second simulation differs from the first only by $Q_1Q_2 \equiv 0$. The (x,a,m)- response with $Q_1Q_2 \equiv 0$, as compared to control, clearly cannot be interpreted to reflect development of membrane hyporesponsiveness.

Of course, pulsatile signals are more realistic than stationary ones in view of *in vivo* conditions. As an approximation to pulsatile signals let us consider square-wave stimula-tion. The latter shall be exemplified by two pulses of equal amplitude and duration. The two pulses shall be separated by a stimulus-free interpulse interval. Figure 7 demonstrates the responses of the resting system (1-5), with Q_1Q_2 not the zero function, to such proto-cols of stimulation. Duration of the first pulse is sufficiently prolonged to move the system near to the equilibrium determined by receptor-kinase activity. When the period of the

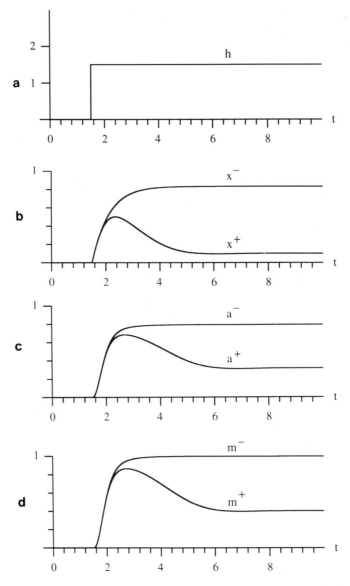

Figure 6. Normalized (x,a,m)- response of (1-4) to a sustained signal of fixed positive amplitude $h > 0$, after previous equilibration of the system to the resting state $(c_3/c_1,0,0,0)$. Triplet (x^+, a^+, m^+) is obtained under the condition of "Q_1Q_2 not the zero-function", i.e. control. Triplet (x^-, a^-, m^-) is obtained under the condition of "$Q_1Q_2 \equiv 0$", representing complete inhibition of receptor-kinase.

stimulus-free interval is short, responsiveness of (1-4) to the second pulse is essentially the same as it was at the end of the first pulse (see Figure 7 A). When the period of the stimulus-free interval is prolonged, responsiveness of (1-4) to the second pulse becomes increased (see Figures 7 B,C). The phenomena illustrated in Figures 7 A,B,C can be interpreted as "non-resensitization", "partial resensitization", and "complete resensitization", respectively.

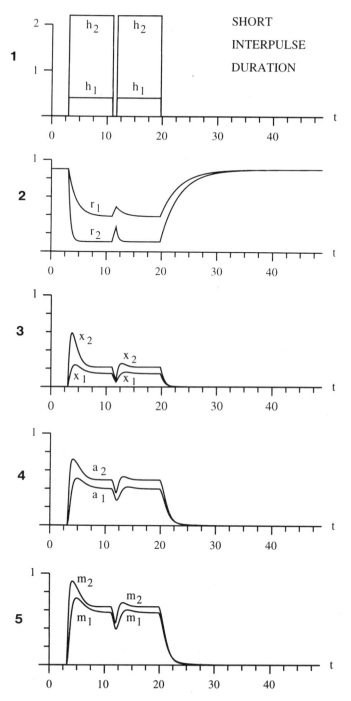

Figure 7.A. Normalized response of the (r,x,a,m)-system to stimulation by two square wave pulses (short interpulse duration) of equal amplitude and duration. Shown are the numerical results corresponding to two distinct pulse amplitudes h_i, $i = 1,2$, with $0 < h_1 < h_2$. Quadruplet (r_i,x_i,a_i,m_i) is obtained with amplitude h_i.

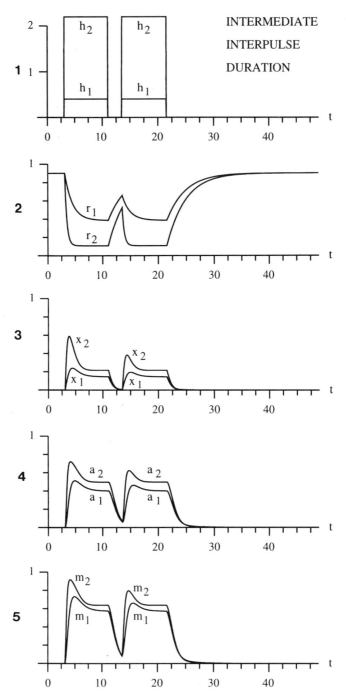

Figure 7.B. Normalized response of the (r,x,a,m)-system to stimulation by two square wave pulses (intermediate interpulse duration) of equal amplitude and duration. Shown are the numerical results corresponding to the pulse amplitudes h_i , $i = 1,2$, underlying Fig. 7.A . Quadruplet (r_i,x_i,a_i,m_i) is obtained with amplitude h_i .

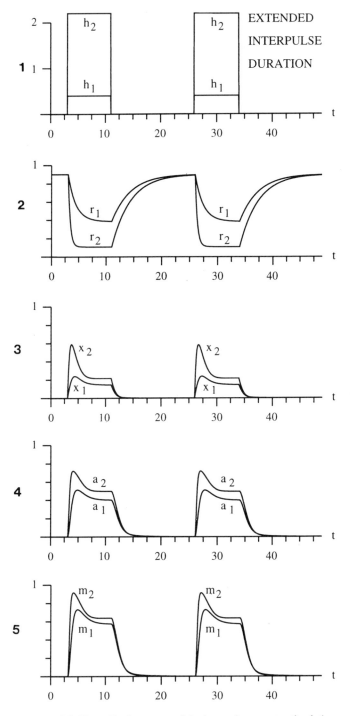

Figure 7.C. Normalized response of the (r,x,a,m)-system to stimulation by two square wave pulses (extended interpulse duration) of equal amplitude and duration. Shown are the numerical results corresponding to the pulse amplitudes h_i, $i = 1,2$, underlying Fig. 7.A. Quadruplet (r_i, x_i, a_i, m_i) is obtained with amplitude h_i.

CONCLUDING REMARKS

The biological findings outlined in the beginning of this work motivate consideration of receptor-kinase action in our mathematical study of cAMP synthesis. The expression $Q_1(a)Q_2(x)$ in equation (2), as already stated, represents action of receptor-kinase in the model (1-4). Introduction of Q_1Q_2 is of phenomenological nature: First, the modeling procedure involves definition of a function ρ_1 relating phenomenologically the kinase-action to the level of active G-protein. Second, the modeling procedure involves definition of a function ρ_2 relating phenomenologically the kinase-action to the level of agonist-coupled, nonphosphorylated, cell surface receptor ("active" receptor). Third, multiplication of ρ_1 by ρ_2 is used to consider synergism between active G-protein and active receptor. The modeling procedure includes rescaling of variables, thus changing $\rho_1\rho_2$ to Q_1Q_2. We currently pursue more elaborate analysis, involving modification of Q_1Q_2, of receptor-kinase action: Adenylate cyclase activation regulated *by* receptor-kinase action is investigated with our attention focused on regulation *of* receptor-kinase action. Our efforts in this regard include consideration of recent experimental results demonstrating regulation of receptor-kinase action by some phospholipids at the plasma membrane (DebBurmann et al., 1996). Examples of such lipids are the phosphoinositides phosphatidylinositol (PI), phosphatidylinositol-4-monophosphate (PIP), and phosphatidylinositol-4,5 bisphosphate (PIP$_2$). In the light of this we take note of the fact that PI,PIP, and PIP$_2$ are constituents of the complex inositol lipid metabolism. PI is synthesized at the endoplasmic reticulum and transported to the plasma membrane via a phosphatidylinositol/phosphatidylcholine-transfer protein. At the plasma membrane, PI is phosphorylated by PI(4)-kinase to PIP. PIP is phosphorylated by PI(4)P5-kinase to PIP$_2$. The latter, for instance, serves as an immediate precursor of several bioactive molecules, but is also involved in the control of numerous cellular processes by its own. The distinct reactions inherent in inositol lipid metabolism are targets of both endogeneous and exogeneous regulators. A well known example of exogeneous regulator is lithium which inhibits inositol phosphatase (Allison and Steward, 1971), thus interfering with production of PI involved in receptor-kinase regulation.

ACKNOWLEDGMENTS

Financial support granted by the Deutsche Forschungsgemeinschaft (DFG) is gratefully acknowledged by the authors.

NOTE

Figures 3,4,5 are reproductions of Figs. 4, 5, 6 of the the article by Nauroschat and an der Heiden (1997). Reproduction-permission has been granted by Springer-Verlag owing copyright to the article.

REFERENCES

Allison, J.H., and Stewart, M.A., 1971, Reduced brain inositol in lithium-treated rats, *Nature Lond.* 233:267.

Borle, A.B., and Uchikawa, T., 1978, Effect of parathyroid hormone on the distribution and transport of calcium in cultured kidney cells, *Endocrinology* 102:1725.

DebBurmann, S.K., Ptasienski, J., Benovic, J.L., and Hosey, M.M., 1996, G protein-coupled receptor kinase GRK2 is a phospholipid-dependent enzyme that can be conditionally activated by G protein βγ subunits, *J. Biol. Chem.* 271:22552.

Hepler, J.R., and Gilman, A.G., 1992, G-proteins, *Trends in Biochem. Sci.* 17:383.

Holmgren, J., 1981, Actions of cholera toxin and the prevention and treatment of cholera, *Nature* 292:413.

Jaber, M., Koch, W.J., Rockman, H., Smith, B., Bond, R.A., Sulik, K.K., Ross, JR., J., Lefkowitz, R.J., Caron, M.G., and Giros, B., 1996, Essential role of β-adrenergic receptor kinase 1 in cardiac development and function, *Proc. Nat. Acad. Sci. USA* 93:12974.

Koch, W.J., Rockman, H.A., Samama, P., Hamilton, R., Bond, R.A., Milano, C.A., and Lefkowitz, R.J., 1995, Cardiac function in mice overexpressing the β-adrenergic receptor kinase or a βARK inhibitor, *Science* 268:1350.

Liscovitch, M., and Cantley, L.C., 1994, Lipid second messengers, *Cell* 77:329.

Mueller, S., Hekman, M., and Lohse, M.J., 1993, Specific enhancement of β-adrenergic receptor kinase activity by defined G-protein β and γ subunits, *Proc. Nat. Acad. Sci. USA* 90:10439.

Nakata, H., Kinoshita, Y., Kishi, K., Fukuda, H., Kawanami, C., Matsushima, Y., Asahara, M., Okada, A., Maekawa, T., and Chiba, T., 1996, Involvement of betadrenergic receptor kinase-1 in homologous desensitization of histamine H_2 receptors in human gastric carcinoma cell line MKN-45, *Digestion* 57:406.

Nauroschat, J., and an der Heiden, U., 1997, A theoretical approach to G-protein modulation of cellular responsiveness, *J. Math. Biol.* 35:609.

Prentki, M., Biden, T.J., Janjic, D., Irvine, R.F., Berridge, M.J., and Wollheim, C.B., 1984, Rapid mobilization of Ca^{2+} from rat insulinoma microsomes by inositol-1,4,5-trisphosphate, *Nature* 309:562.

Robison, G.A., Butcher, R.W., and Sutherland, E.W., 1971, *Cyclic AMP,* Academic Press, New York.

Yu, S.S., Lefkowitz, R.J., and Hausdorff, W.P., 1993, β-adrenergic receptor sequestration - a potential mechanism of receptor resensitization, *J. Biol. Chem.* 268:337.

AMPLIFICATION OF SWITCHING CHARACTERISTICS
OF BIOCHEMICAL-REACTION NETWORKS
INVOLVING Ca^{2+}/CALMODULIN-DEPENDENT PROTEIN KINASE II:
IMPLICATION FOR LTP
INDUCED BY A SINGLE BURST DURING THE THETA OSCILLATION

Hiroshi Okamoto[†] and Kazuhisa Ichikawa

Foundation Res. Lab., Fuji Xerox Co., Ltd.
430 Sakai, Nakai-machi, Ashigarakami-gun,
Kanagawa 259-01, Japan
[†] okamoto@rfl.crl.fujixerox.co.jp

INTRODUCTION

Ca^{2+}/calmodulin-dependent protein kinase II (CaMKII) is involved in a variety of cellular phenomena triggered by Ca^{2+} signalling. For example, CaMKII plays a crucial role in long-term potentiation (LTP) (see references in [1]). Induction of LTP requires transient Ca^{2+} influx into neurons provoked by activation of postsynaptic N-methyl-D-aspatate class of glutamate receptor/channels (NMDA-R/Cs). CaMKII is a very likely target of the Ca^{2+} influx through NMDA-R/Cs because the enzyme is highly concentrated in the postsynaptic regions.

In the previous study [1], we proposed a model for intracellular biochemical-reaction networks describing Ca^{2+}/calmodulin-dependent autophosphorylation of CaMKII versus dephosphorylation of the enzyme. The model was investigated by computer simulation to see how autophosphorylation of CaMKII progresses as a function of Ca^{2+}-signalling pattern. Results obtained showed switching characteristics of the model biochemical-reaction networks: There is a threshold with respect to the intensity of Ca^{2+} signalling; if the intensity is above the threshold, autophosphorylation of CaMKII largely progresses with time (switch-on); if otherwise, the enzyme remains almost dephosphorylated (switch-off). These switching characteristics were further confirmed by a hysteresis loop gained by tracing the degree of autophosphorylation in chemical equilibrium with regard to quasi-static increase and decrease in the intracellular Ca^{2+} concentration ($[Ca^{2+}]$). The hysteresis indicates that the dynamical system describing the model is bistable for $[Ca^{2+}]$ satisfying $\beta < [Ca^{2+}] < \alpha$ with α and β being the threshold and sub-threshold values, respectively, (see Fig. 2A), and it is monostable for $0 < [Ca^{2+}] < \beta$ or $\alpha < [Ca^{2+}]$. The appearance of the discreteness at $[Ca^{2+}] = \alpha$ can therefore be accounted for as a discrete change in the structure of the dynamical system, from bistable to monostable.

It has recently been reported by experimental studies that autophosphorylation results in drastic increase in the calmodulin-binding affinity and full activation of the total activity. These effects probably have significant influences upon CaMKII-related cellular responses to Ca^{2+} signals but were not taken into consideration in our previous study. In this report, we will show that the switching characteristics of the model biochemical-reaction networks are amplified if these effects are included in the model. We will further show that the amplification of the switching characteristics provides a possible explanation for molecular mechanisms underlying a novel form of LTP induced by a single burst during the theta oscillation.

THEORY

Reaction scheme

First, we briefly review regulatory properties of CaMKII and recapture the intracellular biochemical-reaction networks postulated in our previous study [1]. CaMKII is an oligomeric enzyme composed of 8-12 almost identical subunits. The primary structure of CaMKII subunit is characterised by the three domains: a catalytic domain near NH_3-terminus, a regulatory domain in the central portion and an associative domain in the COOH-terminal half. The catalytic domain has potential kinase activity. The regulatory domain interacts with the catalytic domain to block its kinase activity. However, the regulatory domain contains the Ca^{2+}/calmodulin-binding site and if Ca^{2+}/calmodulin binds to this site, the inhibitory interaction between the regulatory and catalytic domains is neutralised, and hence the subunit becomes active.

In the presence of Ca^{2+}/calmodulin, CaMKII rapidly autophosphorylates the threonine residue located in the regulatory domain (Thr[286/287] in the α/β isoform of CaMKII subunit). Once Thr[286/287] is thus phosphorylated, the inhibitory interaction between the catalytic and regulatory domains does not recover even if Ca^{2+}/calmodulin is removed. Therefore, Thr[286/287]-phosphorylated subunit is active also in the absence of Ca^{2+}/calmodulin. This Ca^{2+}/calmodulin-independent activity is called "autonomous activity".

It has long been known that autophosphorylation of CaMKII is an intramolecular reaction. Recent studies have shown that autophosphorylation of Thr[286/287] is an intersubunit reaction [2, 3]. It has also been shown that phosphorylation of Thr[286/287] must be preceded by binding of Ca^{2+}/calmodulin to the 'substrate' subunit [3, 4]. On the bases of these observations, we supposed three kinds of Thr[286/287]-autophosphorylation reactions:

a) A Thr[286/287]-dephosphorylated subunit binding Ca^{2+}/calmodulin phosphorylates another Thr[286/287]-dephosphorylated subunit binding Ca^{2+}/calmodulin;

b) a Thr[286/287]-phosphorylated subunit binding Ca^{2+}/calmodulin phosphorylates a Thr[286/287]-dephosphorylated subunit binding Ca^{2+}/calmodulin;

c) a Thr[286/287]-phosphorylated subunit without Ca^{2+}/calmodulin phosphorylates a Thr[286/287]-dephosphorylated subunit binding Ca^{2+}/calmodulin.

The intracellular biochemical-reaction networks postulated in our previous study consisted of the following processes: Ca^{2+} binds to or dissociates form calmodulin; calmodulin binds to or dissociates from Thr[286/287]-phosphorylated or Thr[286/287]-dephosphorylated CaMKII subunit; CaMKII undergoes Ca^{2+}/calmodulin-dependent autophosphorylation of Thr[286/287]; Thr[286/287] is dephosphorylated by phosphatase activity. The scheme of the intracellular biochemical-reaction networks postulated in our previous study is shown in Fig. 1. This scheme was also used in the present study, except for small changes in control parameters.

The equations describing the time course of the enzyme concentrations are given in Appendix A. To represent autonomous activity quantitatively, we introduced an order parameter, m, defined by

$$m = \sum_{n=0}^{N} n[K_n].$$ (1)

In the previous study, we assumed, for simplicity, that the rate constants for the three autophosphorylation reactions a), b) and c) are the same and that there is no difference between the calmodulin-binding affinities of Thr[286/287]-phosphorylated and Thr[286/287]-dephosphorylated subunits. By the use of the control parameters η and ξ introduced in Appendix A, these assumptions are represented by

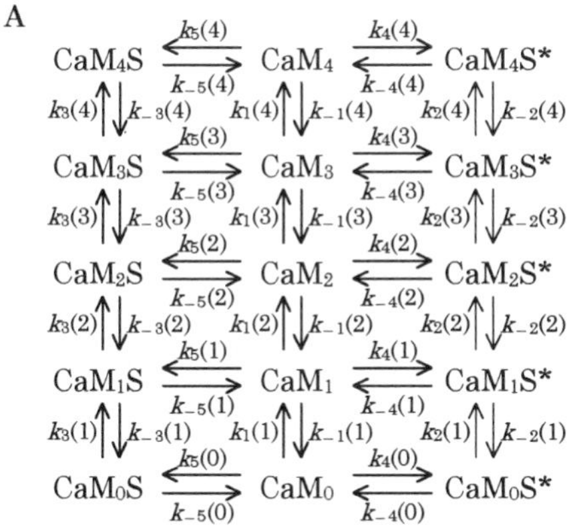

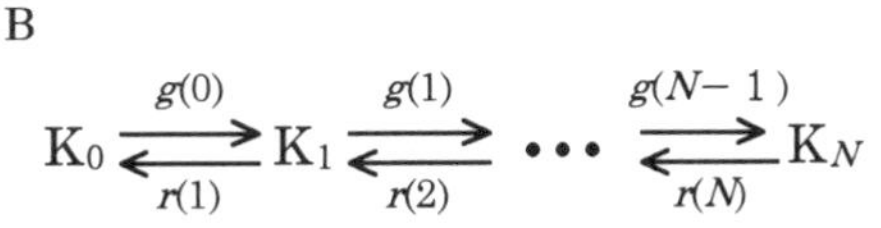

A

B

Figure 1. Reaction scheme of the intracellular biochemical-reaction networks. A, Reactions among Ca^{2+}, calmodulin and Thr[286/287]-dephosphorylated and Thr[286/287]-phosphorylated CaMKII subunits. CaM_i, S and S* symbolise calmodulin binding i calcium ions, Thr[286/287]-dephosphorylated subunit and Thr[286/287]-phosphorylated subunit, respectively. B, Ca^{2+}/calmodulin-dependent autophosphorylation versus dephosphorylation of CaMKII. K_n symbolises CaMKII holoenzyme comprising $N - n$ phosphorylated and n dephosphorylated subunits. The $g(n)$ and $r(n)$ are probabilities per unit time for transitions $n \rightarrow n+1$ and $n-1 \leftarrow n$, respectively, and given by: $g(n) = g_a(n) + g_b(n) + g_c(n)$ with $g_a(n) = k_a p_4^2 (N-n)(N-n-1)$, $g_b(n) = k_b p_4' p_4 n (N-n)$ and $g_c(n) = k_c (1-p_4')p_4 n (N-n)$; $r(n) = nV_D / \left(K_D + \sum_{n'=0}^{N} n'[K_{n'}] \right)$. Here, k_a, k_b and k_c are rate constants for the autophosphorylation reactions a), b) and c), respectively; definitions of p_4 and p_4' are given in Appendix A; V_D and K_D are Michaelis constant and maximal velocity of the dephosphorylation reaction of CaMKII subunit, respectively.

$$(\eta, \xi) = (1, 1) . \qquad (2)$$

In the present study, we again examined switching characteristics of the model under this parameter setting. The results thus obtained will be used as base-line data to see know how the switching characteristics are modified if the above assumptions are altered.

Drastic increase in the calmodulin-binding affinity

Mayer et al. measured the effects of autophosphorylation of Thr^{286} on the calmodulin-binding affinity of α CaMKII using dansylated calmodulin [5, 6]. They found that autophosphorylation of Thr^{286} markedly slowed the release of bound calmodulin; the release time increased from less than a second to several hundred seconds. These indicate that the affinity increases more than 1000-fold after autophosphorylation. If such a drastic increase in the affinity is also the case for naïve calmodulin, it should have significant influence upon CaMKII-related cellular responses to Ca^{2+} signals. Under the assumption that the increase in the affinity is 5000 times, these effects can be taken into our model only by replacing (2) with

$$(\eta, \xi) = (1, 5000) . \qquad (3)$$

Full activation of the total activity

Recently, several experimental studies have provided evidence indicating that autophosphorylation of $Thr^{286/287}$ is essential not only for generation of autonomous activity but also for full activation of the total activity (the activity measured in the presence of Ca^{2+}/calmodulin) [7-9]. It is generally difficult to measure the activity of dephosphorylated CaMKII in the presence of Ca^{2+}/calmodulin and ATP/Mg^{2+} because the enzyme undergoes very rapid autophosphorylation of $Thr^{286/287}$ in the presence of Ca^{2+}/calmodulin and ATP/Mg^{2+}. Fujisawa and his collaborators overcame this difficulty by the use of adenosine 5'-O-(3-thiotriphosphate) (ATP γ S) in place of ATP (see the descriptions in [8, 9]); they found that there is a linear relationship between the total activity and the velocity of Ca^{2+}/calmodulin-independent phosphorylation of an exogenous substrate, and the value for the total activity corresponding to the point of the velocity of zero was approximately 10% of the maximum value (refer to Fig. 3C in [9]). These results indicate that CaMKII initially posses a basal low level of the total activity (about 10% of the maximum) and that autophosphorylation of $Thr^{286/287}$ results in full activation of the total activity. These effects can be included into the model by replacing (2) with

$$(\eta, \xi) = (10, 1) \qquad (4)$$

RESULTS

Amplification of the switching characteristics

The intracellular biochemical-reaction networks contain a lot of kinetic parameters (Fig. 1). We found that, among them, K_D and V_D, Michaelis constant and maximal velocity for the dephosphorylation reaction, respectively (see the caption of Fig. 1), most effectively govern the appearance of switching characteristics with respect to $[Ca^{2+}]$ (data not shown). In the previous study, we verified switching characteristics of the model biochemical-reaction networks for fixed values for K_D and V_D. However, real values for these parameters will scatter owing to variable intracellular conditions. In the present study, therefore, we examined switching characteristics for wide rages of K_D and V_D in order to confirm that the appearance of the switching characteristics are general features, not tied to special selections of K_D and V_D.

For each set of K_D and V_D, evaluation of switching characteristics was performed in the same way as in the previous study: For a fixed value of $[Ca^{2+}]$ that was sufficiently small, chemical equilibrium of the biochemical-reaction networkswas achieved; $[Ca^{2+}]$ was then increased in the quasi-static way; after $[Ca^{2+}]$ had become sufficiently large, then, in turn, it was quasi-statically decreased until it reached the initial value; changes in m were traced along these increase and decrease in $[Ca^{2+}]$.

First, switching characteristics of the model were thus examined under the parameter setting (2). In some cases, we could obtain a hysteresis loop as typically shown in Fig. 2A; in the others, a hysteresis loop was not gained (Fig. 2B) and discrete switching of m could not be seen. In the former, we defined the efficiency of discrete switching characteristics by the area of the region enclosed by the hysteresis loop (indicated by S in Fig. 2A). Table 1A shows the numerically-calculated efficiency of switching characteristics as functions of K_D and V_D.

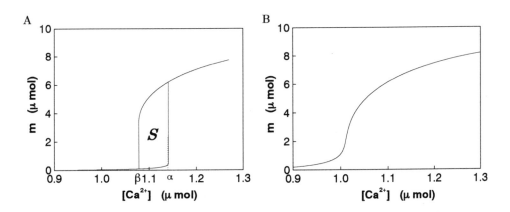

Figure 2. $[Ca^{2+}]$ dependence of m in chemical equilibrium for $(\eta, \xi) = (1, 1)$. A, $(K_D, V_D) = (0.1, 1.0)$. B, $(K_D, V_D) = (1.0, 1.0)$.

A ((μ mol)²)				B ((μ mol)²)				C ((μ mol)²)		
K_D \ V_D	0.01	0.1	1.0	K_D \ V_D	0.01	0.1	1.0	K_D \ V_D	0.01	0.1	1.0
0.1	1.68	1.09	0.14	0.1	2.51	1.85	0.63	0.1		2.47	1.70
1.0	0.60	0.32	–	1.0	1.64	1.13	0.23	1.0	4.26	2.63	0.32
10.0	–	–	–	10.0	0.11	0.002	–	10.0	35.50	7.71	0.23

Table 1. A numerical value in each box is the efficiency of discrete switching characteristics defined by the area of the region enclosed by the corresponding hysteresis loop (indicated by S in Fig. 2A). When a hysteresis loop is gained (Fig. 2B), the box is filled with '−'. A, $(\eta, \xi) = (1,1)$. B, $(\eta, \xi) = (1, 5000)$. C, $(\eta, \xi) = (10, 1)$.

Next, to observe how the switching characteristics were modified if drastic increase in the calmodulin-binding affinity of CaMKII was taken into the model, the same examination but with the parameter setting (3) instead of (2) was performed. The data thus obtained are shown in Table 1B. Comparing these data with the base-line data in Table 1A, one finds that the switching characteristics are amplified.

Finally, Table 1C shows the date obtained under the parameter setting (4). Comparison of these data with the baseline data in Table 1A reveals that the switching characteristics are amplified if full activation of the total activity is taken into the model.

Implication for LTP induced by a single burst given at a peak of the theta oscillation

The dynamical system is bistable for $[Ca^{2+}]$ satisfying $\beta < [Ca^{2+}] < \alpha$ (see Fig. 2A). We observed the tendency that the larger the efficiency of the switching characteristics became, the wider the $[Ca^{2+}]$ window for bistability got (data not shown). Therefore, the data in Table 1A, if compared with those in Table 1B or Table 1C, indicate that the $[Ca^{2+}]$ window for bistability is narrow under the parameter setting (2). The narrowness of the $[Ca^{2+}]$ window for bistability is not a problem if the hysteresis serves only for the discrete switching of m. Occurrence of a discrete change in the structure of the dynamical system from the bistable to monostable states, for which the appearance of a hysteresis loop is necessary, is sufficient for the discrete switching of m, and the efficiency of the switching characteristics itself does not matter.

The narrowness of the $[Ca^{2+}]$ window for bistability, however, makes us difficult to consider the bistable state as a physiologically-realistic state that can persists for a physiologically-significant duration of time. In other words, continuance of the bistable state requires fine-tuning of $[Ca^{2+}]$ to put it within this narrow window for a certain duration, which is physiologically unnatural. However, if either of drastic increase in the calmodulin-binding affinity or full activation of the total activity model, the $[Ca^{2+}]$ window for bistability enlarges. This makes us possible to assume, without tiny-specification of $[Ca^{2+}]$, that the bistable state persists for a physiologically-significant duration of time. Now we will see that this assumption leads to a possible explanation for molecular mechanisms underlying a novel form of LTP induction recently demonstrated by Heurta & Lisman [10-12].

LTP is an activity-dependent form of synaptic plasticity, the property thought to underlie memory formation in brain regions such as the hippocampus and the neocortex. Typically, LTP is induced by high frequency (100 Hz) stimulation of about 1 sec duration. However, a pyramidal neuron does not fire for a full second during brain function because of spike frequency adaptation. Therefore, synaptic stimulation usually required for LTP induction is unlikely to represent the natural trigger for synaptic plasticity. A common pattern of hippocampal activity is repetitive, brief bursts (2-7 spikes at 100-200 Hz) at theta-frequency (5-12 Hz). The theta (θ) oscillation in neuronal activity is a characteristic pattern seen in the hippocampus of an animal during spatial learning. The colinergic system that originates in the medial septum and innervates the whole hippocampal formation is considered to be involved in the θ oscillation in the hippocampus. In fact, bath application of high dose of a colinergic agonist carbachol (CCh) to a hippocampus-alone slice induces a network oscillation in the θ range [10-12].

Heurta & Lisman assessed sensitivity of synaptic plasticity to a very brief burst (4 shocks at 100 Hz) during the θ oscillation produced by CCh, resembling the pattern that occurs naturally in the hippocampus [10-12]. They found that a single burst given at a peak but not at a trough of the θ oscillation can induce homosynaptic LTP (see Fig. 1 of [11]). LTP thus induced was referred to as θ-LTP. It should be noticed that the shape and the amplitude of the field potential they recorded (Fig. 1A of [11]) indicate that the θ oscillation produced by CCh was a subthreshold oscillation of the membrane voltage (V_m) and spike activity was not generated.

First, we hypothesise that [Ca^{2+}] is elevated during the θ oscillation produced by CCh. It is natural to set this hypothesis because various pathways which lead to an elevation of [Ca^{2+}] during CCh application can be considered, as follows.

i) Subthreshold oscillation of V_m is supposed to activate the T-type low-threshold voltage-operated Ca^{2+} channels. The T-type conductance undergoes quick onset followed by a rapid inactivation, but it is deinactivated by modest hyperpolarisation. Therefore, the θ oscillation of V_m promotes inactivation/deinactivation cycles of the T-type conductance, provoking Ca^{2+} influx in waves.

ii) The PI turnover is a major effector system of the metabotropic type of acetylecholine receptors, muscarinic receptors. Therefore, CCh application may result in production of IP_3 which induces Ca^{2+} release from intracellular Ca^{2+} stores.

iii) Heurta & Lisman observed that the amplitude of the θ wave was significantly reduced by the NMDA-R/C antagonist AP5 [11]. Therefore, it may be supposed that, during the θ oscillation, NMDA-R/Cs are repeatedly activated and generates Ca^{2+} inflow in waves.

The next hypothesis is that [Ca^{2+}] elevated during the θ oscillation is in the window for bistability. We can set this hypothesis without fine-tuning of [Ca^{2+}] if the window for bistability is wide enough.

Suppose that autonomous activity is initially switched off, and then a single burst is given at a peak of the θ oscillation. Activation of NMDA-R/Cs by the burst will elicit a brief Ca^{2+} influx. This Ca^{2+} influx is much smaller than that provoked by a full second of tetanic stimulation and by itself cannot achieve a large amount of

CaMKII autophosphorylation sufficient for LTP induction. However, one can suppose that this Ca²⁺ influx triggers CaMKII autophosphorylation to such an extent that the state point only exceeds the potential barrier dividing the phase space into attractor basisns for switched-off and switch-on states. If this is the case, then, even after this brief Ca²⁺ influx is ceased, CaMKII autophosphorylation should continue to grow until autonomous activity converges to the switched-on state. Thus, the very brief synaptic stimulation which is far shorter than a full second can bring out a large degree of autonomous activity of CaMKII which is retained as far as the θ oscillation continues.

We confirmed the above idea by computer simulation of our model. Both of full activation and drastic increase in the calmodulin-binding affinity of CaMKII were postulated in the calculation by setting

$$(\eta, \xi) = (10, 5000). \tag{5}$$

K_D and V_D were choosed as

$$K_D = 0.1\mu\text{mol}, V_D = 1.0\mu\text{mol sec}^{-1}, \tag{6}$$

for which switching characteristics of a large efficiency were gained (data not shown). Details of the modelling of the θ oscillation and a brief Ca²⁺ influx elicited by a single burst are described in Appendix B.

In the simulation, a single burst was given at a peak or a trough of the θ oscillation and the time course of m was examined. The results obtained are shown in Fig. 3. A brief single burst as short as 40 ms can switch autonomous activity of CaMKII on if it is given at a peak of the θ oscillation; by contrast, autonomous activity remains switched off if a single burst is given at a trough.

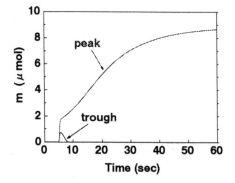 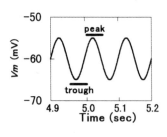

Figure 3. Time course of m during the θ oscillation. A single burst is given at a peak or a trough of the θ oscillation. The single-burst stimulation protocol used in the simulation is depicted on the right of the figure.

The differentiation of the time course of m with regard to the timing of the single-burst stimulation can be explained as follows. A single burst given at a peak of the θ oscillation provokes a relatively large Ca^{2+} influx because the depolarised membrane voltage at peaks of the θ oscillation relieves the voltage-dependent Mg^{2+} block of NMDA-R/Cs. This Ca^{2+} influx is sufficient for moving the state point beyond the potential barrier. However, since the membrane voltage is hyperpolarised at troughs of the θ oscillation, a single burst given at a trough provokes a relatively small Ca^{2+} influx which cannot afford to move the state point beyond the potential barrier.

SUMMARY

We have shown that the switching characteristics of the model for the intracellular biochemical-reaction networks involving CaMKII are amplified if drastic increase in the calmodulin-binding affinity of CaMKII or full activation of the enzyme is taken into consideration. Utilising the enlargement of the $[Ca^{2+}]$ window for bistability associated with the amplification of the switching characteristics, we have also found a possible explanation for why a brief single burst much shorter than a full second, duration usually required for LTP induction by tetanus, can induce LTP if it is given at a peak but not at a trough of the carbachol-induced θ oscillation.

The results of the present study suggest the existence of a novel state of synapses in the hippocampus of living animals: This state emerges when the hippocampus is receiving sustaining cholinergic stimulation from the medial septum; in this state, synapses are greatly sensitised to brief glutaminergic stimulation; that is, synapses are kept ready and can easily change their efficiencies if brief glutaminergic stimulation is timely given. We call this novel state "stand-by state". Since the hippocampus normally requires cholinergic stimulation from the medial septum during spatial learning, we are led to the following hypothesis: Preceding the induction of synaptic plasticity responsible for spatial learning, synapses must wait in the stand-by state.

APPENDIX A

For the reaction scheme in Fig. 1, the time courses of the enzyme concentrations are described by the following equation:

$$\frac{d[K_n]}{dt} = g(n-1)[K_{n-1}] - \big(g(n) + r(n)\big)[K_n] + r(n+1)[K_{n+1}] \tag{A1}$$

$$\begin{aligned}
\frac{d[CaM_i]}{dt} &= k_1(i)[Ca^{2+}][CaM_{i-1}] - k_{-1}(i)[CaM_i] \\
&\quad - k_1(i+1)[Ca^{2+}][CaM_i] + k_{-1}(i+1)[CaM_{i+1}] \\
&\quad - k_4(i)[CaM_i][S] + k_{-4}(i)[CaM_iS] - k_5(i)[CaM_i][S^*] + k_{-5}(i)[CaM_iS^*]
\end{aligned} \tag{A2}$$

$$\frac{d[CaM_iS]}{dt} = k_2(i)[Ca^{2+}][CaM_{i-1}S] - k_{-2}(i)[CaM_iS]$$
$$- k_2(i+1)[Ca^{2+}][CaM_iS] + k_{-2}(i+1)[CaM_{i+1}S]$$
$$+ k_4(i)[CaM_i][S] - k_{-4}(i)[CaM_iS] - \delta_{i4}G + p_i^*R \tag{A3}$$

$$\frac{d[CaM_iS^*]}{dt} = k_3(i)[Ca^{2+}][CaM_{i-1}S^*] - k_{-3}(i)[CaM_iS^*]$$
$$- k_3(i+1)[Ca^{2+}][CaM_iS^*] + k_{-3}(i+1)[CaM_{i+1}S^*]$$
$$+ k_5(i)[CaM_i][S^*] - k_{-5}(i)[CaM_iS^*] + \delta_{i4}G - p_i^*R \tag{A4}$$

$$\frac{d[S]}{dt} = \sum_{i'=0}^{4}\left(-k_4(i')[CaM_{i'}][S] + k_{-4}(i')[CaM_{i'}S]\right) + q^*R \tag{A5}$$

$$\frac{d[S^*]}{dt} = \sum_{i'=0}^{4}\left(-k_5(i')[CaM_{i'}][S^*] + k_{-5}(i')[CaM_{i'}S^*]\right) - q^*R \tag{A6}$$

$$\sum_{n'=0}^{N}[K_{n'}] = \sum_{i'=0}^{4}\left([CaM_{i'}S] + [CaM_{i'}S^*]\right) + [S] + [S^*] = C_K \tag{A7}$$

$$\sum_{i'=0}^{4}\left([CaM_{i'}] + [CaM_{i'}S] + [CaM_{i'}S^*]\right) = C_{CaM} \tag{A8}$$

$$\sum_{n'=0}^{N}n'[K_{n'}] = \sum_{i'=0}^{4}[CaM_{i'}S^*] + [S^*] \tag{A9}$$

$$p_i = \frac{[CaM_iS]}{\sum\limits_{i'=0}^{4}[CaM_{i'}S]+[S]}, \quad p_i^* = \frac{[CaM_iS^*]}{\sum\limits_{i'=0}^{4}[CaM_{i'}S^*]+[S^*]}, \quad q^* = \frac{[S^*]}{\sum\limits_{i'=0}^{4}[CaM_{i'}S^*]+[S^*]} \tag{A10}$$

$$G = \sum_{n'=0}^{N}g(n')[K_{n'}], \quad R = \sum_{n'=0}^{N}r(n')[K_{n'}] \tag{A11}$$

where $n = 0, \cdots, N$ and $i = 0, \cdots, 4$; δ_{i4} in (A3) and (A4) represents Kronecker's delta; C_K in (A7) and C_{CaM} in (A8) are the total concentrations of CaMKII and calmodulin, respectively.

Parameter values used in the present study are listed as follows:

$\eta \times k_a = k_b = k_c = 1.0 \sec^{-1}$; $k_1(1) = k_2(1) = k_3(1) = 2.67 \sec^{-1}$;
$k_1(2) = k_2(2) = k_3(2) = 7.41 \sec^{-1}$; $k_1(i) = k_2(i) = k_3(i) = 20.0 \sec^{-1}$ $(i = 3,4)$;
$k_{-1}(i) = k_{-2}(i) = k_{-3}(i) = 20.0 \mu mol^{-1} \sec^{-1}$ $(i = 1,2)$;
$k_{-1}(3) = k_{-2}(3) = k_{-3}(3) = 600.0 \mu mol^{-1} \sec^{-1}$;
$k_{-1}(4) = 600.0 \mu mol^{-1} \sec^{-1}$; $k_{-2}(4) = \xi \times k_{-3}(4) = 0.06 \mu mol^{-1} \sec^{-1}$;
$k_4(i) = k_5(i) = 0.015 \sec^{-1}$ $(i = 0, \cdots, 3)$; $k_4(4) = k_5(4) = 150.0 \sec^{-1}$;
$k_{-4}(i) = k_{-5}(i) = 5.0 \mu mol^{-1} \sec^{-1}$ $(i = 0, \cdots, 3)$; $k_{-4}(4) = \xi \times k_{-5}(4) = 5.0 \mu mol^{-1} \sec^{-1}$;

$C_K = 1.0 \mu\text{mol}$; $C_{CaM} = 50.0 \mu\text{mol}$; $N = 10$.

Notice η and ξ multiplied to k_a and $k_{-5}(4)$, respectively; they are introduced to take full activation and drastic increase in the calmodulin-binding affinity of CaMKII into consideration in the model (see the text). The ξ is also multiplied to $k_{-3}(4)$ in order to hold 'detailed balance' in the reactions between Ca^{2+}, calmodulin and CaMKII subunit (Fig. 1A).

APPENDIX B

The time course of $[Ca^{2+}]$ during the θ oscillation was defined by the following set of equations:

$$\frac{d[Ca^{2+}]}{dt} = A_1 I_{NMDA} - k_{ext}\left([Ca^{2+}] - [Ca^{2+}]_\theta\right) \tag{B1}$$

$$I_{NMDA} = B(t) G_N (V_m - E_N) Mg \tag{B2}$$

$$Mg = \frac{1}{1 + A_2 e^{-A_3 V_m}} \tag{B3}$$

$$V_m = \frac{V_{dep} - V_{hyp}}{2} + \frac{V_{dep} + V_{hyp}}{2}\sin 2\pi f t \tag{B4}$$

where I_{NMDA} is the Ca^{2+} current through NMDA-R/Cs; $[Ca^{2+}]_\theta$ is the base-line value of the elevated $[Ca^{2+}]$ during the θ oscillation; the second terms in the right-hand side of (B1) represents the force to restore $[Ca^{2+}]$ to $[Ca^{2+}]_\theta$; Mg represents voltage-dependent Mg^{2+}-block of NMDA-R/Cs; V_m varies with time in waves between V_{hyp} and V_{dep}; $B(t)$ represents occurrence of bursting stimulation and

$$B(t) = \begin{cases} 1 & \text{(for 40 ms during bursting stimulation)} \\ 0 & \text{(otherwise)} \end{cases} \tag{B5}$$

Values for the parameters introduced in (B1)-(B4) were set as follows:
$A_1 = 0.2595 \mu\text{mol}\,(\text{fA})^{-1}\text{sec}^{-1}$; $\quad A_2 = 0.244$; $\quad A_3 = 0.0635\,(\text{mV})^{-1}$; $\quad G_N = 70\,\text{pS}$, $k_{ext} = 10.0\,\text{sec}^{-1}$; $\quad [Ca^{2+}]_\theta = 1.2\,\mu M$; $\quad V_{hyp} = 65\,\text{mV}$; $\quad V_{dep} = 55\,\text{mV}$; $\quad f = 10\,\text{Hz}$ (θ frequency).

REFERENCES

1. H. Okamoto and K. Ichikawa, A role of Ca^{2+}/calmodulin-dependent protein kinase II in the induction of long-term potentiation, in: *Computation in Cellular and Molecular Biological Systems*, R. Cuthbertson, M. Holcombe and R. Paton, eds., World Scientific, Singapore (1996).
2. S. Mukherji and T.R. Soderling, Regulation of Ca^{2+}/calmodulin-dependent protein kinase II by inter- and intrasubunit-catalyzed autophosphorylations, *J. Biol. Chem.* 269: 13744-

13747 (1994).

3. P.I. Hanson, T. Meyer, L. Stryer and H. Schulman, Dual role of calmodulin in autophosphorylation of multifunctional CaM kinase may underlie decoding of calcium signals, *Neuron* 12: 943 (1994).

4. S. Mukherji, D.A. Brickey and T.R. Soderling, Mutational analysis of secondary structure in the autoinhibitory and autophosphorylation domains of calmodulin kinase II, *J. Biol. Chem.* 269: 20733 (1994).

5. T. Meyer, P.I. Hanson, L. Stryer and H. Schulman, Calmodulin trapping by calcium-calmodulin-dependent protein kinase, *Science* 256: 1199 (1992).

6. J.A. Putkey and M.N. Waxham, A peptide model for calmodulin trapping by calcium/calmodulin-dependent protein kinase II, *J. Biol. Chem.* 271: 29619 (1996).

7. A.P. Kwiatokowski, D.J. Shell and M.M. King, The role of autophosphorylation in activation of the type II calmodulin-dependent protein kinase, *J. Biol. Chem.* 263: 6484 (1988).

8. T. Katoh and H. Fujisawa, Autoactivation of calmodulin-dependent protein kinase II by autophosphorylation, *J. Biol. Chem.* 266: 3039 (1991).

9. A. Ishida, T. Kitani and H. Fujisawa, Evidence that autophosphorylation at Thr-286/Thr-287 is required for full activation of calmodulin-dependent protein kinase II, *Biochim. Biophys. Acta* 1311: 211 (1996).

10. P.T. Huerta and J.E. Lisman, Heightened synaptic plasticity of hippocampal CA1 neurons during a cholinergically induced rhythmic state, *Nature* 364: 723 (1993).

11. P.T. Huerta and J.E. Lisman, Bidirectional synaptic plasticity induced by a single burst during cholinergic theta oscillation in CA1 in vitro, *Neuron* 15: 1053 (1995).

12. P.T. Huerta and J.E. Lisman, Low frequency stimulation at the troughs of θ-oscillation induces long-term depression of previously potentiated CA1 synapses, *J. Neurophysiol.* 75: 877 (1996).

INFLUENCE OF CALCIUM BINDING TO PROTEINS

ON CALCIUM OSCILLATIONS

AND ER MEMBRANE POTENTIAL OSCILLATIONS.

A MATHEMATICAL MODEL

Stefan Schuster,[1,2] Marko Marhl,[2,3]
Milan Brumen,[2,4] and Reinhart Heinrich[3]

[1]Max Delbrück Center, D-13125 Berlin-Buch, Germany
[2]University of Maribor, Faculty of Education, Koroška cesta 160,
 SI-2000 Maribor, Slovenia
[3]Humboldt University, Institute of Biology, D-10115 Berlin, Germany
[4]J. Stefan Institute, Jamova 39, SI-1000 Ljubljana, Slovenia

INTRODUCTION

Oscillations of the cytosolic calcium concentration have turned out to be an important phenomenon in a variety of living cells and have recently been intensely investigated (Woods et al., 1986; De Young and Keizer, 1992; Dupont and Goldbeter, 1993; Li and Rinzel, 1994; Jafri and Gillo, 1994; Jouaville et al., 1995; Goldbeter, 1996). They play an important role in intracellular information processing (Cuthbertson, 1989; Goldbeter et al., 1990). The risk of overloading the cell with calcium in the process of signalling is believed to be avoided by oscillatory mechanisms instead of adjustable stationary messenger concentrations (Berridge, 1989). Another advantage is that a pulsatile signal can carry two types of information: a digital signal (oscillation or stationarity) acting as a switch, for example, between two modes of metabolism, and an analogue signal which may determine the flux of some metabolic pathway (Cuthbertson, 1989).

In a variety of cells, the oscillatory mechanism is considered to be closely related to the autocatalytic release of calcium out of the endoplasmic reticulum (ER) (Berridge, 1989; Goldbeter et al., 1990; Somogyi and Stucki, 1991; Goldbeter, 1996) or mitochondria (Jouaville et al., 1995). This mechanism is called calcium-induced calcium release (CICR). Frequently, further secondary messengers such as inositol 1,4,5-trisphosphate (IP$_3$) are involved in this process.

Pioneering models for calcium oscillations were developed by Meyer and Stryer (1988, 1991), Goldbeter et al. (1990); Somogyi and Stucki (1991), Cuthbertson and Chay (1991), and Dupont and Goldbeter (1993). More elaborate models have been published recently (Laurent and Claret, 1997; Borghans et al., 1997). Jafri and Gillo (1994) drew

Information Processing in Cells and Tissues
Edited by Holcombe and Paton, Plenum Press, New York, 1998

attention to the fact that calcium oscillations are closely interrelated with oscillations of the electric potential difference across the endoplasmic membrane and that modelling can be improved by including this potential difference as a variable. The question whether the oscillations of this quantity have detectable amplitudes in non-excitable cells is subject to dispute. It is very difficult to measure the potential difference across the ER membrane because it has a very small volume and is a dynamic structure. Any insertion of electrodes causes fragmentation of the ER into small vesicles. So only indirect methods are available so far. They give results indicative of very low values of the transmembrane potential, notably less than 10 mV (Beeler et al., 1981; Dawson et al., 1995). However, some authors seem to be mislead by the assumption that the high permeability of the ER membrane for sodium and chloride ions lead to dissipation of any potential. This is, to our eyes, an oversimplified view because the highly permeable ions simply follow the potential according to the Nernst equation. We believe that modelling the interrelation between calcium oscillations and transmembrane variations is a suitable means to elucidate possible interactions and, thus, to devise new experiments.

Recently, we proposed an electrochemical model for calcium oscillations and their interrelations with ER membrane potential oscillations (Marhl et al., 1997). Elaborating on the models presented by Jafri et al. (1992) and Jafri and Gillo (1994), we considered the binding of calcium to cytosolic proteins. Moreover, we included monovalent anions and cations because they make up a considerable portion of the balance of electric charges (see also Schuster and Mazat, 1993). In order to keep the dimension of the differential equation system small, we made use of the quasi-electroneutrality approximation (Schultz, 1980; Brumen and Heinrich, 1984).

It has sometimes been argued that calcium oscillations can only arise if there is a permanent exchange of calcium between the cell and its surroundings (Goldbeter et al., 1990; Dupont and Goldbeter, 1993; Jafri and Gillo, 1994). However, since the diffusion of calcium in the cytosol is relatively slow (Petersen et al., 1994), we believe that this exchange is of minor importance in comparison to the sequestration by the ER and the calcium-binding proteins. Therefore we have neglected it in our earlier model (Marhl et al., 1997) and do so in the present, refined model.

Experimental findings (reviewed by Smith et al., 1996) show that the velocities of calcium binding differ significantly for different types of proteins. In the present paper, we investigate the effect of the simultaneous presence of different types of calcium-binding molecules. For the sake of simplicity, we consider only two types of sites with high and low binding velocities. We will investigate the transfer of calcium between the different binding sites during oscillations.

DESCRIPTION AND ANALYSIS OF THE MODEL

We consider a cell with the ER functioning as an internal calcium store. Focusing on the ion exchange between the cytoplasm and the calcium store, we neglect the ion exchange with the extracellular medium. Three types of Ca^{2+} fluxes are included: a) the ATP-dependent uptake of Ca^{2+} from the cytosol into the ER, b) the Ca^{2+} efflux from the ER lumen through channels following the CICR mechanism (Dupont et al., 1991; Somogyi and Stucki, 1991), and c) a Ca^{2+} leak out of the ER into the cytosol (Cuthbertson and Chay, 1991). This leak, which can be assumed to be potential-dependent, may flow across the lipid phase of the membrane or through calcium channels or both.

Importantly, the binding of calcium to cytosolic proteins is included. One can roughly distinguish between two classes of binding sites (Pr^I and Pr^{II}). The first class is made up of the binding sites of specific proteins which have a high affinity, but relatively low rate constants of binding and dissociation with respect to calcium. This class contains proteins such as parvalbumin, calbindin, and the C-domains of calmodulin and troponin-C (Smith et al., 1996). They are sometimes called buffering sites. Other molecules such as phospholipids may also belong to this class. The second class (denoted by Pr^{II}) contains sites which bind calcium very fast, but with a much lower affinity than the first class. Examples are provided by the N-domains of calmodulin and troponin-C (Smith et al., 1996). The model system is schematically presented in Fig. 1.

The equations for the kinetics of calcium binding to the two classes of proteins in the cytosol read:

$$\frac{dPr^{I}}{dt} = k_{-}^{I}\,CaPr^{I} - k_{+}^{I}\,Ca_{\text{cyt}} \cdot Pr^{I}, \qquad \frac{dPr^{II}}{dt} = k_{-}^{II}\,CaPr^{II} - k_{+}^{II}\,Ca_{\text{cyt}} \cdot Pr^{II} \; , \; (1a,b)$$

where Pr^{I} and Pr^{II} are the concentrations of the free forms of the slow and fast calcium-binding sites, respectively, and Ca_{cyt} is the cytosolic calcium concentration. The meaning of the parameters in eq. (1a,b) and in all following equations is given in Table 1.

The time dependence of the cytosolic calcium concentration Ca_{cyt} is determined by the fluxes across the ER membrane and by the Ca^{2+} binding to proteins. Thus the following equation can be written:

$$\frac{dCa_{\text{cyt}}}{dt} = J_{\text{ch}} - J_{\text{pump}} + J_{\text{leak}} + k_{-}^{I}\left(Pr_{\text{tot}}^{I} - Pr^{I}\right) - k_{+}^{I}Ca_{\text{cyt}} \cdot Pr^{I} +$$

$$+ k_{-}^{II}\left(Pr_{\text{tot}}^{II} - Pr^{II}\right) - k_{+}^{II}Ca_{\text{cyt}} \cdot Pr^{II} \qquad (2)$$

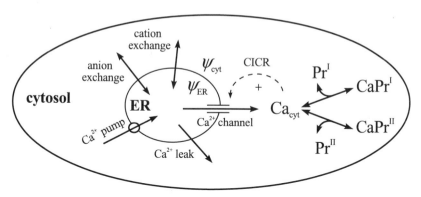

Fig. 1. Schematic presentation of the model system.

We now apply a rapid-equilibrium approximation to the binding sites Pr^{II}; that is, we assume the fast binding reaction to be in equilibrium,

$$\frac{Pr^{II} \cdot Ca_{\text{cyt}}}{CaPr^{II}} = K_{d}^{II} \; . \qquad (3)$$

This is justified by the very high values of rate constants for a number of calcium-binding proteins (Smith et al., 1996). Using the conservation relation $CaPr^{II} = Pr_{\text{tot}}^{II} - Pr^{II}$, this gives

$$Pr^{\mathrm{II}} = \frac{K_d^{\mathrm{II}} Pr_{\mathrm{tot}}^{\mathrm{II}}}{K_d^{\mathrm{II}} + Ca_{\mathrm{cyt}}} \ . \tag{4}$$

The terms corresponding to the fast processes in eq. (2) can be eliminated by subtracting eq. (1b) (cf. Heinrich and Schuster, 1996)

$$\frac{d\left(Ca_{\mathrm{cyt}} - Pr^{\mathrm{II}}\right)}{dt} = J_{\mathrm{ch}} - J_{\mathrm{pump}} + J_{\mathrm{leak}} + k_-^{\mathrm{I}}\left(Pr_{\mathrm{tot}}^{\mathrm{I}} - Pr^{\mathrm{I}}\right) - k_+^{\mathrm{I}} Ca_{\mathrm{cyt}} \cdot Pr^{\mathrm{I}} \ . \tag{5}$$

Inserting eq. (4) into eq. (5) allows us to simplify the differential equation system (1a,b), (2),

$$\frac{dCa_{\mathrm{cyt}}}{dt} = \beta \left[J_{\mathrm{ch}} - J_{\mathrm{pump}} + J_{\mathrm{leak}} + k_-^{\mathrm{I}}\left(Pr_{\mathrm{tot}}^{\mathrm{I}} - Pr^{\mathrm{I}}\right) - k_+^{\mathrm{I}} Ca_{\mathrm{cyt}} \cdot Pr^{\mathrm{I}} \right] \tag{6}$$

with
$$\beta = \frac{1}{1 + K_d^{\mathrm{II}} Pr_{\mathrm{tot}}^{\mathrm{II}} \big/ \left(K_d^{\mathrm{II}} + Ca_{\mathrm{cyt}}\right)^2} \ .$$

A term similar to this factor β has been derived earlier by Smith et al. (1996).

We define $\Delta\psi$ as the electric potential difference between the cytosol and the ER lumen, $\Delta\psi = \psi_{\mathrm{cyt}} - \psi_{\mathrm{ER}}$. The reversal potential for calcium, E_{Ca}, is calculated by using the Nernst equation:

$$E_{\mathrm{Ca}} = \frac{RT}{2F} \ln\left(\frac{Ca_{\mathrm{ER}}}{Ca_{\mathrm{cyt}}}\right) , \tag{7}$$

where R and F have their usual physico-chemical meaning. E_{Ca} is the hypothetical potential which would arise if the actual calcium concentrations corresponded to an equilibrium state. The concentrations Ca_{cyt} and Ca_{ER} are linked by the conservation of calcium in the cell:

$$Ca_{\mathrm{tot}} = Ca_{\mathrm{cyt}} + \rho Ca_{\mathrm{ER}} + Pr_{\mathrm{tot}}^{\mathrm{I}} - Pr^{\mathrm{I}} + Pr_{\mathrm{tot}}^{\mathrm{II}} - Pr^{\mathrm{II}} , \tag{8}$$

where ρ denotes the ratio between the volumes of the ER and cytosol.

The flux through the channels, J_{ch}, is given by (Jafri et al., 1992; Marhl et al., 1997):

$$J_{\mathrm{ch}} = \frac{g_{\mathrm{Ca}}}{2FV_{\mathrm{cyt}}}\left(E_{\mathrm{Ca}} - \Delta\psi\right) . \tag{9}$$

Note that the expression in parentheses in eq. (9) is the electrochemical potential difference of Ca^{2+}. The conductance of the channel is described by a Hill equation with Hill coefficient 2 (Jafri et al., 1992):

$$g_{\mathrm{Ca}} = \tilde{g}_{\mathrm{Ca}} S \left(\frac{Ca_{\mathrm{cyt}}}{K_{\mathrm{Ca}}}\right)^2 \Big/ \left[1 + \left(\frac{Ca_{\mathrm{cyt}}}{K_{\mathrm{Ca}}}\right)^2\right] . \tag{10}$$

For simplicity's sake, the calcium flux J_{pump} into the ER lumen mediated by ATPase is described by a linear dependence on Ca_{cyt} (cf. Jafri et al., 1992),

$$J_{\text{pump}} = k_{\text{pump}} Ca_{\text{cyt}} . \tag{11}$$

A potential-dependent leak flux out of the ER lumen is included, as described by:

$$J_{\text{leak}} = \kappa_{\text{leak}}(E_{\text{Ca}} - \Delta\psi) . \tag{12}$$

We now impose the quasi-electroneutrality condition (Schultz, 1980; Brumen and Heinrich, 1984; Schuster and Mazat, 1993) for the cytosol:

$$C_{\text{cyt}} - A_{\text{cyt}} + 2\,Ca_{\text{cyt}} - 2\,Pr^{\text{I}} - 2\frac{K_d^{\text{II}} Pr_{\text{tot}}^{\text{II}}}{K_d^{\text{II}} + Ca_{\text{cyt}}} = 0 , \tag{13}$$

where C_{cyt} and A_{cyt} are the cytosolic concentrations of monovalent cations (such as sodium and potassium) and monovalent anions (such as chloride), respectively. Applying this condition is justified because the specific capacitance of the membrane is very low, so that a tiny net difference of electric charges is sufficient to generate considerable potential differences. Because of a high permeability of the ER membrane for small ions other than Ca^{2+} (Beeler et al., 1981; Läuger, 1991), the cations and anions involved can be supposed to be in Nernst equilibrium:

$$\frac{C_{\text{ER}}}{C_{\text{cyt}}} = \frac{A_{\text{cyt}}}{A_{\text{ER}}} = \exp\left(\frac{F\Delta\psi}{RT}\right) . \tag{14}$$

Since any exchange of ions with the extracellular medium is neglected, the ion concentrations obey, in addition to eq. (14), the conservation relations

$$C_{\text{tot}} = C_{\text{cyt}} + \rho C_{\text{ER}} , \quad A_{\text{tot}} = A_{\text{cyt}} + \rho A_{\text{ER}} . \tag{15a,b}$$

Eqs. (14) and (15) imply

$$C_{\text{cyt}} = \frac{C_{\text{tot}}}{1 + \rho\exp\left(\dfrac{F\Delta\psi}{RT}\right)} , \quad A_{\text{cyt}} = \frac{A_{\text{tot}}}{1 + \rho\exp\left(-\dfrac{F\Delta\psi}{RT}\right)} . \tag{16a,b}$$

Using the electroneutrality condition (13) for the cytosol and eqs. (16a,b), $\Delta\psi$ can be expressed in terms of the other two model variables, Ca_{cyt} and Pr^{I},

$$\Delta\psi = \frac{RT}{F}\ln\left(\frac{-b - \sqrt{b^2 - 4ac}}{2a}\right) , \tag{17}$$

where $\quad a = \rho(Q - A_{\text{tot}})$, $b = C_{\text{tot}} - A_{\text{tot}} + Q(1 + \rho^2)$, $c = \rho(Q + C_{\text{tot}})$,

with $\quad Q = 2Ca_{\text{cyt}} - 2Pr^{\text{I}} - 2\frac{K_d^{\text{II}} Pr_{\text{tot}}^{\text{II}}}{K_d^{\text{II}} + Ca_{\text{cyt}}} .$

Table 1. The model parameters for which all results are calculated unless otherwise stated.

Parameter	Meaning	Value
Geometrical parameters		
V_{cyt}	Volume of the cytosol	$5.84 \cdot 10^{-8}\,\text{cm}^3$
S	ER surface area	$6.16 \cdot 10^{-3}\,\text{cm}^2$
ρ	Volume ratio ER/cytosol	0.01
Conservation sums		
C_{tot}	Total concentration of cations	$5 \cdot 10^3\,\mu\text{M}$
A_{tot}	Total concentration of anions	$3.65 \cdot 10^3\,\mu\text{M}$
Ca_{tot}	Total concentration of calcium	$45\,\mu\text{M}$
Pr_{tot}^{I}	Total concentration of slow binding sites	$600\,\mu\text{M}$
Pr_{tot}^{II}	Total concentration of fast binding sites	$120\,\mu\text{M}$
Kinetic and thermodynamic parameters		
K_{Ca}	Half-saturation concentration for calcium	$5\,\mu\text{M}$
k_{-}^{I}	Off rate constant of slow binding sites	$0.005\,\text{s}^{-1}$
k_{+}^{I}	On rate constant of slow binding sites	$0.001\,\mu\text{M}^{-1}\text{s}^{-1}$
K_{d}^{II}	Dissociation constant of fast binding sites	$5\,\mu\text{M}$
$\bar{g}_{Ca}$	Maximal channel conductance per unit area	$18\,\mu\text{S}\,\text{cm}^{-2}$
k_{pump}	Rate constant of the calcium pump	$16.7\,\text{s}^{-1}$
κ_{leak}	Rate constant of the leak	$10.0\,\mu\text{M}\,\text{V}^{-1}\text{s}^{-1}$
T	Temperature	$293\,\text{K}$

Our model involves three main variables (Ca^{2+} concentration in the cytosol Ca_{cyt}, concentration of free slowly binding proteins Pr^{I}, and potential difference $\Delta\psi$ across the ER membrane). All other model variables can be expressed in terms of these variables. Note that $\Delta\psi$ is an explicit function of Ca_{cyt} and Pr^{I} (eq. (17)), which means that our model system is actually two-dimensional.

The model parameters used in our calculations are listed in Table 1. Most of them are commented on in our previous paper (Marhl et al., 1997). For the rate constants k_{+}^{I} and k_{-}^{I} of the binding to, and dissociation from, the slow binding sites, respectively, we use very low values to clearly demonstrate the effect of two very different classes of binding sites. For parvalbumin, for example, Smith et al. (1996) report values of $k_{+}^{I} = 6\,\mu\text{M}^{-1}\text{s}^{-1}$ and $k_{-}^{I} = 1\,\text{s}^{-1}$, but mention that k_{+}^{I} is lower under physiological conditions because Mg^{2+} must dissociate before calcium can bind. Diffusional resistance for calcium might also lower these rate constants. The values of the dissociation constants used in our calculations are in qualitative agreement with the values reported by Jafri et al. (1992) and Smith et al. (1996).

Important steps in the analysis of any dynamical system are the determination of steady states and their stability. We can express the steady-state condition for eqs. (2) and (1a) as:

$$\frac{dCa_{cyt}}{dt} = f_1\left(Ca_{cyt}, Pr^{I}, \Delta\psi\left(Ca_{cyt}, Pr^{I}\right)\right) = 0 \ , \tag{18}$$

$$\frac{dPr^{I}}{dt} = f_2\left(Ca_{cyt}, Pr^{I}, \Delta\psi\left(Ca_{cyt}, Pr^{I}\right)\right) = 0 \ . \tag{19}$$

For any given set of model parameters, solving eqs. (18,19) simultaneously and using eq. (17) gives the steady-state values $\overline{Ca}_{cyt}$, $\overline{Pr}^I$, and $\overline{\Delta\psi}$ of the main model variables. An analytical calculation of these values is not straightforward because the variables enter the equations in a non-linear way. This difficulty can be circumvented by formally considering one variable to be known and some parameter as a function of this variable, as explained in Marhl et al. (1997). For the parameters given in Table 1, the steady-state values $\overline{Ca}_{cyt}$ are plotted versus the parameter k_{pump} in Fig. 2. Similar plots are obtained also for $\overline{Pr}^I$ and $\overline{\Delta\psi}$. For certain other parameter values, there occur three coexisting stationary states.

For the calculated steady states a local stability analysis on the basis of the two-dimensional Jacobian matrix has been performed. Care has to be taken to include the quasi-electroneutrality condition appropriately. Analysis of the trace and determinant of the Jacobian, in particular of their signs, allows one to distinguish between different types of steady states (e.g. stable node, unstable focus, etc., cf. Strogatz, 1994; Goldbeter, 1996; Heinrich and Schuster, 1996).

The transition from an unstable focus to a stable focus or *vice versa* is called a Hopf bifurcation. It is determined by the conditions that the trace of the Jacobian equals zero and that its determinant is positive. Hopf bifurcations are supercritical or subcritical according to whether the limit cycle bifurcating from the steady state is stable or unstable, respectively (Strogatz, 1994). Both types of bifurcation have been found in our model, denoted in Fig. 2 by points A and B, respectively. The unstable limit cycle branching off at point B turns stable at a so-called saddle-node bifurcation of limit cycles at a value of k_{pump} which is only a little bit higher than the value at the subcritical Hopf bifurcation. Therefore, if we inserted, in Fig. 2, a curve representing the unstable limit cycle, it would be almost vertical. Whenever a saddle-node bifurcation of limit cycles accompanies a subcritical Hopf bifurcation, the latter is characterized by the phenomenon that the amplitude of the observed limit cycle grows beginning with a finite value. This is sometimes called hard excitation, as opposed to the weak excitation occurring at a supercritical Hopf bifurcation.

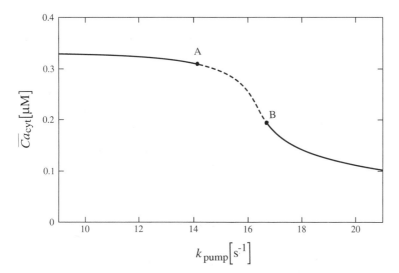

Fig. 2. Dependence of steady-state values Ca_{cyt} on the parameter k_{pump}. Parameter values are as given in Table 1. Stable steady states are indicated by solid lines, while unstable foci are plotted as dashed lines. A and B indicate the supercritical and subcritical Hopf bifurcations, respectively, which occur at $k_{pump} = 14.15$ s^{-1} and 16.73 s^{-1}, respectively.

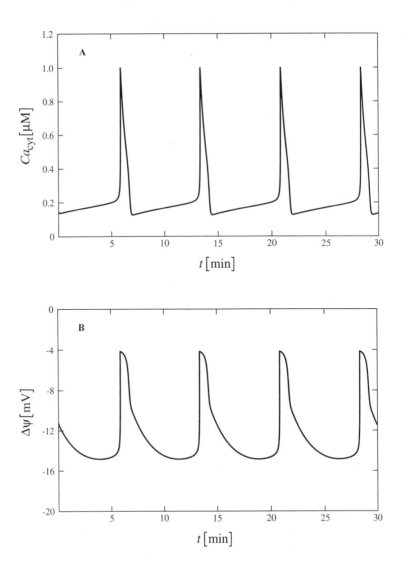

NUMERICAL SIMULATION OF OSCILLATIONS AND MODEL PREDICTIONS

The local stability analysis gives us ranges for model parameters for which oscillations can be expected. To show that self-sustained oscillations in the sense of limit cycles occur in the parameter range yielding unstable foci, numerical integration of the model equations is required. Some results are shown in Fig. 3. The initial conditions were chosen so as to fulfil electroneutrality in both compartments.

Our model reproduces the typical form of calcium spikes frequently observed experimentally (Woods et al., 1986; Berridge, 1989; Rooney et al., 1989). The potential difference oscillates in phase with the calcium spikes. When calcium is released from the ER lumen into the cytosol, the ER membrane is depolarized (i.e., the absolute value of $\Delta\psi$ decreases) and when the calcium is pumped back into the ER, the membrane is repolarized (cf. Läuger, 1991).

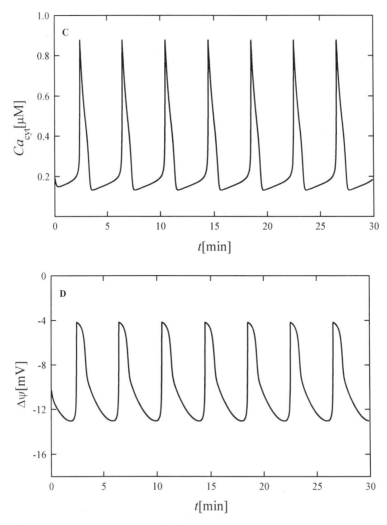

Fig. 3. Oscillations of the cytosolic calcium (A,C) and the potential difference across the ER membrane (B,D) for Pr_{tot}^{II} = 120 μM and $\tilde{g}_{Ca}$ = 18 μScm^{-2} (A,B) and $\tilde{g}_{Ca}$ = 19 μScm^{-2} (C,D) (see Table 1 for the values of other model parameters).

Fig. 3 shows the effect of increasing concentration of the agonist inducing the calcium oscillation (e.g., phenylephrine or vasopressin). The effect of the agonist can indirectly be taken into account by changing the parameter $\tilde{g}_{Ca}$, since this parameter is sensitive to IP$_3$ (cf. Dupont and Goldbeter, 1993; Marhl et al., 1997; and Appendix C in Chay et al., 1995). It can be seen in Fig. 3 that an increase in agonist concentration brings about an increased fequency with nearly unaltered amplitude and spike shape. This is in agreement with the well-known frequency-encoding phenomenon.

It is interesting to consider the interplay of the two classes of binding sites. After Ca^{2+} is released into the cytosol, the signalling proteins bind Ca^{2+} very fast, thus fulfilling their physiological role of receiving the signal. After a short delay, the higher affinity of the slow buffering proteins causes a shift of the bound Ca^{2+} to these proteins. It may be supposed that the physiological function of this transfer of Ca^{2+} from signalling proteins to buffering proteins is to inactivate the spike and enable oscillations. In the next phase, Ca^{2+} is slowly released from the buffering sites and pumped back into the ER.

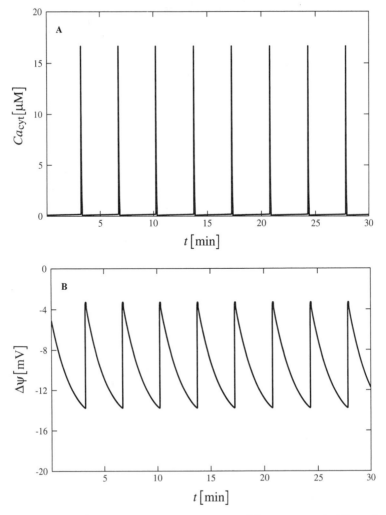

Fig. 4. Oscillations of the cytosolic calcium (A) and the potential difference across the ER membrane (B) for $Pr_{tot}^{II} = 0$ (see Table 1 for the values of other model parameters).

Fig. 4 shows the oscillations for the case where there is only one class of calcium-binding sites, which bind slowly. It can be seen that in comparison to the case of two different classes of binding sites, the amplitudes of calcium oscillations and potential oscillations are higher and lower, respectively. The results shown in Fig. 3 are in better agreement with experimental results (see Discussion).

DISCUSSION

In the present paper, we have developed an extended mathematical model accounting for oscillations of intracellular calcium and the electric potential difference across the membrane of the endoplasmic reticulum. The model describes both the switch between stationary and pulsatile regimes, which serves as a digital signal, and changes in oscillation frequency, which serve as analogue signals. The onset or termination of oscillations can be induced by changing a parameter, for example the rate constant of the ATPase, k_{pump}. Calcium oscillations can indeed be triggered or stopped by addition of specific inhibitors of the ER Ca^{2+}-ATPase (Foskett et al., 1991). Both effects can be simulated by our model when k_{pump} is gradually decreased from a given high value, so that two consecutive Hopf bifurcations occur. For the *in vivo* situation, these transitions are in most cases induced by a change in the maximal channel conductance, $\tilde{g}_{Ca}$, due to an increase in IP_3 concentration, for example (DeYoung and Keizer, 1992; Chay et al., 1995).

The oscillation period obtained is in good accordance with the upper end of the range of experimental values reported. By changing parameter values, we are able to decrease the period down to several seconds, which is at the lower end of the experimental values (see also Marhl et al., 1997). Calcium oscillations as well as other biological oscillations are often referred to as frequency encoded, because their frequencies usually change more significantly than their amplitudes, *e.g.* upon hormone stimulation (cf. Goldbeter et al., 1990). This fact can be reproduced by our model, both in the case of only one (slow) class of calcium-binding sites (Marhl et al., 1997) and in the presence of slow and fast binding sites.

It may be supposed that frequency encoding is particularly well feasible near the subcritical Hopf bifurcation, because frequency there changes more significantly than amplitude upon variation of the bifurcation parameter. This is, in addition, due to the fact that calcium oscillations represent a special type of relaxation oscillations, because they comprise fast movements (spikes) and slow movements (intermediate phases between spikes). Changes in oscillation period are mainly due to an increase or decrease in the duration of the intermediate phase.

The jump-like transition from stable steady states to self-sustained osillations at subcritical Hopf bifurcations corresponds well to experimental observations. For example, challenge of hepatocytes by $0.5\,\mu M$ phenylephrine leads to relaxation to the original steady state after a single calcium spike, while stimulation by $0.6\,\mu M$ phenylephrine triggers oscillations with a period of about $100\,s$ (Woods et al., 1986; Rooney et al., 1989). Interestingly, in the cited experiments, the first spike in an oscillatory pattern after hormone stimulation has almost the same amplitude as the subsequent ones, although Hopf bifurcations need not necessarily imply a very sharp jump because the oscillation amplitude may be attained after a transient.

The shape of the calcium spikes slightly differs in different models. While in some models (Cuthbertson and Chay, 1991; Chay et al., 1995) spikes with a sudden upstroke, an exponential decrease and a constant intermediate phase are found, Dupont and Goldbeter (1993) found more symmetric spikes with a slightly increasing concentration in the intermediate phase. The model of Jafri and Gillo (1994) yields sinusoidal intermediate phases. Our model gives slightly asymmetric spikes where the upstroke is somewhat faster than the decrease. This is in agreement with experimental results (Woods et al., 1986; Rooney et al., 1989), and so is the resting level of about $200\,nM$. The shape of calcium spikes found in our calculations is partly due to inclusion of the membrane potential. During the upstroke, the potential is depolarized, which implies that the driving force of the calcium efflux from the store is diminished both by the decrease in the calcium gradient and by the decrease in the electric gradient. The intermediate phases produced by our model resembles those obtained by Dupont and Goldbeter (1993).

Following the approach of Jafri and Gillo (1994), we have included the effect of other ions. We have not only included cations, but also anions since they contribute to the charge balance by about the same extent. Since the permeability of the ER membrane for these ions is very high, they can be assumed to be in Nernst equilibrium. This allows us to keep the dimension of the differential equation system small, because we need not include differential equations for these ions. The model is likely to be applicable to other intracellular calcium stores, but care has to be taken on whether the equilibrium assumption for monovalent cations remains valid.

The binding of calcium to proteins has turned out to be of essential importance in the generation of membrane potential oscillations. Since free calcium has a very low intracellular concentration, it does not contribute very much to the electroneutrality balance and would not, without this calcium buffering system, make possible remarkable changes in transmembrane potential. While we considered only one class of (slow) binding sites in our previous model (Marhl et al., 1997), we have here extended the model by including a second class with very high binding velocities. This results in values for the amplitude of calcium oscillations of about 1 μM, which are in better agreement with experiment (Woods et al., 1986; Berridge, 1989; Cuthbertson, 1989) than the values of about 17 μM obtained when only one class of proteins is considered, both determined with parameters yielding realistic values of about 10 mV for the amplitude of the transmembrane potential.

The effect of inclusion of a fast binding class of proteins (signalling proteins) besides the buffering proteins on the amplitudes of potential and cytosolic calcium can be rationalized as follows. The additional proteins rapidly bind the calcium flowing into the cytosol, thus decreasing the amplitude of free calcium. The amplitude of $\Delta\psi$, however, does not decrease (or it even slightly increases), because the calcium flux somewhat increases and, hence, slightly higher fluxes of anions in the same direction across the ER membrane and of cations other than calcium in the opposite direction are induced. The presence of slowly binding proteins is still necessary for the generation of oscillations because of the time delay they cause. Indeed, if one included only binding sites with very high binding rates so that rapid equilibrium could be assumed, the effective dimension of the differential equation system would reduce to one, so that oscillations would not arise.

It has sometimes been stated that a permanent calcium entry into, and efflux out of, the cytosol is necessary for oscillations to arise (Goldbeter et al., 1990; Dupont and Goldbeter, 1993; Jafri and Gillo, 1994). In our model, we have neglected any exchange of calcium with the surroundings of the cell (see also DeYoung and Keizer, 1992). The results show that oscillations can arise all the same. This is not, however, in contradiction of the above-mentioned view, since the role of alternating supply and withdrawal of calcium is played, in our approach, by the fluxes of the dissociation and binding of calcium to and from binding sites.

In future extensions of the model, it will be worthwhile also including the binding of calcium to proteins inside the ER. Furthermore, the sites of such proteins that cannot bind calcium but the charges of which contribute to the electric balance may shift the values of $\Delta\psi$. It would also be interesting to include, in addition to the ER, mitochondria as calcium stores. It may be supposed that this would lead to more constant amplitudes of the calcium spikes because any excess calcium is sequestered by mitochondria.

One of the physiological roles of calcium oscillations is likely to be the regulation of protein phosphorylation, which is mediated by the calcium-binding proteins (Cuthbertson, 1989; Goldbeter et al., 1990). Our model shows that the slowly binding proteins have rather smooth time curves, while the fast binding proteins exhibit spikes in their time behaviour (not shown). This difference in behaviour is likely to have an effect on the regulation of enzyme cascades (dynamic encoding), which deserves to be studied in the future.

Acknowledgement

The financial support for a 6 months stay of M.M. at Humboldt University (Berlin) by the Slovenian Science Foundation and for a 3 months stay of S.S. at the University of Maribor by the Ministry of Science and Technology of the Republic of Slovenia is gratefully acknowledged.

REFERENCES

Beeler, T.J., Farmen, R.H., and Martonosi, A.N. , 1981, The mechanism of voltage-sensitive dye responses on sarcoplasmic reticulum, *J. Membrane Biol.* 62:113–137.

Berridge, M.J., 1989, Cell signalling through cytoplasmic calcium oscillations, in: *Cell to Cell Signalling: From Experiments to Theoretical Models*, A. Goldbeter, ed., Academic Press, London, pp. 449–459.

Borghans, J.A.M., Dupont, G., and Goldbeter, A., 1997, Complex intracellular calcium oscillations. A theoretical exploration of possible mechanisms, *Biophys. Chem.* 66:25–41.

Brumen, M., and Heinrich, R., 1984, A metabolic osmotic model of human erythrocytes, *BioSystems* 17:155-169.

Chay, T.R., Lee, Y.S., and Fan, Y.S. (1995) Appearance of phase-locked Wenckebach-like rhythms, Devil's Staircase and universality in intracellular calcium spikes in non-excitable cell models, *J. theor. Biol.* 174:21–44.

Cuthbertson, K.S.R., 1989, Intracellular calcium oscillations, in: *Cell to Cell Signalling: From Experiments to Theoretical Models*, A. Goldbeter, ed., Academic Press, London, pp. 435–447.

Cuthbertson, K.S.R., and Chay, T.R., 1991, Modelling receptor-controlled intracellular calcium oscillators, *Cell Calcium* 12:97–109.

Dawson, A.P., Rich, G.T., and Loomis-Husselbee, J.W., 1995, Estimation of the free $[Ca^{2+}]$ gradient across endoplasmic reticulum membranes by a null-point method, *Biochem. J.* 310:371–374.

DeYoung, G.W., and Keizer, J., 1992, A single-pool inositol 1,4,5-trisphosphate-receptor-based model for agonist-stimulated oscillations in Ca^{2+} concentration, *Proc. Natl. Acad. Sci. USA* 89:9895–9899.

Dupont, G. Berridge, M.J., and Goldbeter, A., 1991, Signal-induced Ca^{2+} oscillations: properties of a model based on Ca^{2+}-induced Ca^{2+} release, *Cell Calcium* 12:73–85.

Dupont, G., and Goldbeter, A., 1993, One-pool model for Ca^{2+} oscillations involving Ca^{2+} and inositol 1,4,5-trisphosphate as co-agonists for Ca^{2+} release, *Cell Calcium* 14:311–322.

Foskett, J.K., Roifman, C.M., and Wong, D., 1991, Activation of calcium oscillations by thapsigargin in parotid acinar cells, *J. Biol. Chem.* 266:2778–2782.

Goldbeter, A., Dupont, G., and Berridge, M.J., 1990, Minimal model for signal-induced Ca^{2+} oscillations and for their frequency encoding through protein phosphorylation, *Proc. Nat. Acad. Sci. USA* 87:1461–1465.

Goldbeter, A., 1996, *Biochemical Oscillations and Cellular Rhythms*, Cambridge University Press, Cambridge.

Heinrich, R., and Schuster, S., 1996, *The Regulation of Cellular Systems*, Chapman & Hall, New York.

Jafri, M.S., and Gillo, B., 1994, A membrane potential model with counterions for cytosolic calcium oscillations, *Cell Calcium* 16:9–19.

Jafri, M.S., Vajda, S., Pasik, P., and Gillo, B., 1992, A membrane model for cytosolic calcium oscillations. A study using *Xenopous* oocytes, *Biophys. J.* 63:235–246.

Jouaville, L.S., Ichas, F., Holmuhamedov, E.L., Camacho, P., and Lechleiter, J.D., 1995, Synchronization of calcium waves by mitochondrial substrates in *Xenopus laevis* oocytes, *Nature* 377:438–441.

Läuger, P., 1991, *Electrogenic Ion Pumps*, Sinauer Associates, Sunderland.

Laurent, M., and Claret, M., 1997, Signal-induced Ca^{2+} oscillations through the regulation of the inositol 1,4,5-trisphosphate-gated Ca^{2+} channel: an allosteric model, *J. theor. Biol.* 186:307–326.

Li, Y.-X., and Rinzel, J., 1994, Equations for InsP$_3$ receptor-mediated $[Ca^{2+}]_i$ oscillations derived from a detailed kinetic model: A Hodgkin-Huxley like formalism, *J. theor. Biol.* 166:461–473.

Marhl, M., Schuster, S., Brumen, M., and Heinrich, R., 1997, Modelling the interrelations between calcium oscillations and ER membrane potential oscillations, *Biophys. Chem.* 63:221–239.

Meyer, T., and Stryer, L., 1988, Molecular model for receptor-stimulated calcium-spiking, *Proc. Nat. Acad. Sci. USA* 85:5051–5055.

Meyer, T., and Stryer, L., 1991, Calcium spiking, *Ann. Rev. Biophys. Chem.* 20:153–174.

Petersen, O.H., Petersen, C.C.H., and Kasai, H., 1994, Calcium and hormone action, *Annu. Rev. Physiol.* 56:297-319.

Rooney, T.A., Sass, E.J., and Thomas, A.P., 1989, Characterization of cytosolic calcium oscillations induced by phenylephrine and vasopressin in single fura-2-loaded hepatocytes, *J. Biol. Chem.* 264:17131–17141.

Schultz, S.G., 1980, *Basic Principles of Membrane Transport*, Cambridge University Press, Cambridge.

Schuster, S., and Mazat, J.-P., 1993, A model study on the interrelation between the transmembrane potential and pH difference across the mitochondrial inner membrane, in: *Modern Trends in Biothermokinetics*, S. Schuster, M. Rigoulet, R. Ouhabi, and J.-P. Mazat, eds, Plenum, New York, pp. 33–44.

Smith, G.D., Wagner, J., and Keizer, J., 1996, Validity of the rapid buffering approximation near the point source of calcium ions, *Biophys. J.* 70:2527–2539.

Somogyi, R. and Stucki, J.W., 1991, Hormone-induced calcium oscillations in liver cells can be explained by a simple one pool model, *J. Biol. Chem.* 266:11068–11077.

Strogatz, S., 1994, *Nonlinear Dynamics and Chaos. With Applications to Physics, Biology, Chemistry, and Engineering*, Addison-Wesley, Reading (Mass.).

Woods, N.M., Cuthbertson, K.S.R., and Cobbold, P.H., 1986, Repetitive transient rises in cytoplasmic free calcium in hormone-stimulated hepatocytes, *Nature* 319:600–602.

A MODEL OF LHRH "SELF-PRIMING" AT THE PITUITARY

Sinéad Scullion, David Brown* and Gareth Leng

Department of Physiology
University of Edinburgh
Edinburgh, EH8 9AG
UK

*Laboratory of Biomathematics
Department of Neurobiology,
Babraham Institute
Cambridge, CB2 4AT
UK

INTRODUCTION

Luteinising hormone (LH) is secreted from specialised cells (gonadotrophs) in the anterior pituitary gland, and regulates ovarian follicular development and ovulation. The secretory pattern of LH changes throughout the reproductive cycle (the menstrual cycle in women, the oestrous cycle in other female mammals). In all female mammals, a large pre-ovulatory surge in the release of LH is the main trigger for ovulation to occur; at other times LH is released into the systemic circulation in small intermittent pulses, the frequency and amplitude of which are critical for follicular development. LH is packaged into large secretory granules within the gonadotrophs, and these granules are released from the pituitary gland in response to the secretion of LH-releasing hormone (LHRH) from neurones in the hypothalamus. As a consequence of electrical activation of LHRH neurones, LHRH is released into blood vessels which link the hypothalamus and pituitary.

However, the pre-ovulatory LHRH surge alone is too small to trigger the pre-ovulatory LH surge (Fink et al., 1982). The LH surge also depends on a marked increase in pituitary responsiveness to LHRH. This is initiated by the pre-ovulatory rise in oestrogen, which

Information Processing in Cells and Tissues
Edited by Holcombe and Paton, Plenum Press, New York, 1998

influences gene expression in pituitary gonadotrophs. However, the very rapid rise in pituitary responsiveness on the day of the surge is due to the "self-priming effect" of LHRH. In pituitaries which have been exposed to oestrogen, initial exposure to LHRH causes a characteristic and dramatic potentiation of subsequent secretory responses to LHRH (Aiyer et al., 1974; Wang et al., 1976).

The magnitude of LHRH self-priming is highly dependent upon exposure to oestrogen, and it varies throughout the reproductive cycle (Aiyer et al., 1973; Byrne et al., 1996). The cellular mechanism of self-priming depends on oestrogen-induced protein synthesis (Curtis et al., 1985), and on the functional integrity of microfilaments (Pickering and Fink, 1979a). The microfilaments appear to be part of the cellular machinery that is important for making secretory granules available for release. The priming effect of LHRH on LH secretion is accompanied by a significant "margination" of secretory granules (Lewis et al., 1984) and an increase in microfilament length and a change in their orientation relative to the plasma membrane in gonadotrophs.

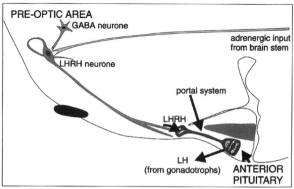

Figure 1. A schematic representation of a para-saggital section through the rat hypothalamus and pituitary. The figure shows the adrenergic projection from the brain-stem, the pre-optic area of the hypothalamus containing LHRH and GABA neurones, the portal system and the anterior pituitary with gonadotrophs releasing LH into the blood.

Inositol phosphate production and mobilisation of intracellular calcium stores, which are involved in LHRH-induced exocytosis (Stojilkovic et al., 1994), are enhanced by prior exposure to LHRH without an alteration of LHRH binding to LHRH receptors (Mitchell et al., 1988). Influx of extracellular calcium is also unchanged. Therefore self-priming involves changes in intracellular signalling and is not simply the result of increased LHRH receptor expression. It is also time-dependent, and occurs within 30-40 minutes of LHRH exposure *in vitro* (Waring and Turgeon, 1983) and within 45-60 minutes of exogenous LHRH administration *in vivo* in the absence of endogenous LHRH release (Fink et al., 1976; Aiyer et al., 1974). This may represent the time necessary for new protein synthesis and modification of existing proteins, as well as for changes in microfilament orientation and margination of secretory granules.

In the original model (Brown et al., 1994) LH release (z) is modelled as a non-linear function of LHRH output (v).

$$\frac{dz}{dt} = p(v) - d_1 z \tag{1}$$

where $p(v)$ is a logistic function of v and d_l is the decay rate of LH. The non-linearity reflects both the fact that sufficiently small fluctuations in v have no effect on z and the pituitary response can be saturated. The model of LHRH actions at the pituitary [equation (1)] was adapted to incorporate the changes that occur in pituitary responsiveness to LHRH throughout the course of the reproductive cycle.

THE MODEL

LH release is based on the Law of Mass Action, the central process being the reversible binding of LHRH, in the vicinity of the gonadotrophs (r), to its free receptor (f), to produce a bound complex (u), $r + f \Leftrightarrow u$.

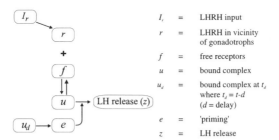

I_r = LHRH input

r = LHRH in vicinity of gonadotrophs

f = free receptors

u = bound complex

u_d = bound complex at t_d where $t_d = t\text{-}d$ (d = delay)

e = 'priming'

z = LH release

Figure 2. A schematic form of the model of LHRH "self-priming" at the pituitary. The central process is the binding of LHRH (r) to its free receptor (f) at the pituitary to form a bound complex (u). LH release (z) is non-linearly related to the current levels of bound complex (u) and the levels of bound complex present some time before the current time (u_d).

The rate constants of the forward and backward reactions are k_1 and k_2 respectively:

$$\frac{dr}{dt} = I_r - k_1 rf + k_7 k_2 u - k_3 r \tag{2}$$

$$\frac{df}{dt} = -k_1 rf + k_2 u \tag{3}$$

$$\frac{du}{dt} = k_1 rf - k_2 u \tag{4}$$

I_r is the rate of LHRH input either from the hypothalamus, or exogenously given in the *in vivo* case, or added to the medium surrounding isolated cells *in vitro*. LHRH is removed at the rate $k_3 r$, in the bloodstream or in the medium. LH release occurs at a rate that is non-linearly linked to the amount of binding of LHRH to its receptor, saturation data for which suggests a single component of binding (Mitchell et al., 1988). This model is an extension of one developed previously for growth hormone-releasing hormone induced growth hormone release (Stephens et al., 1996). The constant of proportionality involves the quantity e, which is a measure of the extent of priming.

$$\text{release rate} = LR = k_6 \phi_n(u)(e + b)k_1 rf \tag{5}$$

where b is a constant and

$$\phi_n(u) = \frac{1}{1 + \exp[-(u - u_{0n})/\delta_n]}$$ (6)

$$\frac{de}{dt} = k_4 \phi_d(u_d) + k_8 u_d - k_5 e$$ (7)

$$\phi_d(u_d) = \frac{1}{1 + \exp[-(u_d - u_0)/\delta]}$$ (8)

$\phi_n(u)$ and $\phi_d(u_d)$ switch between high and low values depending on the level of u and u_d relative to their critical values (u_{0n} and u_0 respectively). The switch can be made more severe by decreasing δ_n or δ.

The extent of priming (e) is dependent, not on the current (time t) levels of bound receptor (u), but rather on u_d, the levels of bound receptor at time t_d, where $t_d = t - d$ and d is a specified time delay between LHRH receptor binding and the onset of priming. Therefore the effect of priming on LHRH-induced LH release is dependent on previous exposure to LHRH as well as the duration of the exposure and the time between exposures. LH concentrations also depend upon its half-life.

MODELLING *IN VIVO* DATA

To investigate the possibility of producing behaviour similar to that seen *in vivo* during LHRH self-priming, several simulations* of the adapted model of LHRH actions at the pituitary (Figure 2, equations 2-8) were run to obtain fits for biological data.

In the *in vivo* experiments of Fink et al. (1976), a continuous infusion, or small frequent injections (pulses) of synthetic LHRH were administered intravenously to rats just before the expected LH surge. During the infusions a total of 75ng LHRH per 100g body weight was administered over 45 or 90 min and the plasma LHRH and LH were measured throughout (Figure 4). Infusion of LHRH for 45 min resulted in a sharp increase in LHRH levels but only a gradual increase in plasma LH. Administration of the same total dose over 90 min resulted in a more gradual increase in plasma LHRH which reached a lower level than seen with the 45 min infusion. The plasma LH also increased gradually for the first 45 min after which there was a rapid increase to a higher level than seen with the 45 min infusion, despite the lower LHRH concentration (LHRH self-priming).

Administration of LHRH infusions, to match those described above, in the model results in a very close fit to the experimental data for the 90 min infusion (Figures 3(a) & 4). Both the plasma LHRH and LH concentrations match throughout the course of the 120 min experiment. For the 45 min infusion data, the model indicates higher expected values of LHRH than were observed, and similarly higher LH values, but a very similar pattern for both.

---

* Simulations carried out with a FORTRAN 77 program using NAG mark 15 numerical routines and Digital Alpha Workstation ®.

Discrete pulses of synthetic LHRH were administered, *in vivo*, in the experiments of Aiyer et al. (1974), [Figure 5(a)]. Two successive injections, each of 50ng LHRH per 100g body weight, were administered intravenously to rats just before the expected LH surge. Pulses were separated by intervals of 30, 60, 120 and 240 min and the plasma LH was measured before and after each injection. The LH response, as measured by the maximal increment in plasma LH, to the second pulse of LHRH was significantly higher than the response to the first for all intervals tested. The response was greatest when the pulses were separated by a 60 min interval, being significantly greater than when separated by 30 or 240 min. The model data [Figures 3(b) & 5(b)] following administration of LHRH pulses show a very similar pattern to the *in vivo* data for the response to the second pulse for all intervals tested. However, the magnitude of the first response is smaller than anticipated.

Figure 3. The effect of varying the LHRH input on priming and LH release in the model described. The panels, from top to bottom, show plasma LHRH concentration, the proportion of bound LHRH receptors, e (a measure of the extent of priming), and plasma LH concentration. Parameter values for these simulations: $k_1=5.0$, $k_2=5.0$, $k_3=2.0$, $k_4=2.5$, $k_5=0.25$, $k_6=7.5$, $k_7=0.01$, $k_8=2.5$, $u_{0n}=0.15$, $u_0=0.2$, $\delta_n=0.0001$, $\delta=0.0001$ and d=30min.

(a)

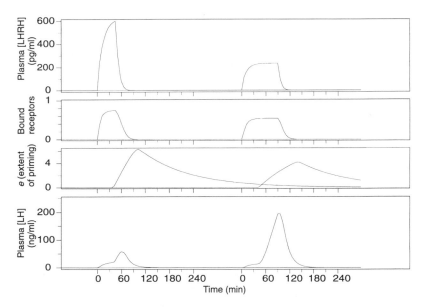

Figure 3(a). The effect of continuous infusions of LHRH in the model. The effects of two infusions are overlaid, the first lasts for 45 min, and the second for 90 min. Infusing LHRH over 45 min results in a sharp increase in LHRH levels and in the levels of bound receptor. While e, the extent of priming, also increases it has no effect on LH release since the proportion of bound receptors has decreased to almost zero at this time. The plasma LH concentration remains low throughout. Administration of the same dose over 90 min results in a more gradual increase in the LHRH concentration and in the proportion of bound receptors, to a lower level than seen with the 45 min infusion. The extent of priming (e) also increases, though to a lower level than in the 45 min infusion. However, with a 90 min LHRH infusion, the increase in e has an effect on LH release since the level of bound receptors remains high after its first appearance. Plasma LH concentrations increase sharply to much higher levels than are seen with the 45 min LHRH infusion.

(b)

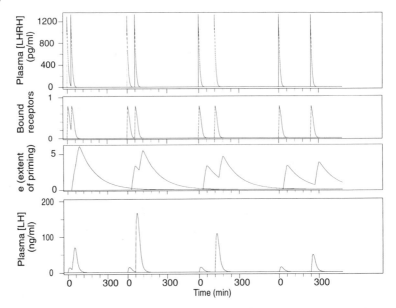

Figure 3(b). The effects of two successive pulses of LHRH, separated by 30, 60, 120 and 240 min are shown. Each pulse of LHRH results in an increase in the proportion of bound receptors and in *e*, the extent of priming. The LH response to the second pulse is always greater than that to the first, and is maximal when the pulses are separated by 60 min.

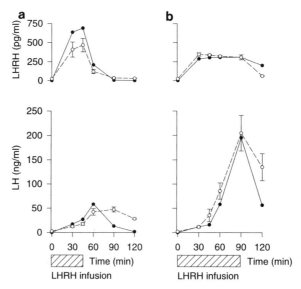

Figure 4. The effect of infusions of LHRH on plasma LHRH and LH levels. The effect of continuous infusions of LHRH on plasma LHRH (top) and LH levels (bottom) *in vivo* (open circles) and in the model of LHRH actions at the pituitary described above (closed circles). In the *in vivo* experiments (data from Fink et al. 1976), 75ng LHRH per 100g body weight was administered to rats just before the expected LH surge over 45 (a) or 90 (b) min. Infusing LHRH for 45 min (a) resulted in a sharp increase in LHRH levels but only a gradual increase in plasma LH. Administration of the same dose over 90 min (b) resulted in a more gradual increase in plasma LHRH to a lower level. The plasma LH concentration also increased gradually, until 45-60 min after the start of the infusion. At this time the LH concentration increased rapidly to a higher level than seen with the 45 min infusion despite the lower LHRH concentration (LHRH self-priming, see text). The model data show a very close fit to the experimental data for the 90 min infusion (b). For the 45 min infusion data, the model indicates higher expected values of LHRH than were observed, and similarly higher LH values, but a very similar pattern for both.

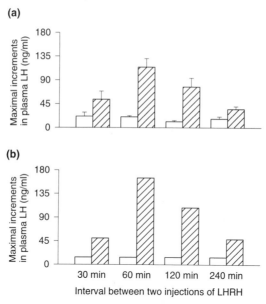

Figure 5. The effect of two successive LHRH pulses on plasma LH. The effect of two LHRH pulses on plasma LH *in vivo* (a) and in the model (b) of LHRH actions at the pituitary described above. In the *in vivo* experiments (data from Aiyer et al., 1974), rats received two intravenous injections of 50 ng LHRH per 100 g body weight, just before the expected LH surge. Injections were separated by 30, 60, 120 or 240 min and the maximal increments in plasma LH were measured (a). The LH response to the second injection (shaded bars) was significantly higher than the first (open bars) for all intervals tested. The response was greatest when the pulses were separated by a 60 min interval. The model data (b) show a very similar pattern to the *in vivo* data for all the intervals tested. However, the magnitude of the first response is smaller than anticipated.

CONCLUSIONS AND FURTHER DEVELOPMENT

In the proposed model [Figure 2, equations (2-8)], the complex changes in pituitary cells that lead to self-priming are modelled as a single component (*e*) which is increased by prior exposure to LHRH. It is a measure of the extent of granule margination, and therefore increases in LH available for release as well as the changes in intracellular signalling that occur during the process of priming. Since LH release is dependent on both the presence of bound receptors at time *t*, as well as *e*, which rises as a consequence of bound receptors at time t_d, LHRH self-priming has no effect on LH release during short infusions. While we do not claim that the changes leading to increased pituitary responsiveness in LHRH self-priming are as simple as the model would suggest, nevertheless we have shown that incorporation of this single parameter is capable of producing results consistent, qualitatively and quantitatively, with *in vivo* results from short term experiments.

The model at present contains no component related to receptor desensitisation, such as proposed by Goldbeter (1996), which occurs when exposure to LH continues over prolonged periods. This would result in a decrease in pituitary sensitivity and LHRH-induced LH release would decrease as exposure continued. There is also no component related to the depletion of LH stores which would also occur should LHRH exposure continue at high levels, or any component reflecting regulated synthesis of LH, all of which may be expected to influence the experimentally observed profile of LH secretion over the longer term. Thus to model the evolving changes in pituitary sensitivity throughout the reproductive cycle the inclusion of components related to these other factors affecting LHRH-induced LH release must be considered. There are several ways to extend the model to incorporate some of these components, but it is not clear that there is sufficient experimental data to resolve them in a useful predictive way. However, by contrast, there are a wide range of data available from a variety of experimental studies (for instance the *in vitro* data of Waring and Turgeon, 1983) for comparing model predictions with acute, short term experimental data.

Thus we note here that substitution of this model of the pituitary release mechanism for equation 1 in the model of the LHRH pulse generator proposed by Brown et al. (1994), may provide a model of the LHRH-LH system that is capable of modelling *in vivo* patterns of LH release throughout the course of the reproductive cycle and is not limited to particular steroid states. Comparison of model predictions with experimental data obtained following acute experimental observations will accordingly provide appropriate tests of the Brown model, though experimental observations reflecting chronic responses will require further elaboration of the model.

Acknowledgements

The authors thank Dr. Colin Brown for careful reading of this manuscript.

Authors' email addresses

s.scullion@ed.ac.uk
d.brown@bbsrc.ac.uk
g.leng@ed.ac.uk

REFERENCES

Aiyer, M.S., Fink, G., and Greig, F., 1973, Sensitivity of the anterior pituitary gland to luteinizing hormone releasing factor (LRF) during pro-oestrus in the rat, *J. Physiol.* 23:32P

Aiyer, M.S., Chiappa, S.A., and Fink, G., 1974, A priming effect of luteinizing hormone releasing factor on the anterior pituitary gland in the female rat, *J. Endocrinol.* 62:573

Brown, D., Herbison, A.E., Robinson, J.E., Marrs, R.W., and Leng, G., 1994, Modelling the luteinizing hormone-releasing hormone pulse generator, *Neuroscience* 63:869

Byrne, B., Fowler, P.A,. and Templeton, A., 1996, Role of progesterone and nonsteroidal ovarian factors in regulating gonadotropin-releasing hormone self-priming in vitro, *J. Clin. Endocr. Metab.* 81:1454

Curtis, A., Lyons, V., and Fink, G., 1985, The priming effect of LH-releasing hormone: effects of cold and involvement of new protein synthesis, *J. Endocrinol.* 105:163

Fink, G., Chiappa, S.A., and Aiyer, M.S., 1976, Priming effect of luteinizing hormone releasing factor elicited by preoptic stimulation and by intravenous infusion and multiple injections of the synthetic decapeptide, *J. Endocrinol.* 69:359

Fink, G., Aiyer, M., Chiappa, S.A., Henderson, S., Jameison, M., Levy-Perez, V., Pickering, A., Sarkar, D., Sherwood, N., Speight, A., and Watts, A., 1982, Gonadotrophin-releasing hormone: Release into hypophyseal portal blood and mechanism of action, in: *Hormonally Active Brain Peptides: Structure and Function*, K. McKearns, ed., Plenum Press, New York.

Goldbeter, A., 1996, *Biochemical Oscillations and Cellular Rhythms: The molecular bases of periodic and chaotic behaviour*, Cambridge University Press, Cambridge.

Lewis, C.E., Morris, J.F., and Fink, G., 1985, The role of microfilaments in the priming effect of LH-releasing hormone: an ultrastructural study using cytochalasin B, *J. Endocrinol.* 106:211

Mitchell, R., Johnson, M., Ogier, S.A., and Fink, G., 1988, Facilitated calcium mobilization and inositol phosphate production in the priming effect of LH-releasing hormone in the rat, *J. Endocrinol.* 119: 293

Pickering, A.J., and Fink, G., 1976, Priming effect of luteinizing hormone releasing factor: in-vitro and in-vivo evidence consistent with its dependence upon protein and RNA synthesis, *J. Endocrinol.* 69:373

Pickering, A.J., and Fink, G., 1979, Priming effect of luteinizing hormone releasing factor in vitro: role of protein synthesis, contractile elements, Ca$\overline{2}$+ and cyclic AMP, *J. Endocrinol.* 81:223

Stephens, E.A., Brown, D., Leng, G., and Smith, R.G., 1996, A model of pituitary release of growth hormone, in: *Computation in Cellular and Molecular Biological Systems*, World Scientific, Singapore

Stojilkovic, S.S., Reinhart, J., and Catt, K.J., 1994, Gonadotropin-releasing hormone receptors: structure and signal transduction pathways, *Endocr. Rev.* 15:462

Wang, C.F., Lasley, B.L., Lein, A., and Yen, S.S., 1976, The functional changes of the pituitary gonadotrophs during the menstrual cycle, *J Clin Endocrin Metab* 42:718

Waring, D.W., and Turgeon, J.L., 1983, LHRH self priming of gonadotrophin secretion: time course of development, *Am. J. Physiol.* 244:C410

COMPUTATION AND INFORMATION
AN INTRODUCTION TO SECTION 2

Ray Paton

Department of Computer Science
University of Liverpool
Liverpool L69 3BX

The first paper in this section is by Lev Beloussov and deals with morphomechanical feedback in embryonic development. Since the development of organisms falls obviously into the category of self-organising processes it requires an essentially non linear feedback between its dynamic components. It is suggested that this feedback can be based upon interaction between passive (imposed from outside) and active (generated within a system itself) mechanical stresses: an externally perturbed part of a developing system moves towards restoration of its initial state of stress, but does this with an overshoot. This is defined as a hyperrestoration (HR) reaction. Some experimental data is reviewed indicating the morphogenetical role of mechanical stresses, which gives a list of the main HR reactions and of the 'standard morphomechanical situations' derived from the latter. An interpretation of a blastula-neurula developmental period of an amphibian embryo as a succession of HR's is presented, in which each next period is derived from the preceding one. The relation of the HR hypothesis to the concepts of positional information and field theories, as well as to morphogenetic functions of chemical inductors is briefly discussed. Morphomechanical feedback as a manifestation of information storage and transfer at a macroscopic level is considered.

The paper by Mark Butler, Ray Paton and Paul Leng considers the properties of non-neural tissues in living organisms and how they can lead to new methods of computation. First a generic characterisation of a tissue is proposed, consisting of properties common to all organisms and properties specific to tissues. Next models of tissues are considered. It is proposed to use a homogeneous network model as it is possible to perform all types of information processing using a mixture of feedforward and recurrent networks. These networks must be able to perform self-modification as information processing in biological systems is causally linked with the capability of the system to modify itself. Models are further complicated by the fact that the networks form modules, linked by both local and global interactions. To clarify these ideas, simple formalisms for self-modification and modularisation in information processing systems are considered. Finally there is a brief discussion of how these will be combined in future tissue models.

Information Processing in Cells and Tissues
Edited by Holcombe and Paton, Plenum Press, New York, 1998

Jerry Chandler's paper looks at issues associated with the semiotics of cells. How are "computational" dynamics of organic components organised to form organisms? The semiotic dilemma is exemplified by noting that the terms organic, organism, and organisation are all derived from the same root! The common semantic origin of these terms strongly suggests the existence of a common perceptual pattern underlying the phenomena of complex pattern generators. Presumably, these generators create cellular "computations." Which mathematical structures are sufficiently rich to support a unified representation of interdependence, of hierarchical degrees of organisation and of nonlinear dynamics? It has been suggested that category theory is appropriate. A notation for a hierarchical structure of the natural sciences is created which is consistent with categorical logic. A notational sequence for composing one - to - one correspondences among the degrees of organisation [O°] of material objects of a cellular system is proposed. Necessary and sufficient conditions for simple and complex systems are described. Causality is also exemplified in terms of the boundary conditions sustaining the system -- bottom - up, top - down, outside - inward and inside - outward. This scientific notation is contrasted with six species of signs of general semiotics.

Ron Cottam, Nils Langloh, Willy Ranson and Roger Vounck take an engineers look at some issues associated with localisation and nonlocality with regard to computation. It is becoming necessary to carefully re-evaluate the meaning attached to the term computation. This paper considers the conditions prerequisite to the implementation of a general description of computation which is formulated in terms of reactions to environmental stimuli, and which can be used to model natural processes and consequently information processing in cells and tissues. The first criterion for survival of an evolutionary entity must be objectivisation of external events, to enable internal computational processes to develop survival-oriented reactions. In its most general form an environmental-reaction processor will operate directly in the interactive domain described by physics, and not in an environment derived from computational preconceptions. The development of an archetypal form for such a processor leads to the suggestion of important relationships between the characteristics to be expected of environmental-reaction computation and electrical and chemical "nonlocal" processes found to operate in the brain.

The paper by Patrik D'haeseleer, Xiling Wen, Stefanie Fuhrman and Roland Somogyi looks at a method for mining the gene expression matrix. Automated measurement of mRNA levels of many different genes simultaneously enables us to take snapshots of the state of a tissue or cell type. As more data of this nature becomes available, the challenge lies in analysing the data in order to infer the underlying gene circuitry. They present a preliminary statistical analysis of the Gene Expression Matrix (*Wen et al.*, 1997), containing expression levels of 112 different genes at nine stages during rat cervical spinal cord development. The data set is analysed for linear and rank correlations, and methods for analysing time series of this type using information theoretic measures are illustrated.

Mike Holcombe and Alex Bell describe a computational model of part of the immune system. The model is novel in respect of it representing the populations of the T cells and other related components of the system as a state-based parallel model. This is then analysed with respect to possible spatial and temporal interactions with the result of a number of simulations presented and some conclusions drawn. One of the interesting aspects of the work is the attempt to construct an integrative model of a complex dynamic system such as the immune system.

Felix Hong notes that mesoscopic processes have been treated as a signal interface between intracellular biocomputing dynamics and macroscopic neural network biocomputing dynamics. In his paper, he shows that such processes exhibit their own rich internal dynamics, in which short-range molecular interactions play a significant role. The control laws governing mesoscopic processes are examined with the following problem in mind. On the one hand, the diffusion-reaction model of intramolecular processes implies biocomputing is excessively plagued by randomness. On the other hand, advocates of classical physical determinism claim that biological determinism is absolute. The dynamics of ion channel activation show that the control laws at the mesoscopic level are not even well defined but are defined probabilistically. At the macroscopic level, however, it reconverges to a well defined control law, as described by the Hodgkin-Huxley theory of action potential and the experimental data they generated. This line of evidence casts a serious doubt on the validity of absolute biological determinism. Nevertheless, a relative degree of determinism is preserved because of the reconvergence of the control law at the macroscopic level. He refrains from the claim of disproving absolute biological determinism since we also realise that absolute physical determinism as enunciated by Laplace is impossible to refute.

A computer system that links gene expression data to spatial organisation of *C. elegans* is presented in the paper by T. Kaminuma, T. Igarashi, T. Nakano and J. Miwa. Using the nematode as a model organism, they have developed a computerised-system that aims to represent embryonic development as the dynamic organisation of a structure, as formed by a cellular aggregate. The system also allows one to link intracellular phenomena, such as gene expression, to a member of the embryonic cell aggregate at a given stage. The system is based on a three layer model: the molecular level (Genomic Space that relates to gene expression, signal transduction pathways, etc.), cellular level (Cellular Space), and organismal level (Developmental Space of three dimensional cell aggregates). The system should be applicable to other multicellular organisms whose developmental processes are well known.

The paper by Ray Paton and Koichiro Matsuno seeks to extend integrative ideas about cellular information processing through a study of verbs, glue and category theory. The discussion is based on the notion of the cellular economy and linguistic metaphors are used to provide a rich framework for exploring relational biology. Ideas are developed in terms of present understanding of the roles of enzymes particularly in relation to a thermodynamic and informational perspective.

Mark Shackleton and Chris Winter present a conceptual architecture for information processing which is based on the processing of molecules carried out by enzymes within biological cells. The architecture is used to create a system which carries out a simple information processing task, namely that of sorting a set of numbers into an ordered list, and results are presented of the time evolution of the system performing this task. A method by which the information processing system can be automatically configured to perform specific application tasks is described, which employs an analogue of protein folding within an evolutionary framework. Features of the architecture are outlined including its applicability to pattern matching problems and improved evolvability versus traditional computer architectures and programming languages.

Roland Somogyi and Stefanie Fuhrman discuss the notion of distributivity as an information theoretic network measure. An important consideration in cellular information processing is the coding and information transmission principles underlying distributed signalling networks. Boolean networks serve as a model system in which these properties can be studied. Based on information theoretic measures, they provide a systematic decomposition of information contributions of individual elements and their respective combinations which determine the state of an output element. They demonstrate this in the analysis of rule tables defining the network architectures. Here they introduce the effective number of inputs of a rule, and its input and output redundancies, important considerations for coding efficiency. Finally, they define distributivity as a quantitative measure for the number of unique information sources controlling an output. While the paper focuses on binary network rules, the information measures introduced are general and can be applied to the analysis of a variety of discrete and continuous data sets.

Alexander Spirov and Maria Samsonova discuss the GeNet database which is designed for the description of genetic networks controlling embryogenesis. GeNet is intended to be a database that provides a new type oriented on representation of results of analysis of genetic networks structure, function and evolution. It contains a quantitative atlas of *Drosophila* segmentation genes expression. Using the information collected in GeNet as an example they discuss mechanisms of regulation of genes controlling early development stages in *Drosophila*. Gap and pair-rule genetic networks are outlined and the results of analysis of these ensembles by means of Boolean network theory are presented.

The paper by Joerg Wellner and Andreas Schierwagen describes the application of cellular-automata-like simulations to dynamic neural fields. Dynamic neural fields show a complex dynamic behaviour which has been investigated successfully by analytical methods so far only for the one-dimensional case. They concentrate our interests mainly to simulations of two spatial dimensions and their different dynamic patterns. In two examples they apply dynamic neural fields to problems in computational neuroscience.

In the concluding paper of this section Jens Ziegler, Peter Dittrich and Wolfgang Banzhaf seek to develop robot control systems with ideas gleaned from prokaryotic information processing systems. Bacteria must be able to detect rapid changes in their environment and to adapt their metabolism to external fluctuations. They monitor their surroundings with membrane-bound and intra-cellular sensors. The regulatory mechanisms of bacteria can be seen as a model for the design of robust control systems based on an artificial chemistry. The capability to process information which is needed to keep autonomous agents surviving in an unknown environment is discussed by means of two examples. (1) is inspired by polymerisation reactions and (2) is inspired by enzyme-substrate kinetics. First experimental results with a real mobile robot are presented.

MORPHOMECHANICAL FEEDBACK
IN EMBRYONIC DEVELOPMENT

L.V.Beloussov

Department of Embryology
Faculty of Biology
Moscow State University
Moscow 119899
Russia

INTRODUCTION

In modern science, a most adequate conceptual framework for treating the behaviour of complex dynamic systems is given by the theory of self-organization (e.g., Prigogine, 1980). The developing organisms may be definitely attributed to self-organizing entities by a number of criteria and, above all, by their capacity for spontaneous breaks of the symmetry order. We define those breaks of macroscopical symmetry as spontaneous which do not imply any definite macroscopical causes (dissymmetrizators), let they be located outside or inside the embryo. As is well established by descriptive and experimental embryology, such symmetry breaks are taking place not only at the level of a visible morphology, but also within the phase space of the developmental potencies. The latter means that embryonic development is always associated with a progressive narrowing and specification of the morphogenetical potencies initially delocalized throughout embryonic space.

We know, on the other hand, that any self-organizing system requires powerful and essentially non-linear regulatory feedback in order to be maintained and evolved. How this feedback may look like in the developing organisms? By the tradition coming from the famous Turing (1952) study, the main feedback in the developing organism is considered to be based upon chemokinetical processes, such as an autocatalysis, autoinhibition and diffusion (e.g. Meinhardt, 1982). Although the processes of this kind may well play a role in establishing, for example, colour patterns on the surface of already created 3-dimensional morphological structures (such as skin integuments, shells, etc) , one can seriously doubt whether they alone are able to create *de novo* these very 3-dimensional structures, bearing in mind the geometrical complexity of the latter and heterogeneity of an entire embryo. Meanwhile, another way for interpreting the formation of complicated morphological structures from more simple (more symmetrical) precursors is to ascribe an important role in the regulatory feedback to mechanical stresses in the embryonic tissues.

EMBRYOMECHANICS (A BRIEF REVIEW)

The main and *a priori* obvious advantage of such a "mechanocentrical" view is, that in the mechanically stressed constructions quite general, uniform and robust relations between the successively established geometries appear automatically, without requiring, from the very beginning, some highly specific prepatterns or signalling mechanisms. At the same time, these geometries themselves may serve as prepatterns for further complication and specification.

The developing organisms (as well as the adult forms) are indeed mechanically stressed constructions. Stresses produced by the cells and tissues can in many cases be traced and more or less precisely measured (Bereiter-Hahn, 1987). According to recent findings, the immediate causes of the most pronounced and morphogenetically important stresses are: (1) turgor pressure within cells, vacuoles, and intercellular cavities produced by a concerted action of ionic pumps and channels; (2) contraction of microfilaments; (3) stretching of cell bodies due to directional assembly of the microtubules and, in some cases, F-actin filaments. The stressed states are stabilized by insertion-resorption of the cell membrane subunits and are largely amplified at a supracellular level, due to coordinated cell movements (see Beloussov *et al.*, 1994; Beloussov, 1997 for more details).

Our present day knowledge of the morphogenetical role of mechanical stresses can be summarized as follows:

- Starting from a state of a germinated egg, all of the developing systems are mechanically stressed to a measurable degree. Initially, stresses are produced by turgor pressure within an egg cell and, later on, within a primary embryonic cavity, the blastocoel. As development proceeds, these stresses are amplified and modulated firstly by the activity of a cortical cell layer and then by a contractile and motile activity of embryonic cells (Beloussov *et al.*, 1975, 1994; Beloussov, 1994, 1996).

- The stresses are arranged according to specific designs, which remain topologically invariable during certain finite developmental periods (those defined in classical embryology as gastrulation, neurulation etc.), drastically changing in between. Definite relations can be established between these stress patterns and the subsequent sets of morphogenetic movements: the latter are usually directed towards relaxation of pre-established stresses, although spending energy by doing this (Beloussov *et al.*, 1975, 1994).

- Experimental relaxation of stresses leads to a number of predictable developmental abnormalities, in particular if done at a blastula - early gastrula stages. The abnormalities are mostly due to an abundant and irregularly located columnarization (increase of length/width ratio) of the relaxed cells (Beloussov *et al.*, 1991).

- By artificially reorienting the dominating stresses, one can also reorient the direction of the subsequent morphogenetic movements. A most direct correspondence has been revealed between the direction of stretching and of a so-called cell intercalation (converging movements of neighbouring cells), playing a leading role in the morphogenesis of many animal groups (Keller, 1987). Actually, cell intercalation have been shown to go on perpendicularly to the stretching direction, thus actively extending the given rudiment just in this direction (Beloussov *et al.*, 1988; Beloussov & Luchinskaia, 1995).

- In some experimental situations, the existence of long-range stresses (those spread over the distances largely exceeding cell diameters' values) was shown to be a necessary (and a crucial) condition for creating regular morphological structures from homogeneous cell populations. This was demonstrated in a mostly straightforward way on the fibroblast cultures (Harris *et al.*, 1984). In our experiments by stretching the explant of a suprablastoporal lip from an early gastrula stage amphibian embryo we were able to show that several minutes stretching at least is required for maintaining a cell integrity of a sample. The samples non-stretched at all have been dissociated into single cells (Beloussov *et al.*, 1988).

More than a decade ago, some first ideas on the mechanical feedback in embryonic development have been suggested. Odell *et al.* (1981) postulated the existence of a "+, +" stretch-contraction feedback within an embryonic epithelia: an epithelial cell, while undergoing an active tangential contraction, stretches a neighbouring cell, firing an active contraction response within a latter. This, in turn, stretches the next cell, etc. In such a way a stretch-contraction relay may go on through large distances. The main deficiency of this model (although reproducing some real situations, see below) was the lack of a negative link which could restrict a continuous relay spreading. Such a negative link, being already included into chemodiffusional models, got its mechanical interpretation in Oster *et al.* (1983) work. The authors implied a coexistence of a short-range positive feedback mediated by a cell-cell adhesion with a negative long-range feedback provided by the tension forces which are generated by the cells themselves and spread along an elastic substrate. A "+, −" short-long range feedback of such a kind was enough for generating Turing-like periodic structures in a purely mechanical way, without the use of a chemical diffusion.

For embryonic epithelia, a somewhat similar model have been suggested by Belintzev *et al.* (1987). Its main idea was that a cell columnarization within an epithelial layer is stimulated by short-range cell interactions (each one columnarized cell induces its neighbour to do the same) while the tangential tension produced by this very process inhibits further columnarization. Hence, we have here again a closed circuit combining a short-range positive feedback with a long-range negative one. As shown by computation, this model was able to generate, in the case of pre-existed tensions, a sharp segregation of an initially homogeneous cell layer into the coherent domains of highly columnarized and extensively flattened cells. This segregation was robust enough (as remaining the same under various initial perturbations) and, most remarkably, scale invariant: the length relations of the both domains remained the same irrespectively of an absolute length of a layer. In such a way, Belintzev's *et al.* model was able to interpret a fundamental phenomenon of so-called embryonic regulations (keeping the same geometry after altering experimentally the amount of embryonic material).

All of the models described demonstrate a heuristic value of the idea of the feedback between passive and active mechanical stresses. We regard as passive those stresses which are created by forces coming outside of a given cell or a tissue piece and which are established in some finite Δt period before a given time moment. On the other hand, the active stresses are those generated within a given tissue piece and within Δt time period. A detection of a spatio-temporal limit discerning the active stresses from the passive ones should be regarded each time as a special empirical task.

Meanwhile, each one of the above models can be applied to no more than some specific morphological situations (see below for more details). Recently we proposed a more generalized construction of such a kind which we have called a hypothesis of a hyperrestoration (HR) of the pre-existed stresses (Beloussov and Mittenthal, 1992; Beloussov *et al.*, 1994). It may be formulated as follows:

- A cell or a tissue piece, after being shifted by an external force (either introduced artificially, or exerted by another part of the same normally developing embryo) from its initial state of stress, develops an active mechanical response, which is directed towards restoring the initial stress value, but as a rule overshoots it to the opposite side (that is, hyperrestores the stress).

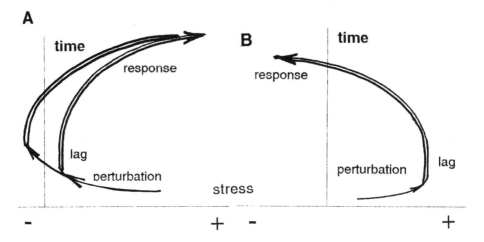

Figure 1. The loops of stresses hyperrestoration. Horizontal axis: stresses (positive are tension stresses while negative are pressure stresses). Vertical axis: time. On this and other pictures single line arrows indicate passive stresses (external mechanical perturbations) while double line arrows show the active stress responses. A: responce to either relaxation (right loop) or to compression (left loop), B: response to stretching.

This concept is illustrated by Fig. 1 A, B. The loops plotted in a stress-time space consist of three main parts: (1) a rather rapid stress shift caused by an external force and thus considered, in relation to the responding structure, as a passive one; (2) a lag-period, required for the structure to elaborate its active response (this period may completely or partly coincide with a continuing action of an external force); (3) an active response itself, which takes, as a rule, much more time than the initial stress shift does and which drives a structure towards hyperrestoration of the initial stress value. The A loop illustrates the response to the relaxation of the preexisted moderate tension, or to the application of come compression force. In this cases, by our idea, the response implies the tension increase overshooting its initial value. The B loop shows a reaction to the external stretching (that is, to the tension increase). Expected now is not only an active tension decrease, but also generation of an internal pressure. This is illustrated by prolongation of the active response branch to the area of the negative (compression) stresses.

The events, which can be considered as HR responses, do not belong entirely to morphogenesis of the multicellular animal embryos: quite diverse kinds of cell activities are stress-dependent and seem to obey similar rules. Such are the responses of plant cells to the turgor pressure shifts (Harold, 1991; Green, 1994); reaction of a cytoplasmic *Physarum* strand to a stretching force (Fleischner & Wohlfarth-Bottermann, 1975); stretch-relaxation dynamics of an isolated neurite (Dennerly et al., 1989); exo- endocytotic activity of the fish zygotes (Ivanenkov et al., 1990), etc. A suggestion very similar to our HR hypothesis has been later on put forward by Banes *et al.* (1995) as analysing the responses of the single cells to loading. In this presentation meanwhile we would like to focus ourselves to the events related to the morphogenesis proper.

HYPERRESTORATION RESPONSES AND STANDARD MORPHO-MECHANICAL SITUATIONS

The main kinds of HR responses relevant for morphogenesis are summarized in Fig. 2. As seen from the Figure, they largely depend upon the border conditions (and, first of all, upon whether the tissue edges are fixed or not). Reaction (1), namely stretch-induced contraction (directed towards reducing an imposed stretch) just corresponds to Odell *et al.* (1981) model. Besides a symmetrical bipolar contraction (the only one shown in Figure) one can also observe, in the case of only one sample's pole being free, a unipolar contraction, that is, a shift of a sample's mass center towards a fixed pole. In such a way the tissue pieces and the individual cells can undergo directed dislocations, which are morphogenetically important and may be defined as a tensotaxis. Reaction (2) illustrates a typical behaviour of a relaxed embryonic cell sheet with free edges. In this case the tension is hyperrestored firstly by curling free edges and then by creating *de novo* an internal osmotically pressurized cavity. The reactions (3) and (4) are going on when the both sample's edges are either fixed or at least semifixed (do not permit spontaneous shrinkage). Such a situation is most of all typical for a normal mophogenesis. Among those, the reaction (3) corresponds to the abovementioned stretch-promoted cell intercalation while the reaction (4) is described by Belintzev's *et al.* model: this is the formation of the columnarized cell domain(s) going on until the tension within a cell layer will not pass over a certain threshold. At the individual cells' level, the reaction (3) is mediated by insertion of the additional portions of a cell membrane (exocytosis), while the reaction (4) is achieved due to resorption of some amount of cell membrane either by endocytosis or by shedding.

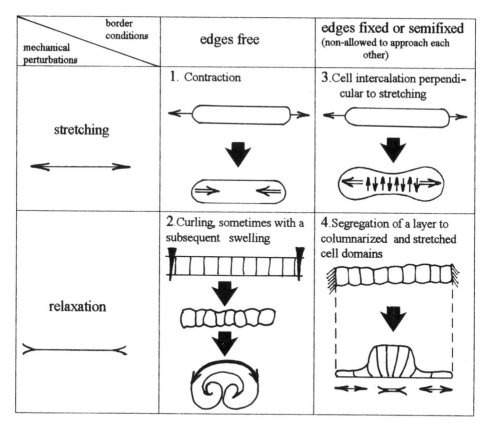

Figure 2. Main reactions of embryonic tissues to the stretching and to the relaxation of stresses. Mechanical perturbations are indicated by single arrows while the active tissue responses by double-line arrows.

Most of these reactions can be perfectly coupled with each other by certain "+, +" feedback circuits. By describing the latter, we come to the formulation of some *standard morphomechanical situations,* SMS, which are repeatedly reproduced during development and provide its advancement (Fig. 3, SMS 1-3). SMS1 is that established within a single planar epithelial layer and couples the reactions (3) and (4). It is easy to see indeed that a stretch-induced cell intercalation (reaction 3) will at least relax (if not compress) the polar regions of a sample, promoting thus the reaction (4), that is cell columnarization (or, more generally, the tangential contraction) to go on in these very regions. This will lead, in its turn, to further stretching of the central region, the stretching will promote again the cell intercalation, and so on. Such a feedback will work until an entire amount of cell material will be involved either into intercalation or in a columnarization. As a result, the shape of a tissue piece will take a dumb-belled shape (SMS1, lower frame). This is what have been exactly observed (Beloussov and Luchinskaia, 1995).

SMS2 should act between two mechanically bound superimposed cell layers, A and B. Suggest that a layer A is initially passively stretched. This will born a stretch-dependent cell intercalation within, transforming the passive stretching into an active similarly directed extension. As a result, the layer B will become passively stretched by A for later on undergoing, in its turn, the stretch-dependent cell intercalation. In such a way the layers will help each other to extend until they become thinned up to a certain extreme.

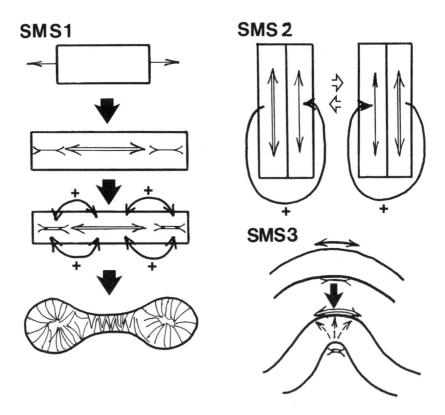

Figure 3. Standard morphomechanical situations. For further comments see text.

SMS3 is expected to be in work when a cell layer possesses a non-zero curvature. Suggest that a layer is firstly bent to some small extent by an external force. As a result, its convex surface will be stretched while the concave one compressed or, at least, relaxed. Consequently, on the convex surface the reaction (3) should be expected (that is, the insertion of an additional amount of a material), while on the concave surface the reaction (4), namely a resorption of some material, should take place. Obviously, the both reactions move a layer towards active increase of a slightly outlined curvature.

DEVELOPMENT OF AMPHIBIAN EMBRYO AS A SUCCESSION OF STANDARD MORPHOMECHANICAL SITUATIONS

Let us demonstrate now how a crucial period of amphibians' development, lasting from a blastula up to a neurula stage, can be represented as a sequence of HR reactions, going on within the SMSs framework. We take as initial conditions: (a) stretching of the blastocoel roof by the turgor pressure within the blastocoel (as shown by a filled oblique triangle, Fig. 4 A), (b) dorso-ventral asymmetry of the embryo, known to be established immediately after fertilization. The first expected HR reaction to the turgor-mediated stretching will be cell intercalation within the blastocoel roof, as shown by the oppositely directed small arrows and by the double-line arrows (Fig. 4 A). This will be enough for initiating SMS1, the latter producing firstly the relaxation and then the active contraction of the peripheral parts of a blastocoel roof (Fig. 4 A, B, single converged double-head arrows). These parts correspond to a so-called marginal zone where at this very stage the gastrulation movements are initiated: a localized contraction is just their starting point. Thus initiated slight invagination is then reinforced according to the SMS3, giving rise to a

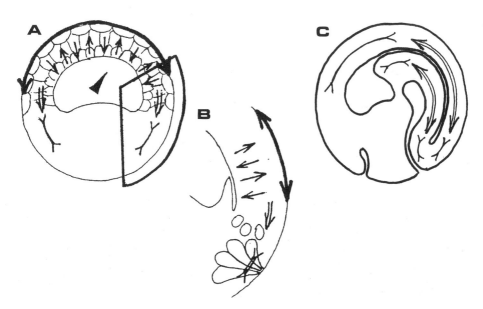

Figure 4. A morphomechanics of amphibian gastrulation. A: blastula stage, B: region framed in A, C: gastrula stage. Designations as in Figs 2, 3.

blastopore. As soon as, due to invagination, even a small piece of an involuted cell layer becomes spread under the pre-involuted one, the both superimposed layers will be involved into the SMS2, being thus extended in the antero-posterior direction (Fig. 4 C). In such a way the internal layer's involution and the external layer's spreading (epiboly) will go on hand by hand, up to the accomplishment of the both processes. At the same time, new relaxation/compression areas will be formed at the posterior and anterior poles of the both layers. Out of these two, the anterior zone will become more pronounced since a posteriorly located compression will be permanently released by the continuing involution movements.

These events create a dynamic basis for a subsequent neurulation. It starts from a passive longitudinal stretching of the dorsal embryo wall, inevitably associated with a Poissonian transversal schrinkage (Fig. 5 A). This stretching is caused by the involution movements of gastrulation. Very soon however (according to SMS1) the passive stretching will be followed by the active, intercalation-mediated extension of the same region, associated by relaxation-compression of the polar regions, and first of all of the anterior one (Fig. 5 B). By the same reasons, an initially passive anterior stress pattern will be soon exchanged by the active one (Fig. 5 C). In such a way an antero-posterior regionalization of a central nervous system can be interpreted. Its transversal regionalization should go on in accordance with the Belintzev et al model: namely, a Poissonian schrinkage should stimulate the formation of a columnarized cell domain along a dorsal embryo midline, this domain becoming flanked, from its ventral sides, by the areas of extensively flattened cells.

A suggested interpretation has been tested by various experiments on relaxing and retensing embryonic tissues. For example, the relaxation of a pre-involuted zone at an early gastrula stage led to the formation of several abnormal clusters of highly columnarized cells (Beloussov, 1994); a relaxation of transversal tensions at the same stage brought into being abnormal ventralwards extension of a columnarized dorsal cell domain, while the retension of the same stage embryos in the different directions produced abnormally located axial organs being elongated in these very directions (Beloussov and Louchinskaia, 1995a).

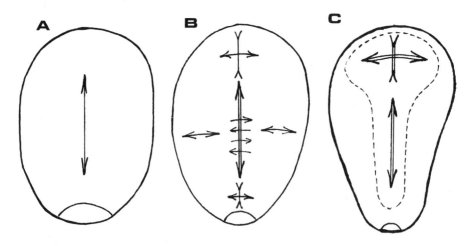

Figure 5. A morphomechanics of amphibian neurulation. Embryos are shown from dorsal sides. Designations as in Fig. 4.

A COMPARISON WITH OTHER CONCEPTS

It is worthwhile to compare the HR hypothesis with the concept of a positional information (PI), firstly in its broadest outlines and then as being applied to some particular cases. The main idea of PI concept (Wolpert, 1996) is that any material element of the embryo, in order to be later on differentiated, at first specifies its position in relation to certain pre-established PI sources and then develops in one-to-one correspondence with the specified positions. One of the deficiencies of this concept (for more detailed analysis see Beloussov, Bereiter-Hahn and Green, 1997) is its non-robustness: any mutual shifts of the postulated PI sources should largely distort the arised patterns. That contradicts to well known and numerous facts of embryonic regulations, that is a capacity of a developing embryo to reproduce normal patterns after extensive mutual displacements of all of their constituent parts. By our idea, meanwhile, in order to participate in a coordinated morphogenesis, any element (say, a cell) should 'know' nothing more than its instantaneous stress value (which is as a rule largely smoothed over an extended embryonic area) and that taking place some time before. Then it compares the both values and makes its 'developmental decision' as a result of such a comparison. In the other words, instead of a detailed initial positional specification, we endow an embryonic element with some kind of a memory about its finite past. The spatial dimensions of an area, with all of their elements sharing the same memory, as well as a 'temporal depth' of the memory may vary substantially in the different developing systems, but always belong to either macro- or mesoscopic realms (that is, are not too small). For amphibians' embryos the corresponding area have the diameters of 10^{-4} - 10^{-3} m orders while the memory duration is up to several hours. Consequently, by our concept (contrary to that of PI) the 'morphogenetic information' is an essentially macroscopic event. This gives to our concept a required robustness.

Let us return now again to the problem of a chemical regulation of development. In no way denying the morphogenetical role of the various chemical factors, including the immediate products of genes' activity, we are inclined to regard them, in general, as smoothly distributed *parameters* of the morphomechanical feedback circuits, rather than the direct *prepatterns* of the morphological structures. Let us address, for example, to an extensively explored problem of so-called embryonic inductors. As suggested by several authors (review: Sasai and Robertis, 1995), they create a chemodiffusional field providing a direct PI for the main embryonic structures. Again, such a view seems to us non-realistic due to its non-robustness. On the other hand meanwhile, it is well known that at least one of the inductive substances, the activin, largely stimulates a cell's capacity to the intercalation movements (Symes and Smith, 1987): this is a typically parametric action naturally incorporated into the above formulated morphomechanical feedback loops. Same idea may be applied to other cases as well. A more precise classification of the known developmental factors according to the categoris of the initial and border conditions, dynamic variables and parameters is extensively required, in order to approach more realistic and integrated views upon embryonic development.

In a broader aspect, we may relate our concept to those field theories which emphasize the ideas of a delocalization and of an implicit - explicit order transformations (Bohm, 1980; see Kortmulder, 1994, for biological applications). The HR hypothesis can be regarded, at least, as a step towards these theories.

The mechanical stresses, which we mentioned in the framework of the HR hypothesis, belong most probably to the hugest and most spatially extended ones. There can be no doubt that within the living cell, a great amount of much more local and short-term stresses are permanently generated and interact with each other. Although the corresponding branch of molecular cell biology (called cytomechanics by Bereiter-Hahn, 1987) is still at the very beginning of its development, some strong evidences of the direct role of intercellular stresses in regulation of gene expression and in modulation of cell signalling pathways already there (reviews: Opas, 1994; Ingber et al., 1994; Traub & Shoeman, 1994) and several attractive hypotheses of this kind have recently been suggested (Forgacs, 1995; Jones et al., 1995).

REFERENCES

Banes, A.J., Tsuzaki, M., Yamamoto, J., Fischer, T., Brigman, B., Brown, T. and Miller, L. (1995). Mechanoreception at the cellular level: the detection, interpretation, and diversity of responses to mechanical signals. *Biochem. Cell Biol.* 73: 349-365.

Belintzev, B.N., Beloussov, L.V. & Zaraisky, A.G. (1987). Model of pattern formation in epithelial morphogenesis. *J.Theor.Biol.* 129: 369-394.

Beloussov, L.V. (1994). The interplay of active forces and passive mechanical stresses in animal morphogenesis. In: *Biomechanics of Active Movement and Division of Cells*, N.Akkas ed. NATO ASI Series H: Cell Biology, V. 84 Springer Verlag Berlin, Heidelberg: 131-180.

Beloussov, L.V. (1996). Patterns of mechanical stresses and formation of the body plans in animal embryos. *Verh. Dtsch. Zool. Ges.* 89: 219-229.

Beloussov, L.V., Bereiter-Hahn, J and P. Green (1997). Mechanical stresses in cells and embryos: expectations of developmental and cell biologists. In: *Dynamics of Cell and Tissue Motion*, W.Alt, A.Deutsch, G.Dunn eds. Birkhäuser-Verlag, Basel (in press).

Beloussov, L.V., Dorfman, J.G. and Cherdantzev, V.G. (1975). Mechanical stresses and morphological patterns in amphibian embryos. *J.Embryol. Exp. Morphol.* 34: 559-574.

Beloussov, L.V., Lakirev, A.V. (1988). Self-organization in biological morphogenesis: general approaches and topo-geometrical models. In: *Thermodynamics and Pattern Formation in Biology* I.Lamprecht, A.I.Zotin eds. W. de Gruyter, Berlin, N.-Y. p. 321-336.

Beloussov, L.V., Lakirev, A.V. (1991). Generative rules for the morphogenesis of epithelial tubes. *J.Theor.Biol.* 152: 455-468.

Beloussov, L.V., Lakirev, A.V., Naumidi, I.I. (1988). The role of external tensions in differentiation of *Xenopus laevis* embryonic tissues. *Cell Diff. Devel.* 25: 165-176.

Beloussov, L.V., Lakirev, A.V., Naumidi, I.I., Novoselov, V.V. (1990). Effects of relaxation of mechanical tensions upon the early morphogenesis of Xenopus laevis embryos. *Int. J. Dev. Biol.* 34: 409-419.

Beloussov L.V. and Luchinskaia N.N. (1995) Biomechanical feedback in morphogenesis, as exemplified by stretch responses of amphibian embryonic tissues. *Biochem. Cell Biol.* 73: 555- 563.

Beloussov, L.V., Saveliev, S.V., Naumidi, I.I., Novoselov, V.V. (1994). Mechanical stresses in embryonic tissues: patterns, morphogenetic role and involvement in regulatory feedback. *Intern. Rev. Cytol.* 150: 1-34.

Bereiter-Hahn, J. (1987). Mechanical principles of architecture of eukaryotic cells. In: *Cytomechanics*, J. Bereiter-Hahn, O.R.Anderson & W.-E. Reif eds. Springer-Verlag Berlin etc, p. 3-30.

Bohm, D. (1980). *Wholeness and the Implicate Order*. London, Routledge and Kegan Paul.

Dennerly, T.J., Lamoureux, P., Buxbaum, R.E. & Heidemann, S.R. (1989). The cytomechanics of axonal elongation and retraction. *J. Cell Biol.* 109: 3073-3083.

Fleischner, M., Wohlfarth-Bottermann, K.E. (1975). Correlations between tension force generation, fibrillogenesis and ultrastructure of cytoplasmic actomyosin during isometric standarts. *Cytobiologie* 10: 339-365.

Forgacs, G. (1995). Cytoskeletal filamentous networks in intracellular signaling: an approach based on percolation. *J. Cell Sci.* 108: 2131-2143.

Green, P.B. (1994). Connecting gene and hormone action to form, pattern and organogenesis: biophysical transductions. *J.Exp.Bot.* 45: 1775-1788.

Harold, F.M. (1990). To shape a cell: an inquiry into the causes of morphogenesis of microorganisms. *Microbiol. Rev.* 545: 381-349.

Harris, A.K., Stopak, D., Warner, P. (1984). Generation of spatially periodic patterns by a mechanical instability: a mechanical alternative to the Turing model. *J. Embryol. Exp. Morphol.* 80: 1-20.

Ingber, D.E., Dike, L., Hansen, L., Karp, S., Liley, H., Maniotis, A., McNamee, H., Mooney, D., Plopper, G., Sims, J., Wang, N. (1994). Cellular tensegrity: exploring how mechanical changes in the cytoskeleton regulate cell growth, migration and tissue pattern during morphogenesis. *Int. Rev. Cytol.* 150: 173-224.

Ivanenkov, V.V., Meschneryakov, V.N., Martynova, L.E. (1990). Surface polarization in loach eggs and two-cell embryos: correlation between surface relief, endocytosis and cortex contractility. *Int. J. Dev. Biol.* 34: 337-349.

Jones, D., Leivseth, G., Tenbosch, J. (1995). Mechano-reception in osteoblast-like cells. *J. Biochem. Cell Biol.* 73: 525-534.

Kortmulder, K. (1994). Towards a field theory of behaviour. *Acta Biotheoretica* 42: 281-293.

Meinhardt, H. (1982). *Models of Biological Pattern Formation.* N.Y., L., Acad. Press.

Odell, G.M., Oster, G., Alberch, P., Burnside, B. (1981). The mechanical basis of morphogenesis. I. Epithelial folding and invagination, *Devel Biol* 85, 446-462.

Opas, M. (1994). Substratum mechanics and cell differentiation. *Int. Rev. Cytol.* 150: 119-138.

Oster, G.F., Murray, J.D., Harris, A.K. (1983) Mechanical aspects of mesenchymal morphogenesis. *J. Embryol. Exp. Morphol.* 78: 83-125.

Prigogine, I. (1980). *From Being to Becoming.* W.H.Freeman and Co, N.Y.

Sasai, Y., De Robertis, E.M. (1997).Ectodermal patterning in Vertebrate embryos. *Devel Biol.* 182: 5-20.

Traub, P., Shoeman, R. (1994). Intermediate filament proteins: cytoskeletal elements with gene-regulatory function? *Int. Rev. Cytol.* 154: 1-104.

Symes K., Smith, J.C. (1987) Gastrulation movements provide an early marker of mesoderm induction in Xenopus laevis. *Development* 101, 339-349.

Turing, A.M. (1952). The chemical basis of morphogenesis. *Phil. Trans. Roy. Soc. L. B* 237: 37-72.

Wolpert, L. (1996) One hundred years of positional information, *Trends in Genetics* 12, 359-364.

INFORMATION PROCESSING IN COMPUTATIONAL TISSUES

Mark H. Butler, Ray C. Paton and Paul H. Leng

Department Of Computer Science
University Of Liverpool
L69 7BX
UK

INTRODUCTION

Biocomputation involves studying organisms using the metaphor of computers so we can discover both new ways of performing computation and defining living systems. An appropriate point to start any discussion of biocomputation is by contrasting the distinct properties of computers and biological systems. Only then is it possible to consider the concept of computation and whether some of the activities of both systems lie within our definition. As computers are man-made their properties are easily described: computers process symbolic information; they solve mathematical problems by algorithms. Physically they are composed of large, regular arrays of interconnected switches so that their component parts can interact in strictly controlled ways. This arrangement, called *structural programmability*, [Con85], gives computers the capability of simulating one symbol processing machine on another. The primary way of performing this simulation is recursion and sequence.

Defining the properties of biological systems is more complicated; a sketchy outline will be considered in this paper. Biological systems are hierarchical collections of semi-autonomous sub-wholes [Koe67]. Their structure and organization complement each other, producing and maintaining system identity [Fle88], keeping the system far from thermodynamic equilibrium [PS85], and permitting creative change over their ontogenetic trajectory [Sal93]. The organism or its components are not defined by their structure at any one point but by their unique life history [AZA88]. [Kam91] has suggested the name component system for systems with these properties and has asserted that they are intrinsically non-algorithmic.

From this simple overview, it is possible to deduce two important differences between the two systems and from this identify two reasons for the importance of biocomputation. Firstly computers are built so their physical structure and the information they process, i.e. programs and data, are as separate as possible to allow for universality. In biology they complement each other and are causally linked. One goal of biocomputation is identifying how we can 'transplant' these properties between the two systems. If we could take universal structural programmability from computers to biological systems then we could realise biological computers. If we could combine universality with structure production we could also realise the molecular assembly devices currently proposed by nanotechnology. Secondly biological systems maintain their identity. Here *identity* refers to the fact that despite the biological system changing over time due to its external environment or its internal state we still recognise it as the same object. Computational systems on the other hand are brittle and error-prone. Internal or external changes can take the system outside its parameters of operation resulting in the halting of the system. Hence maintaining identity is similar to fault-tolerance and adaptation, desirable properties for a computer system.

Many papers have addressed the general area of biocomputation. This paper intends to concentrate on investigating non-neural tissues. A preliminary attempt at characterising the computational power of the liver was made by [PNSBC92]. In this study a number of features were noted. For example there are many possible analogies which can be drawn between computers and the liver. These analogies can be structural or informational. Examples of structural analogies include the liver being viewed as a homogeneous architecture of nodes where the structure of interconnection governs the information processing properties of the system. Alternatively it can be viewed as a heterogeneous distributed architecture, composed of distinct modules which are in communication. The blood supply to tissues often displays fractal organization so can be viewed as using a 'divide and conquer' approach to solving problems. Examples of informational (or instructional) analogies include the liver performing blood sugar regulation being viewed as a SIMD (single instruction stream multiple data stream) system carrying out the same task in parallel. Alternatively the liver processes around 500 metabolites so can be viewed as a MIMD (multiple instruction stream multiple data stream) system where different tasks are carried out in parallel. For example hepatocyte plates may be involved in the simulation of pipelining - a computational paradigm where serially arranged units each complete a different stage of a job. Other analogies can be drawn between the liver and different types of system such as an electrical circuit, an organism or a society.

In this paper we seek to extend this work by providing an integrative perspective on the capacities of tissues, a more general framework of a computational tissue, and a generic design for implementing a computational tissue. In order to fully appreciate the computational capacities of non-neural tissues we must first obtain a general appreciation of their biological capabilities as biological objects in general and tissues in particular. In brief computational processes in non-neural tissues are involved in morphogenesis, tissue maintenance, global pulsing in the heart, pattern formation, hormone production, the antibiotic reaction of epithelial cells and the amplification of hormonal signals [Sta94]. In the first part of this paper we try to summarise some relevant features.

GENERAL PROPERTIES OF BIOLOGICAL SYSTEMS

A useful way of considering the general properties of biological systems is to consider our earlier definition in more depth:

Holons

We defined biological systems as hierarchical collections of semi-autonomous subwholes. One name given to this idea is the holon [Koe67]. A holon is a unit complete in itself which is part of a larger unit. Holons can be considered both individuals with self-assertive tendencies and 'dividuals' with integrative tendencies. Other related ideas include:

- Waddington's homopoiesis / homeorhesis [Dio95]. Here the tissues switches between *homopoiesis*, the satisfying of internal requirements such as the production of energy, cell repair and *homeorhesis*, the satisfying of subsystem requirements such as molecular transformation.

- Rosen's *metabolism-repair systems* [Ros72]. Here the components perform metabolic (external) and repair (internal) functions.

- Dioguardi's idea of *autoisodiastosis* [Dio95]. This means self-same through maintain.

- Eigen and Schuster's idea of *hypercycles* [Eig71], [EGSWO81]. Hypercycles were first proposed as a mathematical description of how RNA like molecules could cooperate in the emergence of life. They involve selection and cyclic coupling. The idea of a hypercycle stems from homeostasis, first suggested by Bernard in cybernetics [Wie53]. A process is said to be homeostatic when it tries to maintain an equilibrium. Biological systems, and particularly tissues, are often trying to fulfill multiple different roles. This is a kind of cooperative homeostasis which can be described by a hypercycle.

Autopoiesis

Our definition also stated "(biological systems) structure and organization complement each other, producing and maintaining system identity". This property is known as *autopoiesis*. Autopoietic systems are self making as they maintain themselves not by properties of the subparts but by relationships among their parts. An autopoietic system also has a distinct identity; this has some similarities with a *regime* in a dissipative structure (see next section). They maintain their identity until large structural changes occur within the organisational pattern of the system. Autopoietic systems also display *closure*, meaning system operations occur within a system boundary. This boundary means the system is *autonomous*. Autopoiesis was first suggested by Maturana and Varela; a useful review can be found in [Fle88] where various properties are defined which the system must have to be autopoietic:

- The system must have an identifiable boundary; this boundary is produced by the interactions and transformations of system components.

- The system must have discrete components; some of these components are themselves products of component interactions and transformations.

- The system is mechanistic i.e. component interactions and transformations are governed by component properties.

Dissipative structures

Our definition also described biological systems "keep far from thermodynamic equilibrium". A dissipative structure is a model for a system far from thermodynamic equilibrium [PS85]. This idea comes from non-equilibrium thermodynamics; it is an attempt to explain how random fluctuations in non-conservative systems can lead to unexplained order such as organized Brownian motion in the presence of a heat source. Biological systems are open systems by necessity; if they reached thermal equilibrium chemical reactions would no longer be possible. Dissipative structures have several identifiable regimes, rather like basins of attraction in dynamical systems. The existence of these regimes can be thought of as a type of identity, similar to autopoietic systems. Dissipation of energy is also important in computers as they rely on non-linear computational components to dissipate energy to avoid errors.

Component systems

Our definition also stated that biological objects "permit creative change over their ontogenetic trajectory". Several people have considered the ability of biological systems to change themselves [Sal93], [Var79]. [Kam91] has proposed the idea of a *component system* to describe this phenomena. Consider a cell - it is highly adaptive and changes its behaviour and structure to respond to different external conditions. In the same way a component system is characterized by a search process which is looking for a self-maintaining system configuration which meets external constraints. If the search process was viewed over the system lifetime, the observer would see a disorderly vibrating pattern of irregular production and decay, punctuated by self-maintaining cycles which are perturbed by external influences.

Streaming

Finally our definition is completed by saying "the organism or its components are not defined by their structure at any one point but by their unique life history". Most research into this has been at the organism, or to a lesser extent, at the tissue level. As well as considering a tissue as a physical object, it is possible to consider it as a history. Zajicek et al. have proposed a phenomena called streaming in the liver where the cells migrate from a region of division. The properties of the cells change as they migrate [AZA88].

GENERAL CHARACTERISTICS OF TISSUES

The broad based analysis of the previous section is important because it captures a rich set of details about biosystems in our computational model. In saying this we note Levins' caveat [Lev70] that no single model in biology can simultaneously optimize for realism, precision and generality. Now it is important to consider what characteristics tissues have which distinguish them from other biological systems. Here we will consider epithelial lining tissues such as those found in the liver or the pancreas. Some of these characteristics have less relevance to other tissues such as the skin. Firstly [Wei77] has proposed several ways in which biological organization produces order from random interactions:

- Most biological units are directed i.e. display polarity.

- Units of a given size are quantitatively reinforced by multiplying the units using a modular design.

- Some tissues use sequences of different processes. This is known as serial design.

- Polar, modular and serial design is combined to form hierarchical design in tree-like structures such as the lungs or glands.

Secondly we can consider a tissue either as a component itself or as being made of components. This is further complicated by our notion of components. A tissue is a fraction of an organ. The term fraction is misleading as most natural objects cannot be divided into equal fractions. They are self-similar or fractal as they can be divided into equal parts which possess the common characteristics of the individual. They are also mosaics of different cell types. Most possess ancillary cell types which originate outside the tissue and invade it either early in its development or during its life. This means for self-similar tissues, there is an absolute limit beyond which a smaller fraction cannot support tissue function. In mixed tissues this fraction must contain all cell types.

Finally there is the structural side of the 'holon' idea discussed before. A tissue can be viewed as a hierarchy as it consists of biological entities on different scalar levels: proteins, ions, sub-cellular structures such as ion channels, cells, supercellular structures such as the acinus, and the tissue itself. It can also be viewed as a mesh-work as although it consists of biological entities on different scalar levels, there is no privileged point of control.

PRELIMINARY MODELS

Now we need to consider how we can build generic models of tissues. Again a suitable point to start is by contrasting computers and biological systems. Computers are physical devices concerned with processing of signals in the informational domain. Biological systems on the other hand exist in several domains. They have an information domain, but also a structural domain, an energetic domain and a genomic domain:

- The structural domain allows the biosystem to modify itself, e.g. the production of enzymes, the production of new cells from a stem cell, the modification of the role of a cell by its neighbours. A computer cannot change its structure (although new types of logic array are being developed which can). In a biosystem information and structure are linked.

- The energetic domain which is concerned with the coherent transfer of energy within the biological system. Energy does not simply dissipate through the biological system; its movement is carefully controlled so the system stays far from equilibrium. This also linked to the informational domain.

- The genomic domain refers to the biosystems ability to transmit information in such a way that it will direct the development and life trajectory of its offspring.

One problem with designing models of these biological systems is these domains are linked. There has been a preponderance of models concerned with only one type of linkage - consider the genomic and informational domains investigated by genetic algorithms. We need to find some way of modeling the linkage between these domains. It must be rich enough to cope with the way biological systems exist on different scalar levels and have global and local interactions. In order to simplify the process it is proposed to only consider informational and structural domains in our models. Initially want to separate the properties of self-modification, modularisation within a hierarchy, and local and global interactions.

Simple models of self-modification

We will begin by considering simple models of self-modification. Consider a standard logic gate but where one set of inputs is connected to a randomised structure altering device. When this set of inputs is presented to the gate it changes the output produced with this input to a random binary value and 'rewires' it self so that it is now triggered by a different random input pattern. We will call such a device the operator M. A possible life trajectory for a 2 input gate can be found in Figure 1.

As we consider the properties of the operator M we realise it is only weakly equivalent to a self-modifying system. It is rather like a neural network which can adjust its weights but not the structure of the system i.e. the number of internal nodes or the number of output nodes. Can we introduce another operator which can do this? Consider an alternative operator N so that when it is invoked by a particular input, it adds or removes either an entire input or an output and updates the truth table accordingly. This means it is possible for a 2 input gate with 4 possible input vectors to be transformed into a three input gate with 8 eight input vectors. However there are many modifications to the system which cannot be made using these operators. Hence the operators N and M are only weakly equivalent to a self-modifying system; for a more in-depth criticism see [Kam94] or [Kam91].

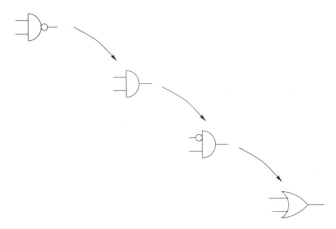

Figure 1. The life trajectory of a simple self-modifying logic gate

Simple models of modularity

So far we have considered life's property of self-modification. However another important property is it's ability to scale itself into levels. How does this work? How does it effect organisms? How can we deduce appropriate design principles?

Here it is proposed to investigate these properties using a network formalism. This will consist of randomly connected feedforward and feedback Boolean networks. The connection between nodes can be represented by a unidirectional graph. Although these networks are random, their properties are described by a schema containing measures such as: the number of input, output or internal nodes; the number of internal nodes in feedback loops; the degree of autonomy i.e. qualitatively whether the network is governed by the state of the internal nodes or the input nodes; the depth of the network as this can be related to the complexity of problem it can solve; and finally the longest state cycle.

In order to investigate scalar levels, this network can then be encapsulated and form a node (a logic unit) in higher level networks. This is shown in Figure 2. Inputs are differentiated into local and global inputs; local inputs connect to a few other nodes whereas global inputs connect to most of the nodes in the network.

Self-modifying modular networks

Making the networks self-modifying is more complicated than making logic gates self-modifying. For example self-modification within a network module can involve the making or breaking of connections (operator M); it can involve the addition or deletion of nodes (operator N). This modifications can either preserve or modify the schema properties of the network module. Some modifications will have a greater effect on the network than others; compare the action of removing a global input to removing a local input.

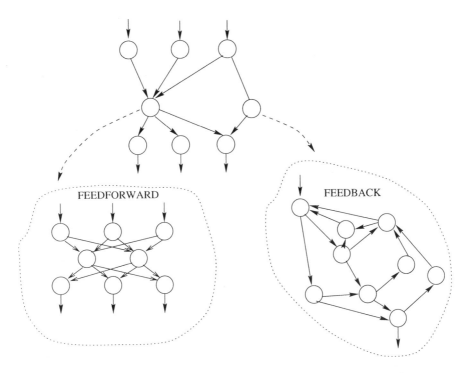

Figure 2. Subnetworks being used as components in higher level networks

These modifications can be triggered by a particular output. Unlike the logic gate these outputs will not be randomised but careful thought must be given to how, if the structure changing output is removed, it can be recovered at a later stage. The modification operators can act at higher or lower level networks thus increasing the complexity of our weak simulation of self-modification. The existence of such networks in biological systems has already been considered; the name given to them is vertical information processing hierarchies [Con92].

CONCLUSIONS

To conclude we have considered some of the properties of biological objects in general and tissues in particular. This has led to the basis of a generic model of tissues based on self-modifying networks. The networks discussed are still random; if we consider the theoretical suggestions made about the systems we are trying to model in the first part of the paper then it becomes clear the networks need some principle to 'guide' them, such as a selectionist scheme in a genetic algorithm, a training rule or a winner takes all strategy in a neural network etc. The rule must be such that the system maintains identity in most cases, until external structural change becomes to alter the system; then creative change commences to place the system in another self renewing configuration.

Another important conclusion is that the models we have discussed must not become more complex; they must be further simplified. We need to refine the questions we are asking and our methods for exploring them.

REFERENCES

[AZA88]   N. Arber, G. Zajicek, and I. Ariel. The streaming liver II: Hepatocyte life history. *Liver*, 8:80–87, 1988.

[Con85]   Michael Conrad. On design principles for a molecular computer. *Communications Of The ACM*, 28(5):464–480, 1985.

[Con92]   Michael Conrad. Molecular computing: The lock-key paradigm. *IEEE Computer*, 25(11):11–20, 1992.

[Dio95]   Nicola Dioguardi. *The Liver, The Hepatone and Its Functioning*. Universita degli Studi di Milano, 1995.

[EGSWO81]   Manfred Eigen, W. Gardiner, P. Schuster, and R WinklerOswatitsch. The origin of genetic information. *Scientific American*, 244(4):78–94, 1981.

[Eig71]   Manfred Eigen. Self-organization of matter and evolution of biological macromolecules. *Naturwissenschaften*, 58:465–522, 1971.

[Fle88]   Gail Raney Fleishaker. Autopoiesis - the status of its system logic. *Biosystems*, 22(1):37–49, 1988.

[Kam91]   George Kampis. *Self-modifying Systems In Biology And Cognitive Science*. Pergamon Press, 1991.

[Kam94]   George Kampis. Life-like computing beyond the machine metaphor. In Ray C. Paton, editor, *Computing With Biological Metaphors*. Chapman And Hall, 1994.

[Koe67]   Arthur Koestler. *The Ghost In The Machine*. Arkana / Penguin Books, 1967.

[Lev70]   R. Levins. Complex systems. In C. H. Waddington, editor, *Towards A Theoretical Biology*, volume 3, pages 73–88. Edinburgh University Press, 1970.

[PNSBC92]   Ray C. Paton, H. S. Nwana, M. J. R. Shave, and T. J. M. Bench-Capon. Computing at the tissue/organ level. In F. J. Varela and P. Bourgine, editors, *Towards a Practice of Autonomous Systems*, pages 411–420. MIT Press, 1992.

[PS85]   Ilya Prigogine and Isabelle Stengers. *Order Out Of Chaos : Man's New Dialogue With Nature*. Flamingo, 1985.

[Ros72]   Robert Rosen. Some relational cell models: The metabolism-repair systems. In Robert Rosen, editor, *Foundations Of Mathematical Biology*, volume 2, pages 217–253. Academic Press, New York, 1972.

[Sal93]   Stanley N. Salthe. *Development And Evolution : Complexity and change in biology*. MIT Press, 1993.

[Sta94]   W. Richard Stark. Artificial tissue models. In Ray C. Paton, editor, *Computing With Biological Metaphors*. Chapman And Hall, 1994.

[Var79]   Francisco J. Varela. *Principle of biological autonomy*. North Holland, 1979.

[Wei77]   E. R. Weibel. The non-statistical nature of biological structure and its implications on sampling for stereology. In *Lecture Notes in Biomathematics: Geometrical Probability and Biological Structures*, volume 23. Springer Verlag, Berlin, 1977.

[Wie53]   Norbert Wiener. *Cybernetics: or control and communication in the animal and machine*. MIT Press, 1953.

SEMIOTICS OF COMPLEX SYSTEMS:
A HIERARCHICAL NOTATION FOR THE MATHEMATICAL
STRUCTURE OF A SINGLE CELL

Jerry LR Chandler

Krasnow Institute for Advanced Study
George Mason University
837 Canal Drive, McLean, VA 22102-1407
JLRChand @ erols.com

ABSTRACT

How are "computational" dynamics of organic components organized to form organisms? The semiotic dilemma is exemplified by noting that the terms organic, organism, and organization are all derived from the same root! The common semantic origin of these terms strongly suggests the existence of a common perceptual pattern underlies the phenomena of complex pattern generators. Presumably, these generators create cellular "computations."

Which mathematical structures are sufficiently rich to support a unified representation of interdependence, of hierarchical degrees of organization and of nonlinear dynamics? It has been suggested that category theory is appropriate. A notation for a hierarchical structure of the natural sciences is created which is consistent with categorical logic. A notational sequence for composing one - to - one correspondences among the degrees of organization [O°] of material objects of a cellular system is proposed. Necessary and sufficient conditions for simple and complex systems are described. Causality is also exemplified in terms of the boundary conditions sustaining the system -- bottom - up, top - down, outside - inward and inside - outward. This scientific notation is contrasted with six species of signs of general semiotics.

Key words: Category theory, complexity, hierarchy, emergence, theoretical biology, biochemistry, semiotics, modeling, information theory, nonlinear dynamics, structuralism, organizational theory, conceptual bindings, E. coli, genetics, health and disease

INTRODUCTION

How are dynamics of the organic components organized to form organisms? The semiotic character of the problem is exemplified by noting that the terms organic, organism, and organization are all derived from the same root! The common semantic origin of these terms suggests the existence of a common perceptual pattern underlies the phenomena of complex pattern generators. If such a perceptual pattern exists, then it should be possible to construct a network of symbols and languages such that the representation of physical, chemical, biochemical, cellular, individual and population processes are meaningfully interrelated. For example, for material objects, a criteria of

Information Processing in Cells and Tissues
Edited by Holcombe and Paton, Plenum Press, New York, 1998

meaning could be a common understanding of the scale of an object, it parts, its forms of birth and death, and its capacities to unite with or separate from other objects (analogous to Thom's archetypal singularities (Thom, 1991)). Meaningful transdisciplinary communication depends on a mutual understanding among the participants, for example, between the cell biologist and the mathematician. To understand the meaning of a semantic term for a material object implies some knowledge of the relationships among the parts, the whole and the embedding system surrounding the object itself (Chandler, 1992). This poses a substantial task for a biological language and notation because of the unique dynamics of each individual organism which are generated by the unique internal and external structures. Since the role of individual DNA base changes in the genesis of specific disease states is an example of "sensitive to initial conditions," a useful notation must represent both the structural form of a complex system as well as its dynamic unfolding.

A crisp representation of causality is also essential to a useful notation because causal pathways are fundamental to the organization of living systems. With the growth of knowledge of such causal pathways as enzyme catalysis, gene expression and neuronal connectedness, the number of potential causal pathways which must be organized within the notation and associated semiotics is large. The notation must record the crucial role of individual components (for example, DNA bases) of a cell without losing the structural specificity in the combinatorial explosion of molecular - biological semiotics. Thus a deep question of cellular information processing is: How does one aggregate unique biochemical structures and variable biochemical dynamics? The formalism of Shannon information theory (which is derived from a priori distributions of symbols) is of little utility for this problem. More importantly, the meaning of the mathematical models should provide a basis for predictions and therapeutic actions. If the scientific models of complex biological systems are not related to clinical practice, my primary goal for studying dose response relationships will not be achieved.

This paper is structured as follows. This introduction has stated the reasons for developing a notation for describing living systems. The next section of this paper introduces a unified view of scientific semiotics in terms of symbolizations, observations, and descriptions. In the third section, a simple hierarchical notation for a cell is introduced; this notation is based on the concept of *degrees of organization* and augmented graph and category theories. A common semantic representation of dynamic behavior at any degree of organization is presented in the fourth section. In section five, concepts from scientific semiotics are used to illustrate a network of directional causes within the degrees of organization of a cell. The discussion contrasts a logician's view of language and general semiotics with this scientific notation.

NEED FOR A COMMON SEMIOTICS SYSTEM OF REPRESENTATION FOR COMPLEX SYSTEMS

Predictability for a complex system requires a congruence among symbolizations, observations, and descriptions, as geometrically illustrated in Figure 1. If science is to become a unified body of knowledge, this congruency must be extensible from simple to complex systems. The requirement for extendibility from one degree of organization to another is absolutely required for the study of living systems. Can the search for congruence among what we observe, what we symbolize and what we describe be viewed as an alternative expression of scientific objectivity? Can this figure also be viewed as a genitive symbol which is independent of the scale, history, or behavior of any particular system? Is it a categorical basis for organizing transdisciplinary communication? Are these questions complementary to one another?

Traditionally, scientists have sought to isolate individual objects or classes of objects for study. Disassembly of natural systems into component parts and analysis of the individual units was a successful strategy for the study of simple natural systems. Such work tends to accent the *independence* of the components. Indeed, physical theories often further accent the notion of independence or autonomy, both in the conceptualization *and the mathematical representation.* Clearly, the nearly universal assumption of independence or autonomy (self-law) is useful for scientific purposes. *However, emergence of a new degree of organization depends on interdependence - a special cooperation among the components such that a vertical genesis process occurs.*

Observations

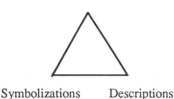

Symbolizations   Descriptions

Figure 1. Inchoate symbol for a common scientific semiotics.

For example, the organic bases within a DNA sequence of a genetic system are not independent, but are tightly linked to each other such that the degrees of freedom form a double helix. This assessment of interdependence among objects was crucial to the logical construction of complexity (the C* hypothesis) in the WESScomm I, II, and III papers, (Chandler, 1991, 1992, 1996) where the interdependence among the logical classes was explicitly incorporated into the description. The description of complexity was built by a system of constraints. For example, the conformation was constrained by the closure, the concatenation was constrained by both the closure and conformation and the cyclicity was constrained by all three concepts - closure, conformation and concatenation (viva infra). Thus, notion of *inter*dependence formed the basis for the organization of the C* theory.

A motivation for these logical interrelationships comes from the analysis of the health of a living organism in an environment, a relationship outside the traditional bounds of physical causality or systems theory (Chandler, 1985). More explicitly stated, causality in biological systems requires an interdependence among degrees of organization:

A genome is a necessary but not a sufficient condition for the existence of an organism.

An ecoment is a necessary but not a sufficient condition for the existence of an organism.

The sufficiency of a specific ecoment to sustain life of a specific organism depends on the genome.

The sufficiency of a specific genome to sustain life depends on the embedding ecoment.

In other words, complex causality may require a specific *inter*dependence between the system and the embedding ecosystem. For example, an ecosystem represents coupled sets of *inter*relationships which sustain the *inter*dependent relationships of reproduction, growth, and senescence of the community of species within a natural ecological web. More abstractly stated, internal and external causes must be fitted together dynamically to create necessary and sufficient conditions for the emergence and sustenance of patterns of complex biological behaviors.

A primary objective of this notation is to express the structure of complexity in symbolic terms such that mathematical relationships can be systematically explored. Which mathematical structures are sufficiently rich to support a representation of interdependence, of hierarchical degrees of organization and of nonlinear dynamics? Ehresmann noted the freedom in creating mathematical structures and that the "theory of categories seems to be the most unifying trend in present day mathematics;... ." Ehresmann and Vanbremeersch (1987), Thom (1991), Rosen (1991), and Baas (1994) have suggested the use of category theory in theoretical biology. The structural simplicity of category theory (objects, morphisms, and compositions) lends itself to logical applications. This simplicity has been exploited in the design of computer algorithms (object oriented programming) (Walters, 1991). Ehresmann and Vanbremeersch have pioneered the use of category theory to describe mental processes by constructing "Memory Evolutive Systems" in a series of papers starting in 1987. A biochemical perspective on complexity (the C* hypothesis) and the Memory Evolutive Systems are closely related theories (Chandler, Ehresmann and Vanbremeersch, 1995). We have published a model of a cell based on category theory and pointed out the critical role of time scales in decision making processes (Vanbremeersch, Chandler and Ehresmann, 1996). Enhanced graph theory, which is somewhat related to category theory, is widely used in the computations of chemical structures. It appears that augmented category and graph theories, along with singularity theory will play a substantial role in the construction of emergent dynamics of complex systems. A robust, constructive, scientific notation is needed to support the mathematical explorations.

PROPOSAL FOR A ROBUST NOTATION FOR ORGANIZED SYSTEMS

Historically, a symbol is used to denote a belief or concept or category or object or a class of beliefs, concepts, categories, or objects. Symbols may also be used to connote a generalized class without specifying the attributes of individual members of the class. A primary objective of this paper is to construct a notation such that a *one to one correspondence* between the languages of scientific observations and the symbolic representation of the degree of organization creates a conceptual basis for the enumeration of complexity. In other words, I choose to select the meaning of the symbols $O°1$, $O°2$, $O°3$, ... such that a construction of one to one correspondences between natural numbers and material objects becomes feasible. Each successive symbol enumerates an increase in the degree of organization of a system. The one to one correspondence between symbols and classes of material objects create the needed ordering relationships among symbols.

$O°1$	subatomic particles
$O°2$	atoms
$O°3$	molecules
$O°4$	biomacromolecules
$O°5$	cells
$O°6$	ecoment
$O°7$	environment

This ordering relationship is was selected for living systems. The atomic table serves to ground the notation in a well-established microscopic ordering relationship. Since it is well-grounded in both the natural numbers and the atomic table, the notation may also be sufficient for some general physical, chemical and engineering purposes. The meaning of these symbols is given in terms of one to one correspondences among the semantic ordering relationships as described in the following paragraphs.

$O°1$, subatomic particles, consist of three material objects: protons (+), electrons (-) and neutrons; other, smaller particles could be added to the class, however, they are not known to play a prominent role in cellular systems. The names of the objects convey a one to one correspondence with the mass and electrical charge of the individual members of the category.

$O°2$, atoms, consists of over one hundred unique objects, composed from the three subatomic particles. The composition of atoms from particles can be enumerated systematically in terms of the natural numbers, preserving the one to one correspondence between particles and particles bound into atoms. The binding operations which creates atoms from subatomic particles are described by specific patterns. Patterns are formed from the principle quantum numbers (p) which designates the number of protons (+) in the nucleus and the other quantum numbers (l,m,s) which designate the patterns of the electrons (-). The non-primary quantum numbers assign a unique role to each and every electron (-) of the atom. Ions are composed from atoms by adding or subtracting charged particles - either positive (+) or negative (-).

$O°3$, molecules, consists of a very large number of different material objects composed from atoms or ions. The binding operations which form neutral molecules from atoms preserve one to one correspondences between atoms and atoms bound into molecules. These binding operations also form geometric patterns. The patterns formed in molecules are created from the organization of the quantum numbers of individual atoms into molecular orbitals. In contrast to the neutral molecules, the binding operations which form electrically charged (either positive(+) or negatively (-)) molecules do not preserve the one to one correspondence between atoms and atoms bound into molecules. Electrically charged particles are either added to or subtracted from the neutral molecule. Since the number of charged particles can be counted, non-neutral molecules can be enumerated in terms of number of subatomic particles and the organization of these particles into atoms and atoms bound into non-neutral molecules.

$O^{\circ}4$, biomacromolecules consists of a large number of material objects, composed from a small number of classes of biochemicals. Biomacromolecules are composed from specific monomers into specific sequences via Boolean operations, guided by a genetic system within an ecoment. The binding operations preserve one to one correspondences between subgraphs of the molecules and the bound subgraphs within the macromolecules.

$O^{\circ}5$, cells, consists of living objects represented as having a boundary sustained by a genetic system. The genetic system is composed of components consisting of $O^{\circ}1$, subatomic particles, $O^{\circ}2$ atoms, $O^{\circ}3$ molecules, and $O^{\circ}4$ biomacromolecules. (In some special cases another (smaller) cell or portion of a cell may be embedded within a (larger) cell.). The ordering relationships among the components of a cell are not completely specified by internal relationships. In general, no one-to-one correspondence exists between a cell and the internal components of a cell; this absence of a simple one to one correspondence is often referred to as the adaptability or plasticity of a living organism. Nonetheless, ordering relations exist among all the essential components of a cell.

$O^{\circ}6$, ecoment, consists of the surrounds of the cell. In natural systems, the surround may include $O^{\circ}1$, subatomic particles, $O^{\circ}2$ atoms, $O^{\circ}3$ molecules, $O^{\circ}4$ biomacromolecules, $O^{\circ}5$ other cells and potentially more highly organized systems. The ordering relationships among the components of an ecoment are not readily specified for natural systems. However, the necessary components of an ecoment are known for many cells and higher organisms - they are named essential nutrients.

$O^{\circ}7$, environment, is the embedding system for the surrounds of organisms.

As suggested by the ordering relationship, the term *ecoment* is introduced to describe the immediate surrounding of the living organism. It is this immediate surroundings which provides the nutrients (the necessary and sufficient conditions) for sustaining life. The term ecoment implies a specific subset of the environment which directly "finances" the economy of a cell and is experienced by the organism. Often, an ecoment is only a small, select subspace of an ecological web.

A sequence of degrees of organization can be designed for other systems. These designs may require either a smaller number or a larger number of degrees of organization, depending on the objectives of the notation. For example, less complex mechanical and / or electrical systems may require fewer degrees of organization than a living organism. Higher degrees of organization can be assigned for more highly organized living systems by composing cells into higher structures.

The conceptual basis of this proposal for a scientific semiotics was the recognition that three degrees of organization are essential to predicting the complete behavior of any cybernetic system and these three O° must form an ordering relationship (Chandler, 1995). In everyday language, any three degrees of organization can be expressed parts, wholes and the surroundings. Co-extensive with these semantics and symbols are the scaling factors -- the parts are smaller than whole, the whole is smaller than it's embedding surroundings.

LANGUAGE OF DESCRIPTION OF A DEGREE OF ORGANIZATION

The next step in designing a scientific semiotics to account for complex behaviors is to list a common set of concepts which can be applied to a specific system. Four terms can be used to describe the behavior of any degree of organization (Chandler, 1996). In order to create and maintain ordering relationships, the meaning of these four terms are mathematically intertwined and must be defined sequentially.

Closure: a domain of discourse, a category, a system, an object, a unity.
Conformation: the components of the closure of a system, the internal patterns of the system, the relationship among the parts of the system, a three dimensional depiction of the internal description of a system, a specific geometric and algebraic description of components (graphs) of the system.
Concatenation: binding parts together, linking changes in the conformation, changes in the internal patterns of a system, the specific linkages between parts of the whole, dynamic processes of the system linking patterns to patterns.
Cyclicity: a pattern of concatenations which sustains the system, the potential cyclic walks or pathways over the conformations (graphs) of the system, the habitual behaviors of the closure.

These four concepts serve as the basis for a linguistic description of simple and complex material systems. *In principle, each concept can be applied to the enumeration of each degree of organization.* When the material composition of a system is known, then these concepts provide a basis for specifying objects and may allow an accounting of the complexity. The following examples illustrate the usage of the symbols and concepts.

Example 1. A carbon atom ($O^{\circ 2}$) is a specific object which will be the domain of discourse to be enumerated. The conformation of a carbon atom is composed from protons, electrons and neutrons (PEN, $O^{\circ 1}$). The quantum numbers (p,l,m,s) specify the concatenation of the of PEN into atomic orbitals -- cyclic behaviors of the electrons and the nucleus, a specific geometric and algebraic pattern of behavior. The principle quantum number, 6, serves to both enumerate PEN and to generate a specific ordering relationship between carbon and all other atoms whose principle quantum number is not equal to 6.

Example 2. A cellular molecule ($O^{\circ 3}$) of Nicotinamide Adenine Dinucleotide, NAD, is composed from $O^{\circ 1}$ and $O^{\circ 2}$. NAD is composed from atoms of hydrogen, carbon, nitrogen, oxygen, and phosphorus with principle quantum numbers of 1, 6, 7, 8, and 15. (This illustrates the role of the quantum numbers for creating ordering relationships among the chemical elements of the atomic table.) The atoms of NAD can be enumerated: it has 26 H atoms, 21 C atoms, 6 N atoms, 15 O atoms and 2 P atoms. The pattern of these atoms are concatenated into a molecular graph or structure (technically, a labeled asymmetric psuedograph, LAP). Geometrically, the conformation of the molecular structure is composed from two almost planar rings of atoms - the Nicotinamide ring and the adenine ring. The conformation of the two rings is nearly coplanar - that is, the Nicotinamide ring and the adenine ring interface with one another. The orbitals of one specific carbon atom ($O^{\circ 2}$) at the 4 position of the Nicotinamide ring ($O^{\circ 3}$) accept or donate electrons ($O^{\circ 1}$) in enzyme catalyzed ($O^{\circ 4}$) oxidation reduction reactions of a cell ($O^{\circ 5}$). Without NAD, a cell cannot sustain itself. Nicotinamide, a vitamin, may imported from the ecoment $O^{\circ 6}$.

Example 3. A macromolecule, alcohol dehydrogenase ($O^{\circ 4}$), is composed from subgraphs of amino - acids ($O^{\circ 3}$) which are composed from atoms ($O^{\circ 2}$) of H, C, N, O, and S with principle quantum numbers of 1, 6, 7, 8, 16. The atoms ($O^{\circ 2}$) are concatenated into one of 20 different patterns (LAPs). The LAPs are concatenated into linear ordering relationships - a specific pattern generated by the genetic system. One portion of the surface conformation of alcohol dehydrogenase specifically complements the surface conformation of NAD and another portion of the surface specifically complements the surface of ethyl alcohol. Transfer of a hydride ion ($O^{\circ 2}$) between alcohol ($O^{\circ 3}$) and the orbitals of the carbon atom ($O^{\circ 2}$) at 4 position of nicotinamide ring of NAD ($O^{\circ 3}$) by alcohol dehydrogenase ($O^{\circ 4}$) is accelerated by more than ten orders of magnitude. ((The class of conformations is termed a <u>co</u>llaborative <u>co</u>nfiguration -- CoCo's) The concomitant generation and acceleration of specific patterns of singularities (births, deaths, scissions and confluences) among atoms, molecules and macromolecules is central to the concept of collaborative configurations. Complex systems can include many collaborative configurations.)

Example 4. Glucose, (a sugar) ($O^{\circ 3}$) may be a component of an ecoment ($O^{\circ 6}$). A yeast cell ($O^{\circ 5}$) may transport glucose within its closure by a specific macromolecule ($O^{\circ 4}$). Within the cell, the LAP of glucose may undergo a specific walk of transformations via a specific sequence of enzymes ($O^{\circ 4}$), eventually generating a molecule of alcohol ($O^{\circ 3}$). The final step of the graphic walk from glucose to alcohol is a transfer of a hydride ion ($O^{\circ 2}$) from NADH to a carbon atom of acetaldehyde, the reverse of the reaction of the previous example. The proton ($O^{\circ 1}$) of the hydroxyl group comes from the aqueous media. The alcohol is expelled from the yeast cell ($O^{\circ 5}$) into the ecoment ($O^{\circ 6}$).

The description of the fermentation of sugars from natural juices illustrates the usage of C* notation at degrees $O^{\circ 1}$, ..., $O^{\circ 6}$ to notate experimental observations. Indeed, the organization of entire internal process of fermentation can be symbolized in terms of pathways among the labeled asymmetric psuedographs and dynamic processes symbolizing the role of collaborative configurations in generating singularities and accelerating the flow of energy. Since fermentation has been cultivated by human cultures since ancient times, these examples can also be extended to higher degrees of organization to describe the role of fermentation in the history of science and law. For example, at the cultural level, the

German laws governing the purity and wholesomeness of beer are among the earliest recorded food safety rules.

CAUSALITY

The examples taken from fermentation illustrate the specific relationships between degrees of organization and the semantic description and enumeration of a complex process. Such specific observations provide the basis for specific enumeration and demonstration of cause - effect relationships in a complex system.

The fermentation process has been described in terms of a set of differential equations. Examination of the equations representing fermentation first illustrate that the cause - effect relationships in complex systems can be deterministic, highly organized, and extremely selective. Of all the possible singularities which could occur among the $O^{\circ}3$, $O^{\circ}4$, $O^{\circ}5$ and $O^{\circ}6$ components of fermentation, the selection of one specific path from the structure (graph) of glucose to the structure of alcohol, among the virtually infinite number of potential singularities, symbolizes a degree of determinism unprecedented in the physical or chemical sciences.

How are we to describe or symbolize this unique set of cause effect relationships? The equations and observations are consistent with the release of energy during fermentation. One description of cause effect relationships (in thermodynamics, for example) is based on energy flow. This will be call this *'bottom - up' causality*. All processes at all degrees of organization seem to involve action and energy flow and hence bottom - up causality. Merely describing energy flow fails to predict the unprecedented determinism observed fermentation and in the growth and behavior of living systems. The examples reveal three additional directions of causality that are implicit to living system.

Secondly, analysis of the fermentation equations show that the cause effects relationships can be ordered in terms of the scale of system. One measure of the scale of a system is the sum of the mass of the components. For any particular hierarchical structure, the sum of the mass will uniformly increase with the degree of organization (since the mass of the whole must be greater than the mass of a one component of the whole.) When one or more higher degrees of organization collaborates with lower degrees of organization, it can be formally defined as *top - down causality*. For example, the enzymatic conversion of glucose to alcohol by an ordered walk from one LAP to the next is symbolized as top down causality because the degree of organization an enzyme ($O^{\circ}4$) is higher than the degree of organization of glucose ($O^{\circ}3$) or alcohol ($O^{\circ}3$). (A non-enzymatic chemical conversion of glucose ($O^{\circ}3$) to alcohol ($O^{\circ}3$) can be described in terms of bottom - up causality without the intervention of the 'top-down' causality expressed within the C* notation.)

Thirdly, examination of the yeast cell fermentation example (that is, the material process of a brewery) indicates the necessary role of the ecoment in the concatenated process. The ecoment ($O^{\circ}6$) both supplies the glucose ($O^{\circ}3$) to the cell ($O^{\circ}5$) and receives the alcohol ($O^{\circ}3$) produced by the cell. In terms of the energy flow of fermentation, the ecoment provides the energy within the glucose LAP and accepts the end products of the process -- heat and the highly reduced / oxidized molecules of alcohol and CO_2, respectively. The ecoment is driving the internal conformation from the outside the closure by financing the internal cycles of metabolisms. This is named *outside - inward causality*. Thus, an ecoment can simultaneously meet the criteria for both top - down and bottom - up causality within the C* notation. This is consistent with causal roles of external objects -- photons, ions, molecules, and so forth.

Finally, the question of biological individuality is related to the concept of a boundary between self and non-self. A cell sustains it boundary, its internal structures and its capacity to divide by responding to an ecoment as a "whole." This cellular response is of a fundamentally different character than the three previous cause - effect relationships. It is termed *inside outward causality*. It emerges from the organization of the first three forms. Its historical and global character are intimately associated with the genetic system of an organism, its ecomental boundary and the embedding environment. Substantial portions of the patterns of inside outward causality are conserved over many generations.

These four facets of causality, viewed from a categorical perspective, determine the directions of energy flows within a cell with respect to the degree of organization. Hence, they are termed 'directional causality'. The bottom - up causality is degenerative in the sense that energy flows tend to be degradative and irreversible. However, outside-inward, inside-outward and top-down causal paths all cooperate to create specific patterns of optional dynamics which can sustain life and can generate new cells. They are generative in character. Since sustaining life depends on collaboration of these three genitive directional causes - life can be viewed as a congenitive thermodynamic system (Chandler, unpublished manuscript, 1993). Both inside outward and outside inward causality function to communicate between the internal organization and the external ecoment. Such tasks are irrelevant to degrees of organization which are represented as semantic closures (open sets) but lack a well-structured boundary, such as particles or atoms.

DISCUSSION

A vast body of observations describe the chemical and biological attributes of cells. In order to amplify the usefulness of this body of knowledge for medical and scientific purposes, it is desirable to organize this loosely woven web of information into a network which is accessible to all disciplines. For example, analysis of dose response relationships requires the contributions of many disciplines - clinicians, physiologists, chemists, logicians, mathematicians, and others. Ideally, after the meaning of each other's languages and symbols is communicated, individual disciplines can contribute to understanding natural phenomena. This notation provides a precise basis for such complex conversations. Usage will determine if it also has a practical utility in public discourse in such areas as scientific research, risk assessment, and physician patient conversations.

The C* notation introduces new constraints on both creating and validating biological and biomedical "models." The nature of these new mathematical constraints are simple. Models expressed in symbols representing a natural language should be validated at both the "natural" $O°$ as well as at lower $O°$ (and upper $O°$ when appropriate.) Finding mathematical structures which fit more than one degree of description can be a non-trivial problem. The Russian logician, Stolyar (Stolyar,1970) notes:

> "A language is well adapted to a precise description of a certain class of objects if in that language two conditions are satisfied: 1. for each object, there is a name for the properties of the object and the relationships between the objects of that class; 2. different objects, their properties and their relationships have different names. If the first of these conditions is not satisfied, the language is poor and insufficient for describing the given class of objects. On the other hand, if the second condition is not met, the language is ambiguous."

Is position of this logician applicable to living systems as well as simpler material objects? It is desirable that the meaning of symbols used in scientific semiotics be unambiguous. Ambiguity at a lower degree begets ambiguity at a higher degree. If ambiguity exists at degree $O°^n$, the ambiguity can be multiplied at $O°^{n+1}$ and may become exponential at the next degree, $O°^{n+2}$. Explication of complex processes in terms of hierarchical symbols organized into categories and graphs should reduce ambiguity by promoting a common understanding of the observations and the discourse. Can the scientific semiotics proposed here meet Stolyar's criteria? If not, why not?

Semiotics is the study of meaning. (The Greek root, *sema*, can be translated as 'mark, sign.') More commonly, the term 'semiotics' is used to refer to the study of the innate capacity of human beings to produce and understand signs of all kinds, but especially linguistic signs. The organizational concepts and notation presented in this paper were developed from scientific observations and languages of the sciences (Chandler, 1991,1992, Lambert and Hughes, 1987). How is a scientific description of natural structure related to general semiotics and normal communication? I seek to put this notational system [O°] within a wider framework by comparing the specific scientific usage with the more common general usage. In "Signs, an introduction to semiotics," the American semiotician, T. Sebeok, asserts that six species of signs are distinct. He asserts that the two parts of the sign are the signifier and the signified. With regard to hierarchy, he asserts: *"It should be clearly understood, finally, that it is not signs that are actually being classified, but more precisely, aspects of signs: in other words, a given sign may -- and* *more often than not does - exhibit more than one aspect, so that one must recognize differences in gradation. But it is equally important to grasp that the hierarchic principle is inherent in the architecture of any species of sign."* The contrast between the views of the logician (Stolyar) and the semiotician (Sebeok) reflects the two cultures. In the following paragraphs, Sebeok's' six species are contrasted with the C* notation developed here.

1. Signal

Semiotic perspective: "The signal is a sign which mechanically (naturally) or conventionally (artificially) triggers some reaction on the part the receiver."

O° systems perspective: The signal is an outside - inward cause which pre-supposes the existence of a poised or anticipatory system. In the absence of knowledge about the ratio of scales of the objects, the nature of the relationship between the signal and the poised system is ill-defined. Signals may or may not trigger an observable reaction, depending on the intensity of the signal, the degree of organization and numerous other variables.

2. Symptom

Semiotic perspective: "A symptom is a compulsive, automatic, non-arbitrary sign such that the signifier coupled with the signified in a manner of a natural link."

O° systems perspective: A symptom can be viewed as a special (abnormal) internal conformation. It can inform the skilled observer on one specific category or on a collection of linkages of a hierarchy of categories.

3. Iconic

Semiotic perspective: "A sign is said to be iconic when there is a topological similarity between a signifier and its denotata." (Sebeok sites Pierce's 1867 paper on 'On a New List of Categories' in which Pierce asserted three kinds of representations -- (a) likenesses (icons) whose relation to their objects is a mere community in some quality' (b.) indices 'whose relation to their objects consist in a correspondence in fact and (c) symbols those the ground of whose relation to their objects is an imputed quality. Subsequently, the category of icon was partitioned into three subclasses, images, diagrams and metaphors by Pierce.)

O° systems perspective: Insofar as one has a likeness (icon, mental image) in mind for each degree of organization, one seeks a topological representation which is consistent within all degrees of organization, not just the immediate one. O° complexity is grounded on observations of many alternative iconic representations for a single closure as illustrated in the examples.

4. Index

Semiotic perspective: "A sign is said to be indexic insofar as its signifier is contiguous with its signified, or is a sample of it. The term contiguous is not to be interpreted literally in this definition as necessarily meaning 'adjoining or adjacent'; thus "Polaris may be considered an index of the celestial pole..."

O° systems perspective: An organized hierarchy require two radically distinctive types of indices - vertical and horizontal. Vertical indices represent the composition of new degrees of organizations with emergent properties. Each degree of vertical organization is described with a novel logical language of dynamic description, often requiring trans-disciplinary terminology. Horizontal indices usually denote ordering relationships within one language of description within one discipline. Horizontal indices can create ordering relationships within one O° -- forming a sequence, a pathway, or a network of causal interrelationships.

5. Symbols

Semiotic perspective:: "A sign without either similarity or contiguity, but only with a conventional link between its signifier and its denotata and with an intentional class for its designatum, is called a symbol."

O° systems perspective: Scientific usage of a symbol requires a specification of meaning in terms of observations (See figure 1). If one views science as a unified body of knowledge, then a specification of meaning is within a hierarchy of natural structures of increasing scale (size, volume). The scientific symbol, "C", representing the concept of carbon, will be used as an example as it is grounded in the same conceptual structures as the O° notation, that is, the natural numbers and the atomic table. Observation of the mass of carbon assign it a value (relative to hydrogen) of close to 12.0. Structural observations of particles indicate a composition of 6 protons, 6 neutrons and six electrons ($O^{\circ}1$). Since both chemical reactivity and quantum descriptions are relative to electrical charge, "C" is assigned the principle quantum number of 6 when viewed as an *autonomous* entity. The symbol "C" in a organic molecule ($O^{\circ}3$) implies a set of atoms bond into a specific geometric form described by a mathematical graph where each carbon atom is one vertex (or node) of the graph. The symbol "C" becomes bound within a new "whole" where it is merely a *constituent* of larger system -- a molecule. When the symbol "C" is used to represent carbon within the structure of a monomer within a biological polymer, the symbol becomes a constituent of a constituent. When the symbol "C" is used in reference to a carbon atom in a monomer within a polymer within a cell, it becomes a constituent of a constituent of a constituent. The relative mass of carbon remains unchanged whether it is an *autonomous* particle or a *constituent*. Dramatic changes in the electrical properties of "C" occur with each genesis of a new degree of constituency. In this example of the scientific usage of the symbol "C", each degree ($O^{\circ}1$, $O^{\circ}2$, $O^{\circ}3$, $O^{\circ}4$, $O^{\circ}5$) is described in terms of a different pattern of organizations with different nouns for the emergent entity. This symbolic linkage of patterns from simple autonomous object to complex cellular object, are composed by iterations of binding operations. Within these symbolic operations, the symbol "C" sustains it identity with respect to the relative mass but the *electrical properties are complexified by the neighboring charges*. For these reasons, the scientific usage of an atomic symbol is remote from the general semiotic usage of a symbol defined by Sebeok.

6. Name

Semiotic perspective: "A sign which has an extensional class for its designatum is called a name. An extensional definition of a class is one given 'by listing the names of the members or by pointing to every member successively." (citing Reichenbach, 1948).
O° systems perspective: A name within a given degree of organization constrains the material composition of the object for degrees one through four. For degrees of organization five, six,... the name marks a historical trajectory but does not necessary mark a specific material composition.

Role of Supra-ordination in Scientific Semiotics and Organizational Theory

Differences between this scientific notation and general semiotics go beyond the meanings of definitions. Sebeok writes of applying the "Law of Inverse Variation", *"The terms, sign, symbol emblem and insignia are here arranged in the order of their subordination, each term to the left being a genus of its subclass to the right and each term to the right being a species of its genus to the left. Thus, the denotation of these terms decreases; for example, the extension of 'symbol' includes the extension of emblem but not conversely."* The Law of Inverse Variation implies the extension of a term may become vanishingly small.

Supra-ordination, rather than sub-ordination, is intrinsic to the C* hypothesis of cellular organization presented here. Supra-ordination is composed by combining lower symbols, $O^{\circ}1$, $O^{\circ}2$, $O^{\circ}3$,..., into specific patterns (mathematical graphs) to form higher categories. In contrast to a decreasing domain of sub-ordinating concepts, this scientific semiotics generates an exponentially increasing domain within the material structure of natural degrees of organization. Supra-ordination follows a historical perspective of the emergence of complexity -- a path of creativity from the simple to the more complex (de Chardin, 1958, Eiseley, 1960, Jantsch, 1981, Miller, 1978, Morowitz, 1992, , 1989, 1996, Chandler, 1991, 1995, 1996). Evolution can be viewed as a sequence of supra-ordinating processes which increase the potential of the system to select among alternative pathways (Chandler, 1996).

This C* hypothesis provides a basis for scientific semiotics which is both richer and poorer than linguistically-based semiotics. O° is richer in the sense that scientific semiotics is grounded within the atomic table of elements; this grounding creates a basis for mathematical consistency within the symbology and semantics of objectivity. It is also richer in the sense that $O°$ provides a basis for the transdisciplinarians to communicate within a common symbolic framework. By the same token, this scientific semiotics is poorer than linguistically based semiotics when attempting to express human desires, values, feelings, and objectives. A simple example illustrates the intellectual challenge. How would the concept of human intentionality be expressed in terms of Stolyar's definitions, Sebeok's six species of general signs or the C* notation? Spiritual traditions, literature, the performing arts, and the natural sciences continue to create ever more highly organized patterns, ever more remote from the simplistic autonomy of mathematically independent particles. The precise role of cellular computations in composing the roles of elementary particles into such highly organized processes is an open question.

Acknowledges

With pleasure, I express my gratitude to Prof. Andree C. Ehresmann for discussions of mathematical philosophy and category theory, to Mr. Jeff Long for discussions of notational systems, and to numerous colleagues in the Washington Evolutionary Systems Society who continue to challenge my imagination.

REFERENCES

Baas, N. A. (1994), Emergence, hierarchies and hyperstructures. Artificial Life III. SFI Studies in the Sciences of Complexity, vol. XVII. Ed.C.G. Langton, Addison-Wesley pp. 515-537.

Chaisson, E. (1996), "Cosmological Complexity on the Grandest Scales," Actes du Symposium ECHO, Amiens Fr., 20-26.

Chaisson, E. (1989), The Life Era. Cosmic Selection and Consciousness. W.W.Norton, NY.

Chandler, Jerry LR (1991), "Complexity: A phenomenologic and semantic analysis of dynamical classes of natural systems," WESScomm 1: 34 - 40.

Chandler, Jerry LR (1992), "Complexity II. Logical constraints on the structure of scientific languages," WESScomm 2: 34-37.

Chandler, Jerry LR. (1995), "Third order cybernetics," Proceedings of 14th International Congress on Cybernetics, Namur, Belgium, p. 489-493.

Chandler, Jerry LR (1996), "Complexity III. Emergence. Notation and symbolization." WESScomm 2: 34-37.

Chandler, Jerry LR, Ehresmann, A.C. Vanbremeersch, J.-P, (1995),

de Chardin, Pierre. T. (1959), The Phenomenon of Man. Harper and Row, NY.

Ehresmann, A. C. and Vanbremeersch, J.-P. (1987), "Hierarchical evolutive systems," Bulletin of Math. Biol. 49: 13-50.

Ehresmann, C. (1966), Trends toward unity in mathematics. Cahier de Topologies et Geometrie Differentielle 8: 759-765.

Eiseley, Loren (1960), The Firmament of Time. Athenaeum, NY.

Jantsch, Erich (1980), The Self-Organizing Universe. Scientific and Human Implications of the Emerging Paradigm of Evolution. Pergamon Press, Oxford.

Lambert, David. M. and Hughes, A. J. (1987), "Keywords and concepts in structuralist and functionalist biology," J. Theor. Biol. 133: 133-145.

Miller, J. G. (1978), Living Systems. McGraw-Hill, NY.

Morowitz, H. J. (1992), Beginnings of Cellular Life. Metabolism Recapitulates Biogenesis. Yale Press.

Rosen, Robert (1991), Life Itself. A Comprehensive Inquiry into the Nature, Origin and Fabrication of Life. Columbia Univ. Press, NY.

Sebeok, Thomas, (1995), Signs. Introduction to Semiotics. Toronto Univ. Press.

Stolyar, A. A. (1970), Introduction to Elementary Mathematical Logic. Dover, NY.

Thom, Rene (1991), Semiophysics Addison Wesley, NY

Vanbremeersch, J.-P, Chandler, Jerry LR and Ehresmann, A.C. (1996), "Are interactions among different time scales a characteristic of complexity?" Actes Du Sym. ECHO Amiens, Fr., 162-167.

Walters, R.F.C. (1991), Categories and Computer Science. Cambridge Press.

LOCALISATION AND NONLOCALITY IN COMPUTATION

Ron Cottam[1], Nils Langloh[2], Willy Ranson[1] and Roger Vounckx

Laboratory for Microelectronics and Technology
Department of Electronics, The University of Brussels (VUB)
Pleinlaan 2, 1050 Brussels, Belgium

INTRODUCTION

It is becoming necessary to carefully re-evaluate the meaning we attach to the term *computation*. This paper considers the conditions prerequisite to the implementation of a general description of computation which is formulated in terms of reactions to environmental stimuli, and which can be used to model natural processes and consequently information processing in cells and tissues.

The first criterion for survival of an evolutionary entity must be objectivisation of external events, to enable internal computational processes to develop survival-oriented reactions. In its most general form an environmental-reaction processor will operate directly in the interactive domain described by physics, and not in an environment derived from computational preconceptions. The development of an archetypal form for such a processor leads to the suggestion of important relationships between the characteristics to be expected of environmental-reaction computation and electrical and chemical "nonlocal" processes found to operate in the brain.

PROCESSING TO STAY ALIVE

A, if not *the*, primary purpose of individual living entities is to survive. Internally-generated reactions to external threatening situations must be available within the time scale dictated by the external process itself. This imposes stringent limitations on the kinds of processing which can be carried out in unexpected contexts. "Real-time" means *slow* processing. Causality is only possible within a domain which exhibits communicational restriction (Prigogine and Stengers, 1984). In the absence of communicative restriction,

---

[1] IMEC university campus researcher
[2] Now with Procter and Gamble, Brussels, Belgium

Information Processing in Cells and Tissues
Edited by Holcombe and Paton, Plenum Press, New York, 1998

197

physical structure is impossible, but universal coherence demands communication *faster* than normal processes.

We are conventionally tempted to choose *between* serial and parallel processing. Serial processing is traditionally viewed as ineffective in the face of extreme demands for speed, and we could conclude that parallel processing would be a better choice for our purposes. Over the last decade, however, the high hopes held out for large scale parallel computation have foundered on the enormous overhead involved in reliably implementing even limited workable schemes. It appears that a more context-dependent compromise is required between serial and parallel processing; but the question is, which one?

SEQUENTIAL REACTIONAL MODELING

We clearly need to have access to simple descriptions of the surrounding environment in order to rapidly compute fail-safe reactions to perceived danger. The difficulty is in integrating these simple models into a scheme where we can afterwards react in a more intelligent manner, given more time for better evaluation of the threat. One solution is to establish a hierarchy of models for a specific phenomenon, ranging from one which is simple but rapidly-processable through a sequence of others which are progressively more complicated.

We arrange for an external stimulus to be projected into the model-assembly as a "query" from the simplest end, and obtain a returning series of sequential "reflections" from the different models (figure 1), these corresponding to stimulus-reactions which take progressively more and more time to appear, but which are progressively more accurate representations of an optimal response (Langloh et al., 1993).

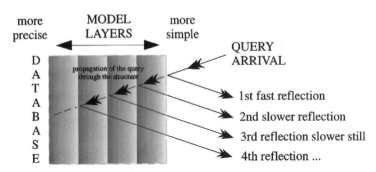

Figure 1. Query-reflection in a hierarchical model assembly.

To a degree, this corresponds to the results of research into the coupling of the visual thalamus to the amygdala (LeDoux, 1992; Davis, 1992), where a similarly multi- (two-) pathway survival-oriented reaction mechanism has been proposed. Each model-layer can be visualised as a self-correlating network, with a number of nodes depending on the complication of representation. As we go further down into the model assembly, the computation required for internal correlation of the model-layers increases, and the propagation velocity of the query is consequently reduced (figure 2), until in the time scale of the external environment it effectively stops.

This corresponds to the imposition of a communication restriction on a causal domain, but here it has more the character of a continuous change between "pure causality" for the simplest model and "pure chaos" for the most complicated representations. Indications of this effect have been observed in the extension of standard quantum mechanics towards numerically large systems (Antoniou, 1995).

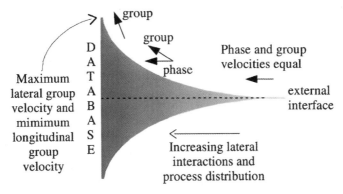

Figure 2. Query slowing-down through a hierarchical model assembly.

NONLOCALITY AND LOCALISATION

We can use this albeit simple picture as the basis for a general model of stimulus-reactional computation in a multi-dimensional "real-world" phase space which defines and describes the environment in which the interactions take place. Symbolically representing a single real-world parametric dimension by a single structure of this kind, we can multiply up the number of represented dimensions if we notice that, in order to sustain universal inter-dimensional correlation, the most complicated representation of each dimension must take into account *all* of the other dimensions, and be associated with *the same* common database (figure 3).

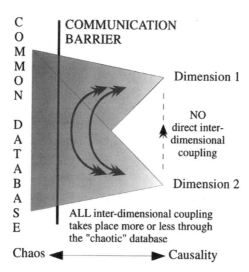

Figure 3. The general representation reduced to an illustration of two dimensions.

Inter-dimensional coupling now calls for a progressive increase in the fractal nature of all dimensions from the "causal" right hand side of the assembly shown in figure 3 towards the "chaotic" left hand side; but more than that, it requires increasing fractality between *all* of the dimensions simultaneously. Mathematically this necessitates a departure from classical probability as a way of characterising the parameter-reducing nature of chance towards a probabilistic description which is closer to that of Dempster (1967) and Schafer (1976), where a defined value of probability is replaced by the upper and lower limits of a uniformly valid probable band. An increase in inter-dimensional coupling then corresponds to an increase in the degree of recursion of a Dempster-Schafer-style described fractality. We can describe this coupling as the degree of *diffuseness* of the system associated with its modeling level.

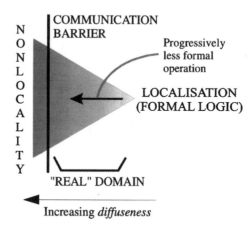

Figure 4. Aquarium as a general model for computation.

We can now provide a simple image for computation in terms of the degree of diffuseness of the current model. Complete diffuseness, as indicated at the left hand side of figure 4, corresponds to the "classical" idea of nonlocality or pure causal chaos, and zero diffuseness, at the right hand side, is equivalent to complete localisation, or perfect determinism. This is consistent with the description of nonlocality as perfect communication, or rather the *capability* for perfect communication, and localisation as non-communicative isolation.

DATA-ZONE BARRIERS AND CONFLICT RESOLUTION

For a processor to be able to deal successfully with the arrival of circumstantially or inherently conflicting data, we should expect the development of partial communication barriers within the processing structure itself, in order to isolate conflicting data within regions associated with, for example, specific kinds of sensory inputs. It would be unlikely, however, for regions of this kind to *remain* completely segregated; to do so would presuppose the development of a kind of schizophrenia. It is difficult to see how any large degree of re-correlation across segregating barriers could be safely carried out in a context which is capable of requiring rapid fail-safe reactions to unforeseen environmental conditions. It would be less risky to isolate first of all the complete system from the consequences of any possibly initially "faulty" imposed inter-regional communication, and only then to allow the nonlocality to increase between previously segregated regions in order to facilitate a required re-correlation.

The global electrical activity of the brain shows up patterns which depend on the degree of our wakefulness and activity, and we can expect them to indicate whether activity is purely local, or whether there is a correlation between the different cerebral regions. Electrical signals generated during the deepest periods of sleep evidence a degree of phase coherence across the complete neural assembly (figure 5). Does this correspond to the conditions required for nonlocal cross-barrier conflict resolution? It is difficult to say at this point, but it is worth noting that electrical phase coherence between different entities is evidence of a high degree of communication, and therefore of a form of "electrical" nonlocality.

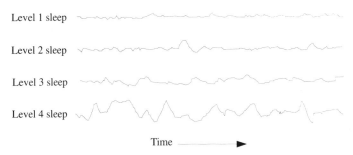

Figure 5. Typical globally measured electrical brain-wave patterns associated with the different levels of sleep.

The conventional engineering approach to software module relocation or re-linking is to first take the machine off-line, perform the necessary maintenance, and then carry out simulations of normal operation in a "close-to-real-conditions" but still off-line environment, which will enable us to see if the changes which we made successfully resolve pre-recognised problems without creating new and possibly worse ones. If we record the progression of periods of sleep and dreaming during a typical night, we see a continual change of sleep-level between level 1, where it is difficult to detect electrical activity different from that occurring during periods when we are awake, and level 4, where a high degree of nonlocal phase coherence can be observed.

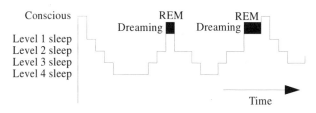

Figure 6. Sleep levels and dreaming during a typical night.

Are dreams equivalent to computational re-correlation testing by simulation? Again, this is difficult to answer, but dreams *do* appear during the periods of lightest sleep, as evidenced by experimentation into rapid eye-movement (REM) sleep (figure 6), when the brain is close to "normal" localising consciousness and the muscles are in a paralytic "off-line" state. And between dream periods the brain *does* apparently descend to a more phase-coherent and therefore more nonlocal state.

CHEMICAL NONLOCALITY

Increasing evidence is coming to light of communication which takes place between different elements of the cerebral system but outside the traditionally considered pathways of rapid electrical neural activity. Research into the recognition of recognisable but variable visual facial characteristics (Rummelhart et al., 1986) has suggested that a good way of storing information of this kind is to associate the functions of a number of physically adjacent neurons. Galley et al. (1990) have proposed that synaptic messengers may diffuse outwards from their presupposed zone of operation and influence the function of other physically adjacent synapses. Is this evidence for the proposition of development of a kind of "chemical" nonlocality?

Recently, communication has been observed (Newman and Zahs, 1997) in real tissue between adjacent glial cells. Is their function in this context merely to remove neurotransmitter chemicals from the synapses, or is this removal itself part of slower and more long-range "nonlocal" communicative processes similar to those we might expect to find towards the left hand side of the reactive computational structure described in figure 4? As Newman (1997) suggests, "... a more intriguing possibility is that they (glial cells) may be able to influence or modulate neuronal activity...". Does the brain resort to diffusive chemical nonlocality because of its inability to successfully implement an information-correlative scheme using the kind of diffuseness we describe here?

ACKNOWLEDGEMENT

This work has is supported by the Inter-University Microelectronics Center (IMEC), Leuven, Belgium.

REFERENCES

Antoniou I., 1995, Extension of the conventional quantum theory and logic for large systems, *Einstein Meets Magritte*, VUB Press, Brussels.

Davis M., 1992, The role of the amygdala in fear and anxiety, *Ann. Rev. Neurosci.* 15, pp 353-375.

Dempster A.P., 1967, Upper and lower probabilities induced by a multivalued mapping, *Annals Math. Stats.* 38, pp 325-339.

Galley J.A., Montague P.R., Reeke G.N. and Edelman G.M., 1990, The NO hypothesis: possible effects of a short-lived rapidly diffusible signal in the nervous-system, *Proc. Nat. Assoc. Sci. US.* 87:9, pp 3547-3551.

Langloh N., Cottam R., Vounckx R. and Cornelis J., 1993, Towards distributed statistical processing, *ESPRIT Basic Research Series, Optical Information Technology*, pp 303-319, S. D. Smith and R. F. Neale, ed., Springer-Verlag, Berlin.

LeDoux J.E., 1992, Brain mechanisms of emotion and emotional learning, *Curr. Opin. Neurobiol.* 2:2, pp 191-197.

Newman E.A. and Zahs K.R., 1997, Calcium waves in retinal glial cells, *Science* 275:5301, pp. 844-847.

Newman E.A., 1997, In J. Kenen, Glial brain cells seen passing messages, Reuter (6 February), Washington.

Prigogine I. and Stengers I., 1984, *Order Out of Chaos: Man's New Dialog with Nature*, p 17, Flamingo-Harper Collins, London.

Rummelhart D.E., Hinton G.E. and Williams R.J., 1986, Learning representations by back-propagating errors, *Nature* 323:6188, pp 533-536.

Schafer G., 1976, *A Mathematical Theory of Evidence*, Princeton University Press, Princeton.

MINING THE GENE EXPRESSION MATRIX:
INFERRING GENE RELATIONSHIPS FROM
LARGE SCALE GENE EXPRESSION DATA

Patrik D'haeseleer,[1] Xiling Wen,[2] Stefanie Fuhrman,[2] and Roland Somogyi[2]

[1]Computer Science Department
University of New Mexico
Albuquerque, NM 87110
WWW: http://www.cs.unm.edu/~patrik
E-mail: patrik@cs.unm.edu
[2]Laboratory of Neurophysiology
NINDS, NIH
Bldg.36/Rm.2C02
Bethesda, MD 20892
WWW: http://rsb.info.nih.gov/mol-physiol/homepage.html
E-mail: <linglinglsfuhrman>@codon.nih.gov, rolands@helix.nih.gov

INTRODUCTION

In order to infer the logical principles underlying biological development and pheno-typic change, it is necessary to determine large-scale temporal gene expression patterns. To quote Eric Lander, "The mRNA levels sensitively reflect the state of the cell, perhaps uniquely defining cell types, stages, and responses. To decipher the logic of gene regula-tion, we should aim to be able to monitor the expression level of all genes simultaneously ..." (Lander, 1996). One method for accomplishing this involves the use of reverse tran-scription polymerase chain reaction (RT-PCR) to assay the expression levels of large num-bers of genes in a tissue at different time points during development, with a standard proto-col. The relative amounts of mRNA produced at these time points provide a gene expres-sion time series for each gene.

The Gene Expression Matrix presented in Wen et al. (1997), contains expression lev-els of 112 different genes at nine stages during rat cervical spinal cord development. Earlier analysis of this data set used Euclidean distance and information theoretic measures to cluster the genes into related expression time series (Somogyi et al., 1996; Wen et al., 1997). A significant problem with this approach is the variety of measures that can be used. Each measure produces a unique clustering of gene expression patterns, and measures must be selected in part according to whether they capture the types of interactions we would

Information Processing in Cells and Tissues
Edited by Holcombe and Paton, Plenum Press, New York, 1998

expect to find in a biological system. For example, a cluster of positively correlated gene expression patterns suggests that all these genes may share the same input(s), but ignores the possibility that other genes may be inversely regulated by those same inputs. Another important consideration is the extent to which the measures capture the total amount of information contained in the gene expression data.

This paper focuses on determining significant relationships between individual genes, based on linear correlation, rank correlation and information theory. Such relationships may be used to form hypotheses concerning potential pathways of information flow between the genes in question, or from an unobserved source towards the genes. Alternatively, the relationships between individual genes can be combined to infer a model of a gene network, in order to form hypotheses about the behavior of such a network (Kauffman, 1993; Somogyi, 1996). These hypotheses can then be tested experimentally by perturbing a specific gene and observing the effect on gene expression levels at different time points following the perturbation.

LINEAR CORRELATION

Looking at the gene expression matrix (Wen et al., 1997), we see that many genes seem to be highly correlated. Indeed, we find positive (Pearson) correlation coefficients of up to 0.992 (between 5HT1b and GRa4, Figure 1 a,b), and negative correlation coefficients of up to -0.986 (between aFGF & IGH II, Figure 1 c,d). Using a two-tailed t-test with n-2=7 degrees of freedom, we can reject (at the 1% significance level) the null hypothesis that the

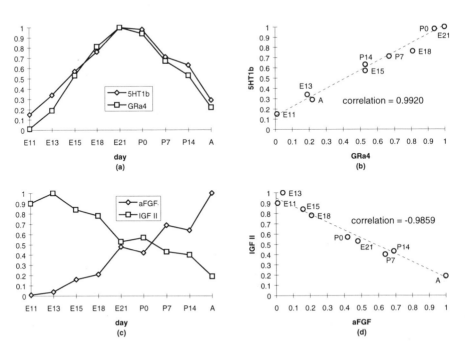

Figure 1. High positive correlation between gene expression patterns 5HT1b and GRa4 (a, b); high negative correlation between aFGF and IGF II (c, d). Expression levels for each gene are normalized with respect to the maximum expression level for that gene.

population correlation coefficient ρ is zero, if the absolute value of the sample correlation coefficient |r|>0.798. Out of the 112x111/2 = 6216 pairs of expression time series, we find 806 pairs that have a significant correlation. Whereas this is much better than having to examine all 6216 pairs of genes, it would still take an immense amount of work to check each of these significant correlations in the laboratory. Furthermore, at the 1% level we would expect about 62 such significant correlations even from independent time series. Because of the large number of gene pairs, we will get a fair number of false correlations even at this significance level.

If we want to further restrict the number of relationships to be examined, we might want to test which correlations are not only significantly different from zero, but also significantly larger than a certain value. For instance, to find those relationships in which at least 50% of the variance is explained by the correlation, i.e. $\rho^2 \geq 0.5$, we need $|r| \geq 0.96$ to reject at the 1% significance level the null hypothesis that $|\rho| < 0.7071$ (we use a transformation from r values to an approximately normal distribution to calculate the significance level). The 33 pairs of time series that have such a high estimated correlation are listed in Table 1.

Table 1. 33 pairs of time series with $|r| \geq 0.96$ ($\rho^2 \geq 0.5$ at the 1% level).

Gene 1	Gene 2	r	Gene 1	Gene 2	r	Gene 1	Gene 2	r
MAP2	SC6	-0.960	ChAT	aFGF	0.982	GRa4	5HT1b	0.992
NFM	mGluR1	0.971	ChAT	IGF II	-0.977	GRa5	mGluR3	0.967
NFM	NMDA2A	0.985	ChAT	IP3R2	0.966	GRa5	mGluR7	0.972
NFH	GRg1	0.971	ACHE	5HT1c	0.981	GRg1	cyclin B	-0.965
S100 beta	GRg1	0.983	ACHE	statin	0.965	GRg3	5HT2	0.976
S100 beta	cyclin B	-0.964	ODC	IGF II	0.975	mGluR1	NMDA2A	0.962
GAD65	preGAD67	0.964	ODC	cyclin B	0.965	mGluR3	NMDA2B	0.970
GAD67	GRg3	0.974	GRa1	GRg1	0.971	mGluR5	NMDA1	0.980
GAD67	5HT2	0.988	GRa2	GRa5	0.972	bFGF	aFGF	0.968
GAT1	SC6	-0.963	GRa2	mGluR3	0.974	bFGF	IGF II	-0.974
ChAT	ODC	-0.962	GRa2	statin	0.970	aFGF	IGF II	-0.986

There are a number of ways to visualize these results. Earlier work on this data (Somogyi et al., Wen et al., 1997) relied on hierarchical clustering of the time series into a dendrogram using the Fitch-Margoliash (1967) algorithm as implemented in the PHYLIP package (Felsenstein, 1993). If we define a distance measure based on the residual variance between any two time series (i.e. $d=1-r^2$; $d=0$ if perfectly correlated, $d=1$ if uncorrelated), we can construct a similar dendrogram based on linear correlation. The resulting tree is quite different from the one based on Euclidean distance (data not shown). Another, more intuitive way to visualize this data is to use multidimensional scaling, which maps time series into a two-dimensional plane, while attempting to preserve the specified distances between them. Figure 2 shows a multidimensional scaling of the 34 genes involved, where the distance measure used was the residual variance between any two time series. Multi-dimensional scaling of the entire set of 112 genes shows these 34 as very tight clusters amidst a more loosely connected set of genes.

Lastly, we might want to ask the following question: given that there is a certain amount of noise in our measurements, at which point can we say that time series A cannot really be distinguished from a linear transformation of time series B? In order to improve the accuracy of the expression data, each value v in the Gene Expression Matrix is really an average over up to three separate individuals. The standard error $s_{\bar{v}}$ on this average is $s_v/\sqrt{n}$, where s_v is the estimated standard deviation of the individuals and n the number of individuals. If we assume the time series are homoscedastic, i.e. the standard error is the

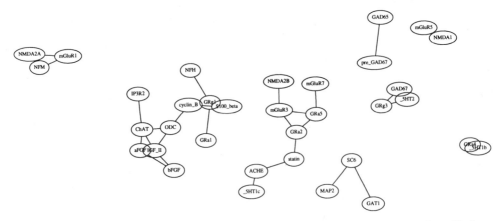

Figure 2. Multidimensional scaling of 34 time series with high correlation. Distance represents the residual variance, $1-r^2$, with those pairs of genes for which $|r|>0.96$ linked by an edge.

same for each time point of a time series A, we can estimate $s_{\bar{A}}$ by the average over the standard errors at each time point. This $s_{\bar{A}}$ is a measure of how much noise there is in the entire time series. If we find that the variance $s^2_{A \cdot B}$ of the residue of the regression of time series A on time series B (i.e. the amount of variation in A that cannot be explained by B) is smaller than $s^2_{\bar{A}}$, we can say that A cannot be distinguished from a linear transformation of B. We may then choose to reduce the data set, by replacing each such pair by a single time series. We find 15 pairs of time series (listed in Table 2) where this occurs.

Table 2. 15 pairs of time series for which the variance of the residues is smaller than the noise on one of the time series.

Gene A	Gene B	$s^2_{A \cdot B}$	$s^2_{\bar{A}}$
GAD65	pre-GAD67	0.079	0.083
GAD67	5HT2	0.058	0.069
GRa2	GRa5	0.092	0.108
GRa2	mGluR3	0.091	0.108
GRa2	statin	0.095	0.108
5HT1b	GRa4	0.040	0.063
CNTFR	IGF I	0.045	0.045
cyclin A	NFH	0.058	0.072
cyclin A	MOG	0.069	0.072
cyclin A	GRa1	0.066	0.072
cyclin A	GRg1	0.061	0.072
cyclin A	IGFR2	0.068	0.072
H2AZ	MOG	0.052	0.061
H2AZ	mAChR4	0.053	0.061
H2AZ	CRAF	0.046	0.061

The approach above is rather simplistic because it relies on homoscedasticity of the time series, and does not give any significance levels. There is a more rigorous approach to answer the same question. We can represent time series A and its best linear fit $A' = a + bB$ as points in a nine-dimensional space, with each dimension corresponding to a specific time point in the series, and the error in each dimension being the standard error on the value for that time point. We can then use a multivariate test to check whether A and A' are significantly different.

NONLINEAR CORRELATION

One obvious way to capture nonlinear relationships between time series is to fit the data to a nonlinear model. If we are interested in monotonic relationships we could suggest a sigmoidal model for example. For a very general sigmoid model one would want to include at least parameters for scaling and offset along both axes. However, with only nine points to fit for each pair of genes, a four-parameter model might be excessive. Pearson correlation, whether using a linear or nonlinear model, also assumes an approximately Gaussian distribution of the points, and may not be very robust for non-Gaussian distributions. For these reasons, we have chosen a more general measure for monotonic relationships: the Spearman rank correlation r_s. The Spearman correlation coefficient is equivalent to the Pearson correlation coefficient calculated over the ranked values. The 1% significance level for r_s for short time series can be derived from tables (Olds, 1938; for n>10 we can use a modified t-test). For n=9, we can reject the null hypothesis that the population rank correlation ρ_s is zero at the 1% significance level if the sample rank correlation $|r_s| \geq 0.833$. Using this test, we find a total of 491 pairs of expression time series, involving 98 genes, which have a significant r_s, ranging from -0.979 to 0.996. 119 of these pairs do not have a linear correlation that is significantly different from zero at the 1% level (i.e. $|r| < 0.798$). Figure 3 shows one such time series pair which has a high rank correlation but low linear correlation.

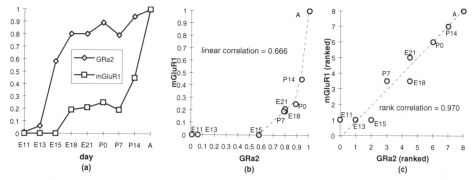

Figure 3. High rank correlation but low linear correlation between mGluR1 and GRa2.

As in the previous section, even if the null hypothesis were true (uncorrelated time series), at the 1% level we would expect an average of 62 pairs where we wrongly reject the null hypothesis and think we see a significant correlation. Again, one might want to find those rank correlations that lie above a certain value. However, unlike linear correlation, it is hard to find a plausible explanation of the meaning of such a level. Also, since there is no adequate mathematical model for the distribution of r_s for small n, the 1% significance level has to be derived from tables or calculated exhaustively (Olds, 1938).

Because of ranking, we may lose a significant amount of information present in the data. For instance, the maximum information content for a time series of nine data points with values ranging between 0.00 and 1.00 (with a .01 resolution, i.e. 101 possible expression levels) is $9\log_2(101) \approx 60$ bits. The maximum information content for a ranked nine-point time series without ties in the ranking is $\log_2(9!) \approx 18.5$ bits (slightly larger if ties are allowed). In effect we may be ignoring up to 70% of the information content of the data.

INFORMATION THEORY

An even more general method to detect association between gene expression time series can be derived from information theory. The mutual information $M(A,B)$ between two information sources A and B represents how much information A contains about B (and vice versa). Formally, if $H(A)$ and $H(B)$ are the (Shannon) entropies of sources A and B respectively, and $H(A,B)$ the joint entropy of the sources, then $M(A,B) = H(A) + H(B) - H(A,B)$ (Shannon, 1948). Mutual information is defined both for continuous and discrete distributions, but the discrete form is much easier to use. To apply this technique we will therefore first discretize the time series by partitioning the expression levels into bins. Some regulatory genes exhibit a close approximation to on/off behavior, with several orders of magnitude difference between basal and induced expression levels. In such a case, the gene expression levels can be discretized without loss of information. However, if part of the regulatory activity of the gene depends on small fluctuations superimposed on the on/off behavior, then this will not be captured by a discretized model. Similarly, the expression levels of some genes mirror continuously varying environmental parameters and have a regulatory effect over their entire range. Discretization of expression levels of such genes will lead to a loss of information.

The fewer bins we use to discretize the data, the more information about the original time series we ignore. On the other hand, too fine a binning will leave us with too few points per bin to get a reasonable estimate of the frequency of each bin, especially when calculating the joint entropy. Given that our data only contains nine time steps, three bins is the most we should use.

As with rank correlation, we can get an estimate of how much information we may ignore in doing this binning. A time series of nine data points between 1 and 3 has a maximum entropy of only $9\log_2(3) \approx 14$ bits, as opposed to approximately 60 bits for values between 0.00 and 1.00 (see above), so we may be ignoring up to 76% of the information content of the data. This is of course a worst-case calculation, and although it casts doubt on the usefulness of this method for the amount of data we currently have, the generality of this approach may outweigh the information loss for larger data sets. This section then should be regarded more as an example of how we could use information theoretic measures once we have more data to work with, rather than as an overview of significant results on the current data set.

Most of the properties covered in this section can be summarized in a single graph, a detail of which can be found in Figure 4. For the complete graph, see the web site at http://rsb.info.nih.gov/mol-physiol/IPCAT/patrik/mgraph.html.

Some time series map to the same discretized series. For two independent time series, the probability of mapping to the same series is $3^{-9} \approx 10^{-4}$. However, due to the coarseness of the binning, identity of binned expression time series is of course a less accurate predictor for association than a small Euclidean distance or large correlation. In total, from 112 unique continuous-valued time series we get 91 discretized time series. Table 3 shows the discretized series that correspond to more than one gene. In Figure 4, these genes are listed within the same node.

Next, we can find those pairs of time series for which there is a one-to-one mapping by permuting the bin numbers, i.e. for which $H(A)=H(B)=M(A,B)$. For instance, in Table 3 we notice that there is a one-to-one mapping for S 100 beta (row 6) and GAD67 (row 7) by permuting bins 1 and 2. In total, we find 17 pairs of unique time series where this occurs. The probability that two independent but unique time series have a one-to-one mapping is approximately $(3!-1) \times 3^{-9} \approx 10^{-3}$. Series for which there exists a one-to-one mapping are equivalent with respect to their information properties, so we can replace such time series by one single series, leaving us with a set of 77 unique, non-equivalent time series. These cor-

Table 3. Gene expression time series that map into the same discretized series.

E11	E13	E15	E18	E21	P0	P7	P14	A	genes
0	0	2	2	2	2	2	2	2	MAP2, pre-GAD67, GAT1
0	0	0	0	0	0	0	1	2	NFM, mGluR1, NMDA2A
0	0	0	0	1	1	1	2	2	NFH, ChAT
0	0	1	1	2	2	2	2	2	synaptophysin, ACHE
0	1	2	2	2	2	2	2	2	neno, trkC
0	0	0	1	1	1	1	2	2	S100 beta, GRg1
0	0	0	2	2	2	2	1	1	GAD67, mGluR5, NMDA1
2	2	2	2	2	2	2	1	1	ODC, CRAF, cyclin A
0	0	1	2	2	2	2	2	2	GRa2, GRa5, statin
0	0	1	2	2	2	1	2	2	mGluR3, NMDA2B
0	0	2	0	0	0	0	0	0	nAChRe, trk
2	2	2	2	2	2	2	2	2	CNTFR, PDGFa, IGF I, H2AZ, TCP, actin

respond to the nodes in Figure 4. Time series which have a one-to-one mapping to each other are grouped in rectangular nodes, with the node compartments corresponding to the different permutations of the bin numbers. Note that any kind of permutation of the bins suffices, which can lead to very discontinuous and not biologically plausible one-to-one mappings for larger numbers of bins.

When there is no one-to-one mapping, we could use various symmetric measures for the information sharing between two time series, such as $M(A,B)/\max(H(A),H(B))$, or $M(A,B)/H(A,B)$. However, the asymmetric measure $R(A,B) = M(A,B)/H(B)$ has shown to be more informative. This relative mutual information measure reflects what fraction of the information in time series B can be derived from A. For some pairs of series, $R(A,B) = 1.0$, which means that all the information about time series B is contained in time series A, i.e. A

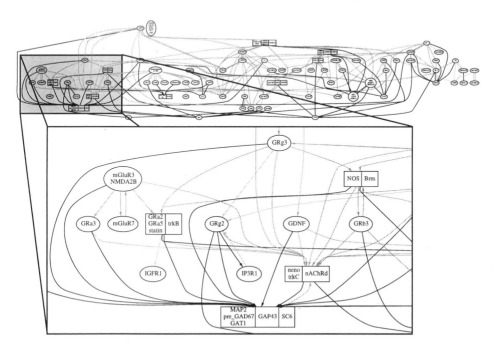

Figure 4. Detail of graph summarizing information theoretic relationships between time series.

maps into *B*. Relative mutual information might be used as an indicator for a possible causal relationship between gene *A* and gene *B*. In Figure 4, $R(A,B) = 1.0$ is represented by a black arrow from node *A* to node *B*. (Note that nodes corresponding to low-entropy time series will tend to have many incoming arrows, whereas high-entropy nodes will have few black arrows coming in). If we cannot find a time series that will map exactly into a given time series *B*, we can still find the series *A'* that carries the most information about *B*, i.e. that maximizes $R(A',B)$. In Figure 4, those nodes *B* with no incoming black arrows will have a gray arrow coming from time series *A'* that carry the most information about *B*.

OTHER METHODS

All the above approaches look only at the relationship between two time series. For those series which do not have a clear one-to-one relationship with any other time series, we might want to look at multi-gene interactions, for instance using multiple regression. However, it turns out that we have too many possible models to match the amount of data in our current data set. Suppose we are given a time series *A*, and we want to find a function f and two other time series *B* and *C* such that $A \approx f(B,C)$. Note that there are $111 \times 110/2 = 6105$ pairs of time series (*B*,*C*) to choose from. Considering that series *A* contains only nine time steps, even randomly generated data will tend to generate a large number of supposedly good models of the form $A \approx f(B,C)$.

Most of the problems we encounter with this data set stem from the fact that we have so few time steps. We can get around this problem somewhat by interpolating between the values. Of course, interpolation does not in fact give us more data to work with, rather it may allow us to get more out of the data we already have. In the information theoretic approach for instance we might be able to use a larger number of bins and still get a reasonable frequency estimation. Another interesting approach might be to plot the data for each of the three individuals that were sampled at each time point. These three individuals will generally be at slightly different stages in their development, because there is a certain degree of error on the estimate of the age of each rat embryo, as well as slight variations in the speed of the developmental process. This data is more noisy, but it might be possible to extract more information out of it using the correct statistical tools.

We can also use continuous information theoretic measures, which requires estimating the continuous joint entropy distributions. Such estimates can be achieved based on interpolation between the time points and the use of kernel estimators to represent the amount of variance on the measurements.

So far, we have not yet taken advantage of the timing information in the time series. It might be possible to infer causal relationships based on the timing with which changes percolate through the system, similar to the work done by Arkin et al. (1995, 1997) on reconstructing reaction pathways. Similarly, hysteresis effects could indicate the ordering of genes within a regulatory chain. However, the time resolution may have to be of the same order of magnitude as the response time of the individual genes. With the current time resolution however (up to one time step per day), we may only be able to infer the global state trajectory of the system and larger-scale interactions between multi-gene subsystems.

CONCLUSION

Linear correlation can be used very effectively to detect linear relationships. Because linear correlation is scale and translation invariant, it will also detect relationships not captured by Euclidean distance, such as high negative correlations. However, it only detects relationships with a large linear component and assumes Gaussian distribution of the expression levels. Many standard statistical techniques using correlation coefficients are available, making it an excellent tool. For instance, Path Analysis can be used to determine the strength of interaction between correlated genes for a given causal scheme (Wright, 1921).

Rank correlation can be used to detect non-linear relationships. It is much more robust with respect to the distribution of expression levels. Some statistical tools for linear correlation may still be applicable. However, rank correlation assumes monotonic relationships and some information may be lost because it only uses the relative ordering of the expression levels.

Information theory can be used to detect genes whose (binned) expression patterns share information. It will detect *any* mapping from time series *A* to *B*, but this also means that it will not distinguish between biologically plausible or implausible relationships. It is an excellent tool for discrete-valued data. However if continuous data needs to be discretized first, a lot of information may be lost.

Of course, the methods presented in this paper could also be used on expression data from a single cell type in culture, where we do not have to worry about any spatial development patterns. These methods are also not limited to time series. In fact, the time resolution available so far might be too coarse to infer cause and effect. If so, sampling states that are temporally unrelated (e.g., under different growth conditions) may tell us more about underlying circuitry because they allow us to cover a larger region of the entire state space.

These are only the first steps towards deciphering the logic of gene regulation based on large-scale gene expression data. As more of this data gets produced, and especially as more data per gene becomes available, we can expect techniques such as these to become even more powerful. We think it is important to start looking now for the tools that we will need tomorrow.

ACKNOWLEDGMENTS

This research is supported in part by grants from the National Science Foundation (grant IRI-9157644) and the Office of Naval Research (grant N00014-95-1-0364). The Santa Fe Institute is to be credited for originating this collaboration.

REFERENCES

Arkin, A., and Ross, J., 1995, Statistical construction of chemical reaction mechanism from measured time-series, *J. Physical Chemistry*, 99:970-979.

Arkin, A., Shen, P., and Ross, J., 1997, A test case of correlation metric construction of a reaction pathway from measurements, *Science*, 277:1275-1279.

Felsenstein, J., 1993, PHYLIP (Phylogeny Inference Package) version 3.5c, distributed by the author, Department of Genetics, University of Washington, Seattle.

Fitch, W.M., and Margoliash, E. , 1967, Construction of phylogenetic trees, *Science* 155:279-284.

Kauffman, S.A. , 1993, *The Origins of Order: Self-Organization and Selection in Evolution*, Oxford University Press, Oxford.

Lander, E.S., 1996, The new genomics: global views of biology, *Science* 274:536-539.

Olds, E.G., 1938, Distributions of sums of squares of rank differences for small numbers of individuals, *Annals of Mathematical Statistics* 9:133-148.

Shannon, C.E., 1948, A mathematical theory of communication, *Bell Sys. Tech. Journal* 27:379-423, 623-656.

Somogyi, R., Sniegoski, C.A., 1996, Modeling the complexity of genetic networks: understanding multigenic and pleiotropic regulation, *Complexity* 1(6):45-63.

Somogyi, R., Fuhrman, S., Askenazi, M., and Wuensche, A., 1996, The gene expression matrix: towards the extraction of genetic network architectures, *Proc. of the Second World Congress of Nonlinear Analysts* (WCNA96). Elsevier Science, in press.

Wen, X., Fuhrman, S., Michaels, G.S., Carr, D.B., Smith, S., Barker, J.L., and Somogyi, R., 1997, Large-scale temporal gene expression mapping of CNS development, *Proc. Natl. Acad. Sci.*, in press.

Wright, S., 1921 Correlation and causation, *Journal of Agricultural Research* 20:557-585.

COMPUTATIONAL MODELS OF IMMUNOLOGICAL PATHWAYS

Mike Holcombe, Alex Bell

Department of Computer Science
University of Sheffield
Sheffield, UK

INTRODUCTION

The immune system in higher animals, particularly humans, is a complex and only partially understood system of enormous significance for medicine and which is essential for survival in an extremely hazardous world. As in many other areas of biology, a vast amount of detailed knowledge of the system is available. Our knowledge of the components parts and workings of the immune system is vast and far greater than any other similarly complex anatomical system, such as, for example, the brain. This information deals with many aspects of the system, its architecture and the structure of its components; the subsystems involved at many different levels; the biochemical behaviour of the components and many more. As in other biological contexts, it is very difficult to bring all these features together into a coherent and complete model which is both realistic and tractable (in the sense that we can analyse it in a useful way and extract new information and understanding from it). Yet, this is perhaps the most pressing challenge that we face if we are to truly understand how these complex systems behave and respond to their world.

The system includes the many mechanisms which exist for fighting disease and infection by alien organisms and substances. It is a combination of specialised cells, proteins, specific structures, organs and tissues which interact together in many subtle and complex ways. The system changes in both space and time, it has the capacity of adapting to new and unusual circumstances, most typically, caused by the appearance in the organism of previously unseen pathogens and diseases. It can also "remember" previous invaders and is able to destroy these in order to protect the organism from further damage.

Central to this system is the role of the T cells and the B cells. These cells can exist in one of several different states and the state influences what they are able to do. The history of the invasions of the organism by alien bodies, bacteria, chemicals, viruses and other damaging substances, determines what state some of these T cells can be in. The behaviour of these cells, their interactions with other parts of the system form the focus of our investigations.

Despite the wealth of detailed knowledge that immunologists have obtained, the overall picture is far from clear. We have much local knowledge and have to make many assumptions about the way it all fits together. In this paper we attempt to build a series of novel models that try to examine the broader picture, analyse them with the help of a modern simulation tool and draw some tentative conclusions about the overall behaviour of such a system. From this knowledge we can identify those areas where more experimental data is needed in order to construct more sophisticated models. We are also exploring the feasibility of this approach, can it deliver a practical way of understanding complex, parallel, system behaviour of this type?

Thus, we attempt to utilise recent advances in modelling complex, hybrid systems which share many of the abstract system features of the immune system and to evaluate the feasibility of developing a powerful dynamic model of the immune system so that the model integrates much of the important specific information about different aspects of the system into one mathematical structure. This structure can then be analysed to provide information about, for example, the stability of the overall system and used as a basis for simulation to establish a way of examining the influences of a variety of both external and internal events on the overall performance and destiny of the system. The immune system normally exists in a balanced state of quiescence but it is subject to considerable perturbations caused by environmental influences and has to be able to adapt to these and "learn" from the experience.

IMMUNOLOGICAL PATHWAYS

The immune system is a complex dynamical system which operates on the levels of organs, cells and biochemicals. A good deal is known of the cellular taxonomy of immunity and about the processes carried out by individual cells, organs and chemicals in isolation, from in-vitro experiments and other observations, but little is known about how these processes affect one another from a more system-oriented context.

The behaviour of the system is strongly dependent upon the spatial arrangement of its components within and between the organs of the system. This arrangement leads to specificity and delays in the communication between sections of the system, which are affected by the physical distance, barriers and chemical gradients. The overall dynamics are also affected by temporal considerations. The strength and effect of a particular process is affected by the time at which it occurs since there is delay between the activation of a process and the process carrying out its function. There is also the issue of immunological memory. There are billions of lymphocytes (T and B) in the system but each have their own unique receptors for antigen. Thus on entry of antigen into the system, only very few from those billions can recognise the antigen and be provoked to respond. This is why the first response must be one of clonal expansion (to increase the size of the effective "fighting force"). Once this has been achieved the cells can differentiate to produce effector cells that will cause the elimination of the antigen. However, the system also normally sequesters part of the expanded cell number into a "memory pool". Thus, if and when we see the antigen again (and the likelihood is that if the system has seen something it recognises once it is likely to see it again) we have more cells ready and able to respond. They also do so more efficiently. This is the basis of immunological memory.

The various stages of the immunological process can be described in terms of a series of process that occur in different locations at different times. These processes combine in both sequential and parallel ways to form a complex dynamic system. We are presenting a cellular level model and the processes involve cellular interactions with different molecules and local environmental conditions, the cells being influenced to change their state and their role in the system by a combination of these interaction events and other, probabilistic influences.

A number of approaches to the mathematical modelling and analysis of these types of system have been tried (McLean, 1994; Sonada, 1992; Marchuk et al., 1991). Classical control theory models involve the solution of differential and integral equations relating system variables with environmental inputs and outputs. Stochastic analysis also plays a major role in some models. Often the mathematical analysis requires the linearisation of the equations to make them soluble or the use of sophisticated numerical strategies to obtain approximate solutions. The number of interdependent variables that can be handled is limited. The control is modelled using continuous analytical variables and system states are identified by these variables. In many biological systems, however, there is a discrete dimension to the state and to the environmental interface. These may be exemplified by the existence of a cell in one of a limited number of states which then may interact with a discrete number of individual subsystems - such as a T cell interacting with an antigen presenting cell. The discrete control models, however, can only model the situation in a somewhat superficial way since the relationships between the discrete behaviour and the continuous processing and environmental effects is also only possible through crude approximations.

Other writers have used connectionist models (Farmer 1986,1990). These require considerable simplification of the system being modelled, often dealing with only one or two components of the system directly, leaving others to be inferred from these. The internal processing of these models is hidden from the user, thus making analysis and modification difficult, and their validity has been questioned (De Boer and Hogeweg, 1989).

Control theoretic models of hybrid non-linear systems have also been proposed, for example the models based on the recent work of Holcombe et al.,(1994; 1995a; 1995b). These hybrid models are able to model both continuous and discrete aspects of the control of the system in an integrated way. They have been developed and analysed in connection with examples from the steel making system where there is important interaction between the discrete events, the continuous processing and stochastic events both within the system and in its immediate environment. These models were obtained by generalising the discrete models and combining with them the non-linear differential equation models as part of the internal state behaviour. Thus for example, while the system is in one state a particular set of differential equations applies to the continuous variables of the system; on reaching a particular leaving condition – perhaps determined by a specific variable reaching a particular value, or by an external discrete event - a state change ensues to a different state where a different set of differential equations hold. The problem with this approach is it is currently computationally unfeasible to try to simulate a sizeable system this way or to reason about it using a suitable formal logic.

One approach to producing discrete models of biological systems has been described in Holcombe (1990), Holcombe (1991), Holcombe (1994), Bell and Holcombe (1996) and we believe that a similar approach involving the integration of continuous, stochastic and discrete models into a hybrid model will be fruitful. In this work we are trying to use a discrete parallel model with a fine temporal grain and with a suitable discretisation of the continuous variables. We have chosen a specific area, namely immunological pathways, to provide a detailed context for the modelling.

We have used the following, highly simplified model of T2 response to a single antigen.

A number of different possibilities can occur in the biological situation:
- Antigens (Ag) which combine with an antibody, become neutralised and marked for pickup by an antigen presenting cell (APC).
- Ags encounter a B cell thereby activating it (its state changes to active)
- Ags encounter an active B cell together with interleukin 5 (I15) causing plasma cells to form.
- Plasma cells produce antibodies of a specific type dependent on the lymphokine it encounters (Interferonγ (IFNγ), I110, IL13, I14 produce different Ab classes)
- B^m cells transform into memory B cells if a plasma cell arrives in a specific time.
- Type 2 helper cells (Th2), when activated by a DC* and Th2* produce I14,5,10,13 etc.
- If Th2* receives I12 or 4 it proliferates. If not it becomes a $Th2^m$ cell.
- The effect of I12 and 4 is increased by I11 and inhibited by IFNγ

THE BASIC COMPUTATIONAL MODEL

The underlying basis of the model is one of a collection of subsystems that can exist in one of a number of possible states. The state of a subsystem will change as a result of interactions with its environment, which includes other parts of the subsystem as well as specific input stimuli, for example bacteria. In our case one part of the system will be a population of T cells which can be in a number of internal states. As the system runs the population profile will change and this will, in turn, affect and be affected by, the activity of other subsystems and the environment.

We are left with the task of constructing a hierarchical model involving a number of levels of system, each comprising a set of communicating subsystems, each of which can be operating concurrently, changing state, producing outputs which affect other parts of the system and the environment. A further complication is the fact that some of this activity is localised in specific tissue and that real communication, for example in the case of cells and chemicals, involves motion through the blood supply system form one site to another.

We will model this by using a graphical language called Statecharts (Harel et al., 1996). One advantage of this approach is that a sophisticated modelling and simulation package (Statemate^{TM}) is available.

The central concept is of a machine and of a *superstate*. The machine is a system that can exist in a number of possible states. The system/machine is influenced by events, these may be specific occurrences of a particular substance in the environment of the machine, or in the elapsing of a certain period of time, and the system will then respond by changing state and perhaps, causing an action to occur, this again might be the production of some output. The state might actually represent a machine or set of machines itself and this is what we mean by a superstate. So that on entering such a superstate it might be the case that several submachines start up in parallel, interacting to events and causing their internal states to change and actions/outputs to occur.

We represent much of this in a graphical way. The following diagram illustrates a simple machine with 3 states.

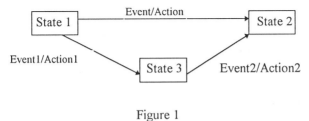

Figure 1

Here, if the system is in State 1 and Event occurs then the system stage changes to State3 and produces the Action. However if Event 1 occurs then the machine goes into State 2 and Action1 occurs. Now State 2 is a superstate, it comprises 2 parallel machine which start up together:

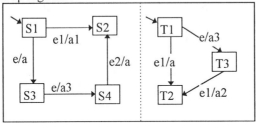

Figure 2

When the system enters the superstate it enters state S1 and T1 simultaneously, if event e occur then the next states are S3 and T3 with actions a and a3 occurring together. If, however event e1 occurs then the next states are S2 and T2 with actions a1 and a occurring.

We need to relate the idea of a state-based system to the idea of an immunological pathway. In the immune system there are many different cells each in a specific state, these cells are subject to a wide number of influences that can cause them to change their state, they are also at a number of different physical positions in the system which also affect their status and behaviour. We develop a simple notation to describe this. It has a similar notation as the "flow chart" model of [1] except that this model was concerned with the molecular rather than the cellular and intracellular level.

At this level vertices are replaced by rounded boxes, representing registers. The interactions are represented by edges between the registers and a polyhedral box representing a function. An arrow or line to a function box represents the input of the availability (A product of abundance, affinity to the reaction and a distance relation based on either the separation of the source and reaction site or on a concentration distribution) of the item represented by the register it originates from. The function in a rectangular box may be the normally accepted velocity equation for that reaction, or it may use the smallest input availability to determine the rate of the reaction, that is how much of each substrate or precursor is used, and product created in a given time based on stochastic considerations. A function box may indicate that the type as well as the quantity of the product of the reaction will vary depending on the nature of the function's inputs. An arrow to a function box also denotes that the result of the function is subtracted from the register (substrate or precursor type behaviour). A line (with no arrowhead) to a function box denotes that the register is unchanged by that function (enzyme, or facilitator type behaviour). An arrow from a function box denotes that the result of the function is added to the register.

Using this notation we produce a representation of the T2 immune response, which we intend to model.

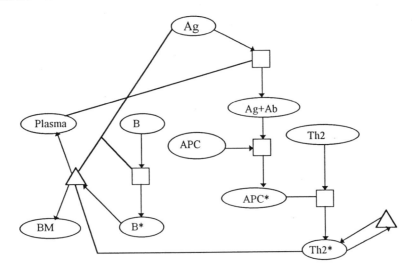

Figure 3

In the flow diagram the rounded register boxes represent populations of cells which are stored as real or integer numbers.

In the state chart these are stored as single variables of real or integer type and are called by those functions to which they are connected in the flow diagram.

Each register in the flow diagram is represented by a state of the same name in the state chart. An arrow into a junction box in the flow diagram means there is a state change in the state chart. (A headless line does not) which is taken after some delay or threshold that is detailed in the function box. The transition goes from the current state to the state named after the register to which the arrow out of the appropriate function box attaches.

If a function box has more than one output arrow, then instead of a pair of transitions which would be non-deterministic we create a superstate containing the new paths allowing them to run in parallel.

In the case of a decision box this is unnecessary since we can label the transitions with conditional guards to ensure that the choice is deterministic.

Function boxes contain details of any delays or transport functions, rate functions or threshold conditions that apply to the transitions and states.

Rate equations in a function box associated with a state change are represented as actions throughout the target state. If a state change occurs through a function box that contains the same rate equation as was already in effect, then a superstate may be created to contain the two states, with the rate equation throughout the superstate.

If there is some equation that is executed throughout all states, then it is placed at the level of the topmost super state that contains all of the states.

The delay information is held in the function boxes in the flow diagram and is represented in the state chart by a dummy state between the two states involved in the change and a time-out condition on the transition from the dummy state to the next state, (e.g. tm(en(DCX),9)). These delay states are not strictly necessary, but can be used without affecting the processing provided the transitions are properly handled.

The initial states of the flow chart are represented in the flow model by those registers with no input arrow. The paths followed from the initial states are contained within separate areas of the top level superstate which contains all states, and run in parallel.

If two paths should combine for some reason, then the areas must merge. This could mean having 2 or more start states in a path that is illegal, so the two paths are contained in a superstate in which they are separated and run in parallel. The transitions that cause the merging of the original path are now a single transition out of this superstate.

The process is terminated by a transition from the topmost superstate to a termination marker after a predetermined delay.

The various states in the representation each contain simple functions as follows:

WHOLE: /AG:=(AG*1.1225)-PC

DCX: /if AG<100 then APCX:=
  Else if AG>1100 then APCX:=1000
  Else APCX:= AG-100 endif endif
DCXX: /if AG-(0.7937*APCX)<100 then APCX:=0
else if AG-(0.7937*APCX)>1100 then APCX:=1000
else APCX:= AG-(0.7937*APCX)-100 end if end if

THX: /TX := T*(APCX/1000)

THP: /T := (1.1225*TX) + T

BX: /BC := 10*AG

P: /PC := 1e-05*BC*TX

All variables are of integer type with minimum set to zero, except APCX which must be of real type in order to avoid error in it's division by 1000 when calculating THX.

The meanings of the states;
DC, TH and B are the initial states, PRE… denotes a delay state.

DCX: The APC population has been activated by antigen.

DCXX: The APC population remains active, but there is desensitisation of part of the population over time.

THX: Th2 cells are active

THP: Th2 cells are proliferating at a rate which doubles the population every 6 time units (24 hours).

BX: B cells are activated.

P: B cells have become plasma cells and are producing antibodies.

Now we can introduce a detailed model of this using the statechart method. Here we identify 3 concurrent subsystems and describes how they behave under the environmental conditions described above.

The simulation we have produced is for a simplified model of the T2 response to an antigen which doubles its population every 24 hours. Because of the modular nature of the modelling tool we can add components or modify behaviours easily, to take into account such things as the T1 response, non-specific response and the systemic effects of cytokine production. At this stage we have not taken into account the effects of physical structure and separation within and between organs. We have hidden the production and effects of cytokines in simple communications between cell populations that take one time step to occur. The model has been kept deliberately

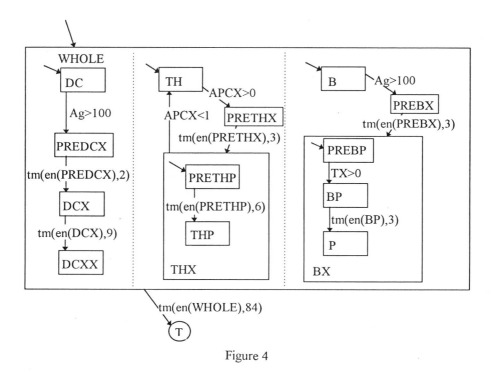

Figure 4

simple in order to make the initial checking and testing of the modelling technique straightforward. Our intention was to produce a modelling technique that can be used to produce models from the bottom up, allowing modifications and expansions to be made easily and accurately in an intuitive way.

The relations we have used are based on results from various in-vitro experiments in the literature (Coffman and Carty, 1986; Fleshier et al., 1988; Langhoff and Steinman, 1989; Steinman, 1991). Reaction rates and response times are simplified, but this was for simplicity in testing the modelling technique and can be made more realistic when proper simulations are required. Units are relative, but are based on the cell concentration (units per ml) information given in the experimental papers.

We have not considered spatial arrangements as yet, but we shall in future models.

Because the processing is necessarily stepwise in a machine simulation, errors arise in the simulation of the quasi-continuous processes between and within populations. We need some way of measuring these errors in order to either keep them within certain limits, or at least to know the limits of our model. We can reduce these errors by reducing the lengths of the time steps used. Since the data may be of integer type there may be a limit on the size of time steps before rounding errors become significant. This means that an acceptable balance between the above mentioned sources of error must be found. State changes are normally considered to be instantaneous reconfiguring of the systems of equations. This is not a real limitation since state changes which are not instantaneous can be replaced by a sequence involving a state representing the period of the state change with an instantaneous beginning event and an instantaneous end event.

In synchronous statemate simulations, state changes take up one time step during which processing within the affected states is suspended, so we must be careful how we treat them. Where there is a delay between the receipt of some signal, and the reaction to that signal, we can safely use state changes provided the maximum size of the time step is half of the lowest common denominator of the lengths of all the delays in the model. If a state change occurs without any delay, we must be careful to ensure that any processing which should not be halted is to be explicitly required as part of the transition step.

The simulation we are presenting here has a very simplistic way of dealing with these delays. After the transition has occurred, the number of the cells which are transformed is dependent on the current level of the activating signal rather than the level which actually caused the activation before the delay period. This leads to responses that may be too rapid or too slow depending on the behaviour of the system since the delay began. We can avoid this problem by using a set of variables to keep track of the recent history of the signals in question.

Using the formalisms of state charts or hybrid X-machines we are also able to perform mathematical analysis of models. We are able to use the usual methods of systems of differential equations, which can produce good information about the general dynamics, but lose more specific numerical information and tend to need serious simplifications in order to make them tractable. We may also be able to use approximation and constraint methods on the systems of algebraic equations in our models in order to produce information about the ranges within which the model can operate. We may treat the model as a stochastic problem dealing with probabilistic interactions between populations distributed in space, time and "activation". This mathematical analysis, while not producing such large amounts of numerical data as machine simulations, avoids some of the errors brought about by the discontinuous nature of such simulations. It will also allow us to develop some notion of the accuracy of our simulations and how much reliance we can place upon them.

These four charts show the results of using an initial level of Ag = 70 and T = 10 000. Other charts show the simulation for Ag = 5 000 and Ag = 15 000 where we see different behaviour. The first chart is dominated by the BC and TX values which are removed in the second to show the behaviour of the other three variables more clearly.
The second pair of charts show longer term behaviour for the Ag = 70 simulation with data taken every 6 steps (ie. 24 hours) instead of every 1 step (4 hours).

Ag: antigen;
APCX: active APC;
PC: Plasma cell;
BC: Active B cells;
TX: Active T cells.

The charts show
For Ag = 70 a slow onset of response leading to a long period of oscillations in cell populations sizes until Ag is cleared in 28 simulated days.

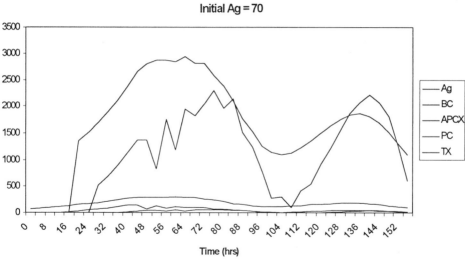

Figure 5

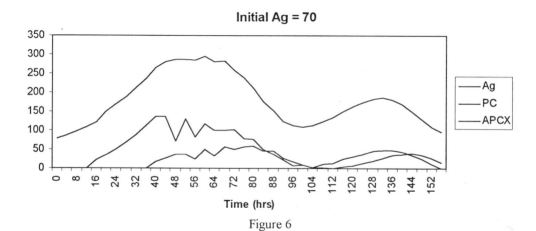

Figure 6

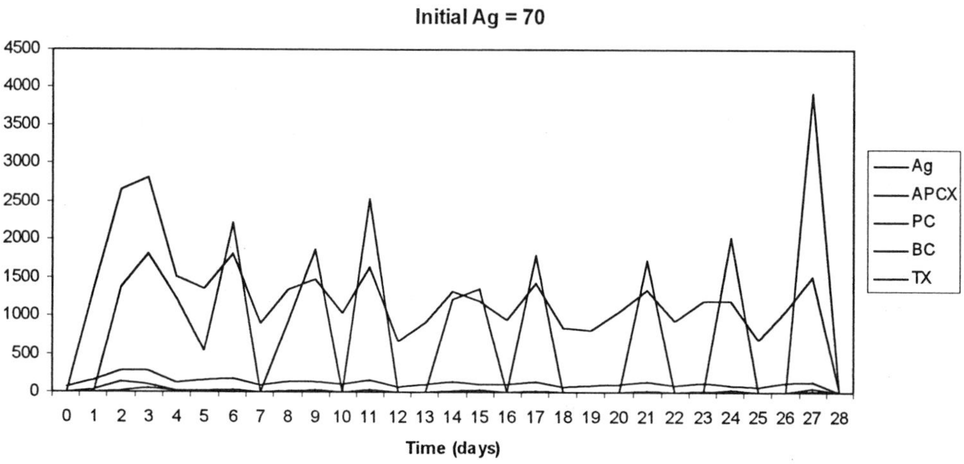

Initial Ag = 70

Figure 7

For Ag = 5 000 a rapid response and clearance in a few days.
For Ag = 15 000 a rapid response leads to a clearing in a larger time than for 5 000 but shorter than for 70.

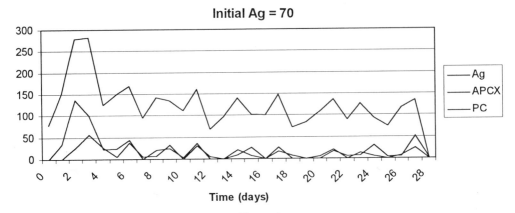

Figure 8

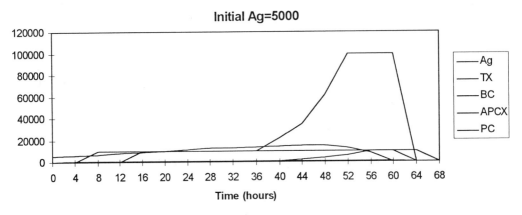

Figure 9

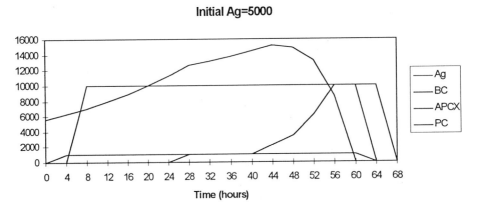

Figure 10

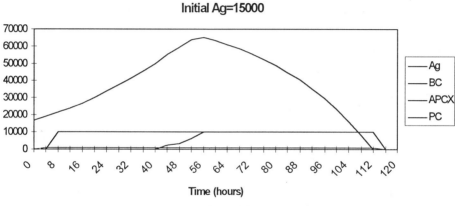

Figure 11

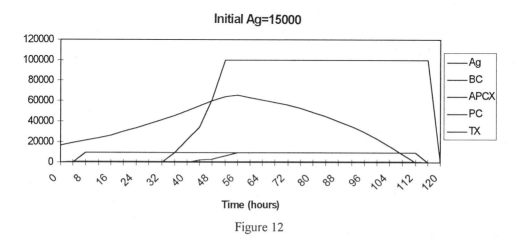

Figure 12

The results of a number of simulations are shown, in each case only the initial dose of antigen was changed. The results lack some subtlety due to the large time steps and the simplicity of the model, but clearly show some complex features. For larger doses of antigen, increasing the dose increases the size and speed of the response, as well as the length of time required to clear the infection, if such clearance is possible at all. For low doses, however, we see an increase in the length of the period of infection with a later, weaker response.

The oscillatory behaviour evident in the results for lower levels of initial antigen dose are probably an artifact of the discrete timesteps, but the long term behaviour appears to be qualitatively unaffected by varying the length of the time steps.

All doses of antigen eventually cause a response, because we have not included any of the body's other defences, or complications such as dormant antigens and so forth.

CONCLUSION

We have described a novel method for modelling and producing simulations of immunological responses in a simple but formal way.

Our aim was to investigate the feasibility of this method, rather than to introduce an accurate model. While our model lacks the subtlety of the real system, we have provided a basis for further refinement.

Integrating information from a number of facets of the system into a complete, or nearly complete, model which addresses the interactions of the subsystems is a challenging task, but our conclusion from this work is that more realistic models of this complex system are possible using these ideas.

REFERENCES

Bell, A., and Holcombe, M., 1996, Computational models of cellular processing, in: *Computation in Cellular and Molecular Biological Systems*, R.Cuthbertson, M.Holcombe and R.Paton, eds., World Scientific, Singapore.

Coffman, R.L., and Carty, J., 1986, A T-cell activity that enhances polyclonal IgE production and its inhibition by IFNγ, *Journal of Immunology* 136: 949-954.

De Boer, R.J., and Hogeweg, P., 1989, Unreasonable implications of reasonable idiotypic network assumptions, *Bulletin of Mathematical Biology* 51: 381-408.

Farmer, J.D., Hackard, N., and Perelson, A.S., 1986, The immune system, adaptation and machine learning. *Physica D* 22: 187-204.

Farmer, J.D., 1990, A rosetta stone for connectionism, *Physica D* 42: 153-187.

Flechner, E.R., Freudenthal, P.S., Kaplan, G., and Steinman, R.M., 1988, Antigen specific T lymphocytes efficiently cluster with dendritic cells in the human primary mixed – leukocyte reaction, *Cellular Immunology III*: 183-195.

Harel, D., and Naamad, A., 1996 The statemate semantics of statecharts, Technical Report, I-Logix.

Holcombe, M., 1994, From VLSI through machine models to cellular metabolism, in: *Computing with Biological Metaphors*, R. Paton, ed., Chapman & Hall, London.

Holcombe, M.,1991, Mathematical models of biochemistry, in: *Molecular Theories of Cell Life and Death*, S.Ji, ed., Rutgers Univ. Press, New Jersey.

Holcombe, M., 1990, Towards a formal description of intracellular biochemical organisation, *Comp. Math. Applic.* 20:107-15.

Holcombe, M., Duan, Z., and Linkens, D., 1994, Timed interval temporal logic and the modelling of hybrid systems, *Pro. Eur. Simulation Multi-Conference*, Barcelona, 534-541.

Holcombe, M., Duan, Z., and Linkens, D., 1995a, Modelling of a soaking pit furnace in hybrid machines, *Systems Analysis, Modelling, and Simulation* 18:153-157.

Holcombe, M., Duan, Z., and Linkens, D., 1995b, Specification of a soaking pit system in parallel hybrid machines, in: *Proceedings of the 21 Euromicro Conference, IEEE/CS*, 241-247, Como, Italy.

Hunt, K.J., 1993, Classification by induction: applications to modelling and control of non-linear dynamical systems, *Intelligent Systems Engineering*, 2(40) 231-245.

Langhoff, E., and Steinman, R.M., 1989, Clonal expansion of human T lymphocytes initiated by dendritic cells, *Journal of Experimental Medicine* 169: 315-320.

Marchuk, G.I., Petrov, R.V., Romanyukka, A.A., and Bocarov, G.A., 1991, Mathematical models of antiviral immune response I, II. *Jour.Theor.Bio.*151: 1-70.

McLean, A.R., 1994, Modelling T cell memory, *Jour.Theor.Bio.* 170: 63-74.

Sonada, T.,1992, Formation and stability of a memory state in the immune network, *J.Phys.Soc.Japan*, 61: 14908-1424.

Steinman, R.M., 1991, The dendritic cell system and its role in immunogenicity, *Annual Review of Immunology* 9: 271-296.

CONTROL LAWS IN THE MESOSCOPIC PROCESSES OF BIOCOMPUTING

Felix T. Hong

Department of Physiology
Wayne State University
Detroit, Michigan 48201 USA

INTRODUCTION

Biocomputing in a multi-cellular organism is implemented in computational networks arranged in a hierarchical fashion (Conrad, 1990a). In addition to the obvious network formed by interconnecting neurons, numerous biochemical processes in the cytoplasm and the nucleus can be regarded as a distributed network. Conrad (1984) has characterized biocomputing as a hierarchy of *microscopic-macroscopic (m-M) network dynamics*, in which microscopic intracellular processes are vertically coupled to macroscopic interneuronal processes by means of second messenger signaling — an intermolecular process commonly referred to as signal transduction. Signal transduction takes place in the membrane and its vicinity, the dimension of which is too large to be considered microscopic but too small to be considered macroscopic. Information processing at the membrane level is thus referred to as a mesoscopic process.

Mesoscopic processes are not mere coupling links between macroscopic and microscopic dynamics. Mesoscopic processes have their own rich dynamics which often involve short-range molecular forces such as hydrogen bonding, electrostatic, hydrophobic and van der Waals interactions. This article follows the point of view of m-M dynamics, and examines the control laws that relate the inputs to the outputs of mesoscopic processes.

A conventional digital computer utilizes strictly deterministic control laws in computation, in which errors enter only at the stages of input and output and are not tolerated in the intermediate steps — a natural consequence of digital processing. In a comparison of the mode of operation between a digital computer and the human brain, Conrad (1989) pointed out that the predominant mode of biological information is not switch-based but shape-based: biosystems utilized the *"lock-key" paradigm* in molecular recognition. He proposed a self-assembly model based on the "lock-key" paradigm (Conrad, 1985, 1992). However, biological information processing at the microscopic level is characterized by random diffusion of biochemical intermediates and by incomplete biochemical reactions because of the almost universal presence of back reactions. As Conrad (1992) pointed out that the chances that two macromolecules would ever find a complementary fit through random collision is negligible, and biocomputing based on "lock-key" paradigm would be slow and inefficient. On the other hand, information processing at the macroscopic network level is rather specific and somewhat deterministic. To bridge this gap, Conrad (1990b, 1992, 1997) proposed a mechanism of electronic-conformational interactions in which electronic wave functions interact with nuclear wave functions when nuclei undergo dynamic motions during conformational changes. He speculated that this interaction allows the superposition of electronic states to speed up the docking process and, thereby, to speed up the input-output transform (*quantum mechanical parallelization*).

In this article, actual examples will be given that illustrate the speeding up of docking process can be understood in terms of mesoscopic processes based on well known short-range intermolecular interactions, which of course are also of quantum mechanical origin. It is not necessary to speculate a novel mechanism to account for the apparent speed of intracellular biocomputing processes. Many mesoscopic processes are not as random as suggested by a simple diffusion-reaction paradigm. We shall review several pertinent examples which demonstrate how electrostatic interactions are configured to control and curtail randomness in mesoscopic processes.

In considering the systems performance of a living organism, it is a pertinent question to examine how deterministic the control laws, which govern the input-output transform, can become. Answering this question will automatically address the problem of how intracellular biocomputing copes with the randomness inherent in molecular diffusion and biochemical reactions. It will also address a problem at the other extreme, namely, the common "belief" that the physical world is absolutely deterministic. If one is willing to assume that biocomputing can be explained solely on the basis of physics and chemistry of the organized collection of cells in a living organism, the extreme form of physical determinism will inevitably lead to the denial of the existence of "free will," (Schrödinger, 1936; Ruelle, 1991) or its conflict with determinism (e.g., Goldman, 1990). Although free will and physical determinism are problems beyond the scope of our present discussion, it is of interest to examine determinism at the level of control laws of biocomputing (biological determinism). Two opposing views called *compatibilists* and *incompatibilists* will be examined. Incompatibilists claim that free will is incompatible with physical determinism whereas compatibilists claim that the conflict does not exist. Implications to machine intelligence and molecular electronics will be discussed.

"GRAY SCALE" OF DETERMINISM IN BIOCOMPUTING

In order to facilitate the subsequent discussion, we shall define here a "gray scale" of determinism. At one end of this scale lies complete randomness if the output has absolutely no correlation with the input (infinite dispersion). At the other end of this scale lies *absolute determinism*, in which, given a known input, the control law prescribes a unique output with no dispersion (complete correlation). This definition of absolute determinism excludes the error arising from the imperfection of observation or measurement processes. Although it is almost impossible to prove or disprove, the view of absolute determinism claims that the outputs are uniquely determined although they may not be accurately observed or measured. In this article, we shall argue that biocomputing probably never achieves absolute determinism and seldom resorts to complete randomness. Instead, biocomputing practices a weaker form of determinism, *relative determinism*, for lack of a better term. Relative determinism is not indeterminate because the dispersion is finite and the computing process is not completely random. On the gray scale of determinism, relative determinism occupies a wide range except the two extremes. It can approach nearly absolute determinism or nearly completely random. Partial randomness in biological information processing is a necessary condition for the inherent intelligence of biological organisms, since a completely deterministic machine only reflects the intelligence of its "creator."

MEMBRANE FLUIDITY AND ELECTROSTATIC EFFECTS

The biological membrane is not merely a cellular boundary and diffusion barrier. Nor is it a mere transit station for coupling the microscopic intracellular processes to the intercellular macroscopic signaling processes. The membrane itself is the substrate of a two-dimensional network of information processing. Two unique features are important for mesoscopic processes: the membrane fluidity and the small but finite thickness of a membrane.

The fluid mosaic model of biological membranes portraits functional proteins as a collection of molecular machinery being anchored in the "sea" of a phospholipid bilayer membrane. These protein components are capable of undergoing both rotational and lateral diffusion in the fluid environment of the lipid bilayer. Interactions or biochemical reactions among these protein components are thus similar to the diffusion-reaction scheme usually associated with intracellular processes in the solution phase of cytoplasm.

The small thickness of a membrane as compared to the dimension of a cell establishes a unique environment for biocomputing. A biomembrane usually carries negative fixed charges at the membrane surfaces because of the presence of phospholipid polar head groups. Electrostatic interactions of the surface charges with the aqueous mobile ions in the adjacent aqueous solution (known as the diffuse double layer) lead to the well known effect of *charge screening* (Verwey and Overbeek, 1948; Blank, 1986). The electric force within the double layer is intense but short-ranged: as the distance from the membrane surface increases, the force falls off faster than that predicted by the inverse square law. However, the electric field also reaches the membrane interior and the aqueous phase beyond the *opposite* membrane surface (e.g., Hong, 1987). In brief, the electrostatic effect of membrane surface charges is limited to the membrane phase and the two adjacent double layers — a mesoscopic process.

MESOSCOPIC BIOCOMPUTING IN BIOENERGETICS

The photosynthetic thylakoid membranes and mitochondrial inner membranes are membrane-based biomachines utilized by green plants and by animals, respectively, for converting energy captured from the environment to chemical energy suitable for cellular metabolism (photosynthetic and oxidative phosphorylation, respectively). Light energy and chemical energy of nutrients are extracted via a sequence of electron transfer (redox) reactions among several redox components. The green plant photosynthetic membrane consists of three spatially separated supramolecular complexes, Photosystem I, Photosystem II and the cytochrome b_6-f complex (e.g., Lee, 1991, Barber, 1983) (Fig. 1). Each supramolecular complex consists of several peptide subunits and prosthetic groups (redox centers). The mitochondria inner membrane consists of four supramolecular complexes formed by peptides and metal-containing redox centers (e.g., Lee, 1987). Electron transfers within a given supramolecular complex follow electron paths that are somewhat fixed ("hard-wired"). Electron transfers between complexes or between the two Photosystems are mediated by mobile electron carriers within the membrane phases, such as ubiquinones and cytochrome c in mitochondria and plastoquinones in photosynthetic membranes.

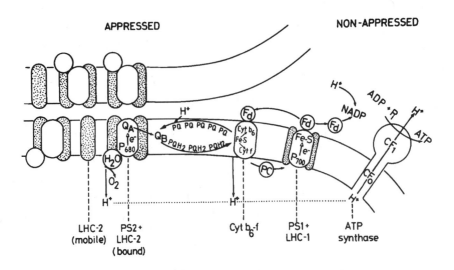

Figure 1. Schematic diagram showing the two photosystems of a green plant. The segments of the thylakoid membrane are shown to be juxta-positioned in the appressed region or widely separated from each other at the non-appressed region. The Photosystem I (PS1) together with its antenna complex (LHC-1, light-harvesting complex I) are shown in the non-appressed region, whereas the Photosystem II (PS2) and its antenna complex (LHC-2) are shown in the appressed regions. Plastoquinone (PQ) is the mobile membrane-bound charge carrier that shuttles electrons between PS2 and the cytochrome b_6-f complex, and shuttles protons from the stromal space (at the top) to the thylakoid space (at the bottom). Ferredoxin (Fd) and plastocyanin (PC) are mobile aqueous electron carriers. The ATP synthetase (with two subunits CF_0 and CF_1) are shown in the non-appressed region. Photophosphorylation of LHC-2 leads to the state 1-state 2 transition. (Reproduced from Barber, 1983)

The net result of transmembrane electron transfer is the formation of a transmembrane proton gradient, which stores converted energy temporarily. This gradient is subsequently utilized to power the synthesis of ATP by ATP synthetase, which also resides in the membrane (membrane level phosphorylation). The constituent biomachines form a loosely connected two-dimensional network because of membrane fluidity. However, the loose network connections are not established by completely random collisions of the complexes with mobile charge carriers. Rather their connections are assisted by short-ranged electrostatic interactions commonly known as salt-bridges. These salt-bridges help to stabilize correct spatial alignment of reactive groups in what Conrad (1985, 1990a) dubbed as "shape-based" information processing ("lock-key" paradigm). For example, cytochrome c, a mobile electron carrier, possesses several positively charged lysine groups, which cytochrome c utilizes to "dock" with complementary negative charges on the b-c_1 complex and cytochrome oxidase — supramolecular complexes that are "upstream" and "downstream," respectively, on the electron transport chain (Smith et al., 1981).

Salt bridge formation helps establish a sufficiently long "dwelling" time for electron transfer between the two encountering supramolecular complexes to consummate. Since the active site of a macromolecule often occupies a small fraction of its total surface area, a chemical reaction between the two encountering parties requires a stringent mutual orientation. Without special provisions, two encountering macromolecules might approach each other by random diffusion and collision and might deflect from each other as rapidly as they came together. The resulting short "dwelling" time of encounter would make reaction nearly impossible to proceed. Short-range intermolecular forces provides a energy potential well at the molecular surfaces for trapping the encountering molecules so as to make possible the search of a specific orientation and match of complementary charge groups. In this

230

way, a three-dimensional random search is transformed by these short-range intermolecular forces into a much faster two-dimensional search (dubbed "reduction-in-dimensionality" effect by Max Delbrück).

In the case of green plant photosynthesis, the coupling of the two photosystems via plastoquinone is also regulated by light via an electrostatic mechanism. Photophosphorylation of the light harvesting protein complex II (LHC-II) increases the negative surface charges and leads to decoupling of the two systems. This regulatory mechanism, called the State 1-State 2 transition (Myers, 1971), thus alters the structure of the two dimensional network of photosynthetic apparatuses and allows for dynamical allocation of resources in accordance with the lighting conditions.

There is a long-standing controversy about the bioenergetic process of membrane-level phosphorylation. The coupling of the transmembrane proton gradient to ATP production requires lateral movement of protons from the electron transfer site to the phosphorylation site. Yet the chemiosmotic theory (Mitchell, 1966) implies that the photon energy conversion process allows protons to be dumped into the vast extracellular space. The major fraction of transported protons appears to be dispersed to the bulk solution phase. Thus, an unacceptable loss of converted free energy may be incurred. This objection against the so-called delocalized chemiosmotic theory is usually referred to as the "Pacific Ocean" effect, and has been very much debated in the past (e.g., Williams, 1978).

A rival "localized" chemiosmotic theory (Williams, 1978) demands that protons be preferentially channeled to the phosphorylation site either through an internal membrane route or through a pathway provided by a proton network on the membrane surface. Kell (1979) suggested that the interfacial region (Stern layer) is separated by a diffusion barrier that keeps most of the converted protons in this adjacent layer and thus preferentially diverts protons to the phosphorylation site. Such preferential proton lateral mobility has subsequently been demonstrated by Teissié and his colleagues (Teissié and Gabriel, 1994).

These authors detected rapid lateral movement of protons on a phospholipid monolayer-water interface. They found that the proton conduction along the surface is considerably faster than proton conduction in bulk phase (2-3 minutes vs. 40 minutes for a comparable distance in their measurement setup). They indicated that the enhanced lateral proton movement occurs along a hydrogen-bonded network on the membrane surface in accordance with a mechanism previously proposed by Onsager for proton conduction in ice crystals. As a result, protons move considerably faster in water than by classical diffusion. In fact, a similar mechanism has been used to explain proton diffusion in bulk water. So what factors make the lateral proton movement on the membrane surface so much faster than proton diffusion in bulk water?

At least two explanations can be proposed. First, the proton movement at the membrane surface is two-dimensional whereas the proton movement in bulk water is three-dimensional (again, reduction-in-dimensionality effect). The second reason is the increased surface concentration of protons as compared to the bulk concentration caused by a negative surface potential. The negative surface potential is primarily due to the net negative surface charges of phospholipid polar head-groups. Teissié's group has shown that there is indeed a steep gradient near the membrane surface (2 pH units in less than 1.5 nm). However, these authors also observed similar facilitated lateral proton movement with monolayers formed from neutral and zwitterionic phospholipids. Therefore, the effect of a negative surface potential is probably secondary and not essential.

The special mechanism underlying proton lateral mobility serves the purpose of "containment" of randomness. Protons which are pumped across the membrane need not search for the site of ATP synthetase randomly and waste time by wandering "purposelessly" into the bulk solution phase (aqueous phase beyond the Debye length).

The above-described examples demonstrate that mesoscopic biocomputing takes place in a two-dimensional network which is partly "hard-wired," in terms of supramolecular complexes, and partly loosely connected, via two-dimensional diffusion in the fluid lipid bilayer. The control laws governing these processes occupy an intermediate position of the "gray scale" of determinism: it is neither completely random nor completely deterministic. In addition, the control laws can be adjusted dynamically via electrostatic interactions.

MESOSCOPIC SWITCHING PROCESSES: ENZYME ACTIVATION

While "shape-based" molecular recognition is ubiquitous in biology, switch-like processes are also important. Many enzymes are activated by phosphorylation, as are many crucial steps of "bifurcation" in biological information processing. Membrane surface charges are often involved in a molecular switching process. For example, the activation of protein kinase C (PKC), a ubiquitous protein in signal transduction, requires the presence of negatively charged phosphatidyl serine (PS) in the plasma membrane (for review see Newton, 1993) (Fig. 2). The mechanism is not purely electrostatic however. Initially, an electrostatic interaction between PKC and an acidic membrane (with negatively charged phospholipid headgroups) in the presence of Ca^{2+} leads to a global conformational change of PKC. In the second stage of activation, the requirement of diacylglycerol and phosphatidyl serine (a negatively charged phospholipid) is rather specific. Substitution with another acidic phospholipid will not work. The specific non-electrostatic interaction induces a local conformational change that exposes the pseudosubstrate domain (with multiple positive charges). The pseudosubstrate domain is further stabilized by binding to any acidic lipid (electrostatic interaction). Thus, electrostatic interactions are invoked twice in PKC activation: first as the homing device to help PKC target acidic membranes, and second as part of the allosteric mechanism to produce the catalytic activity. It is worth noting that the first stage converts a three-dimensional random search to a two-dimensional one (reduction-in-dimensionality effect). In effect, the second stage of the PKC activation constitutes an heuristic search as opposed to a random search which occurs in the first stage.

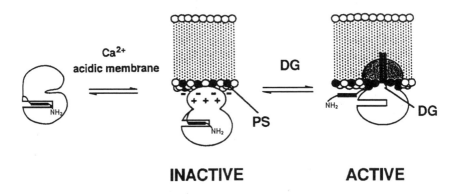

INACTIVE   **ACTIVE**

Figure 2. Model for the interaction of protein kinase C (PKC) with lipid headgroups. In the first step, electrostatic interactions drive PKC to bind to acidic lipids (middle). This binding results in a conformational change that exposes the hinge region of PKC. Diacylglycerol (DG) promotes cooperativity and specificity in the protein-lipid interaction and causes a marked increase in the affinity of PKC to phosphatidyl serine (PS), and thus activates the PKC by causing a PS-dependent release of the pseudosubstrate domain from the active site. (Reproduced from Newton, 1993)

The first stage of PKC activation represents a rather general mechanism shared by the association of coagulation proteins with anionic membranes as well as binding of synapsins (a protein mediating exocytosis of neurotransmitter molecules) to an acidic membrane. We have previously postulated a similar homing mechanism for visual transduction (Hong, 1989): a photoinduced positive surface potential triggers the binding of a G-protein (transducin) by photoactivated rhodopsin and a subsequent phosphorylation of rhodopsin triggers binding of arrestin. The difference here is that the surface charges are photoinduced on membrane-bound protein instead of on phospholipid, which is not light-sensitive.

STOCHASTIC PROCESSES IN ION CHANNEL ACTIVATION

In the examples cited above, the control laws governing the biocomputing processes yield well defined (mean) outputs although the outputs usually contain errors (variances). Were it not for the intervention of short-range interactions, the variances would be even greater. There are however certain mesoscopic processes in which the control laws do not give rise to a well defined mean output value. Instead, the output values are governed by a probabilistic distribution function (power law). A good example is provided by the patch-clamp data of a rat muscle, which shows "fluctuation" of Na^+ channels (Fig. 3).

Neural excitation is generally characterized by a switching event known as an action potential (nerve impulse). An action potential is preceded by a localized change of the membrane potential (known as depolarization) that eventually exceeds a certain threshold value. The cross-over of the threshold is caused by a positive feedback process that starts with increased conductance of Na^+ channels which in turn allows an increased influx of Na^+ into the cytoplasm, leading to further depolarization (Hodgkin and Huxley, 1952). In brief, the switching event is critically depend on a rapid transient increase of Na^+ conductance followed by a slower return to its basal value (activation and inactivation of Na^+ channels, respectively). Both these events have a well defined time course.

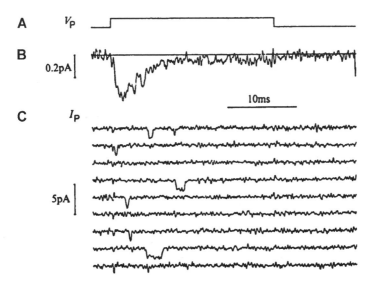

Figure 3. Patch clamp records of sodium ion currents. Tetrethylammonium is used to block K^+ channels that may be present in a small patch of rat muscle membrane in response to a 10 mV depolarization (trace A). The traces in panel C show responses to nine individual 10 mV depolarization. The trace in B is the average of 300 individual responses. (Reproduced from Sigworth and Neher, 1980)

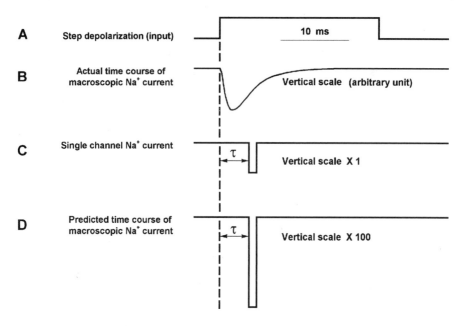

Figure 4. Schematic showing the time course of the sodium ion current in the presence and in the absence of "ion channel fluctuation." Trace A is the step depolarization of the membrane potential that triggers activation of sodium channel. Trace B is a typical macroscopic sodium ion current measured by the voltage clamp technique with a conventional intracellular glass pipette electrode. Trace C is a typical current from a patch clamp recording. Trace D is the hypothetical macroscopic sodium ion current when the "fluctuation" were eliminated and some 300 channels turned on and off in unison. Note the difference in the vertical scale.

The time course of macroscopically measured Na^+ conductance changes in response to a sudden step depolarization is shown schematically in Fig. 4B. Such a macroscopic control law governing the time-dependence and voltage-dependence of Na^+ conductance reflects the collective behavior of thousands of tiny Na^+ ion channels.

At the mesoscopic level, the control law governing the opening and closing of an in dividual Na^+ ion channel shows no resemblance to the corresponding macroscopic control law which prescribed the time course shown in Fig. 4B. Shown in Fig. 3 are measurements of ionic currents flowing through individual ion channels in a small patch of the rat muscle membrane in response to 10 mV depolarization (Sigworth and Neher, 1980). Record C in Fig. 3 shows nine separate measurements in which the opening and closing of an individual ion channel is observed. The time course appears erratic and not reproducible from trace to trace. None of these traces directly suggest the time course of the macroscopic Na^+ conductance changes as shown in Fig. 4B. However, the channel activity is not completely random.

As is evident from Record B in Fig. 3, when 300 individual measurements are averaged, their collective behavior shows a well defined time course that is characteristic of activation and inactivation of Na^+ channels as originally observed by Hodgkin and Huxley (1952) (cf: Figs. 3B and 4B). While the conductance of an individual Na^+ channel is voltage-independent, its opening and closing in response to a given depolarization is governed by a voltage-dependent probabilistic control law. In fact, the channel is more likely to open and close repeatedly during the first few milliseconds after the initiation of depolarization and the activity gradually tapers off over several milliseconds. The probability is enhanced by depolarization. The control law shown here is apparently an intrinsic property of the

Na$^+$ channels; the K$^+$ channels exhibit an entirely different probabilistic control law. As explained in quantitative detail by Hodgkin-Huxley theory, the time course of an action potential critically depends on the macroscopic time course of activation and inactivation of Na$^+$ channels, and that of delayed activation of K$^+$ channels. For example, suppression of delayed activation of K$^+$ channels by a K$^+$ channel blocker prolongs the duration of the action potential.

LAPLACE'S ARGUMENT FOR ABSOLUTE DETERMINISM

Since neural excitation is one of the central events in biocomputing, the above-demonstrated probabilistic control law of ion channel operation suggests that biocomputing does not practice absolute determinism. However, any argument against absolute biological determinism must also confront the notion of absolute physical determinism unless one wishes to invoke vitalism. Therefore, a skeptical inquirer and/or an advocate of absolute physical determinism will probably raise the famous objection once made by Laplace: "any belief that things come about randomly, or by chance, is simply due to our ignorance of the causes" (e.g., see p. 266, Cottingham, 1996) There is some truth in this claim, and it has been supported by new discoveries numerous times in science history.

It is not known presently what causes the ion channels to "fluctuate" in their time sequence of opening and closing. It is quite conceivable that in the future sufficient understanding of the detailed molecular dynamics along with the detailed knowledge of the channel structure will enable investigators to elucidate the factors that lead to the fluctuating ion channel opening and closing event. It may then be possible to control these factors so that all Na$^+$ channels will respond to a sudden depolarization with the same well defined time course, namely, opening abruptly *in unison* at a fixed time interval after the sudden depolarization, staying open for a fixed duration, and then closing abruptly, as shown in Fig. 4C. Under such conditions, the macroscopic Na$^+$ conductance will have exactly the same time course as that of an individual channel but will have a much greater amplitude and will look like a (mathematical) delta-function (Fig. 4D). Therefore, the collective behavior of these channels will not exhibit the characteristic time course of activation and inactivation of Na$^+$ channels as shown in Fig. 4B. It is apparent that the excitable membrane will no longer function properly under such a circumstance. Full control of factors responsible for the apparently irregular and unpredictable sequence of channel opening and closing will also render the ion channels inoperative and unsuitable for the generation of "normal" action potentials. That the "abnormal" action potential would not work can be made clear by the following consideration.

Ventricular cells in the cardiac muscle have an unusually long plateau phase of depolarization (about 100 ms). Such a prolonged depolarization phase serves a vital purpose; it permits voltage-gated Ca^{2+} channels to stay open for a sufficiently long period in order to raise the intracellular Ca^{2+} concentration to a sufficient level for the subsequent contraction of muscle fibers to take place (excitation-contraction coupling). A delta-function like action potential would make excitation-contraction coupling in cardiac muscles literally impossible unless the entire contractile machinery were radically "redesigned."

It is quite obvious that (partial) randomness was recruited by Nature through evolution to serve a vital biocomputing purpose. In so doing, evolution apparently recruited excess dispersion for biocomputing instead of sticking with processes that are more deterministic.

The source of randomness in ion channel fluctuation is unknown. It is generally assumed to arise from conformational changes between various channel states (at least three states for Na$^+$ channels: closed, open and inactivated) (Bezenilla and Stefani, 1994). However, Lev et al. (1993) have demonstrated that pores through non-biological synthetic polymer membranes also show fluctuation between high and low conductance states. They proposed a mechanism based on ionization of fixed charges within a channel or at its mouth of opening (Pasternak, 1997; Bashford, 1997). Here, the main source of partial randomness in channel kinetics is quite local and is not a consequence of intrusion by unwanted and unrelated external agents. Of course, fluctuation in ionization, being a thermal phenomenon, has a distinct contribution from the environment. But a bioorganism has been "designed" to operate in a noisy environment anyway.

The above example suggests that absolute biological determinism is unlikely. However, advocates of absolute physical determinism would probably argue as follows. While the control laws appear to contain variance or even probabilistic in nature, the "hidden" control laws that determine the variance or determine the probability might still dictate the individual occurrence with absolute determinacy. Furthermore, the ultimate understanding the "hidden" control laws does not necessarily mean that one can manipulate the factors so as to eliminate the variance or the "fluctuation." In any case, the "hidden variable" argument remains valid.

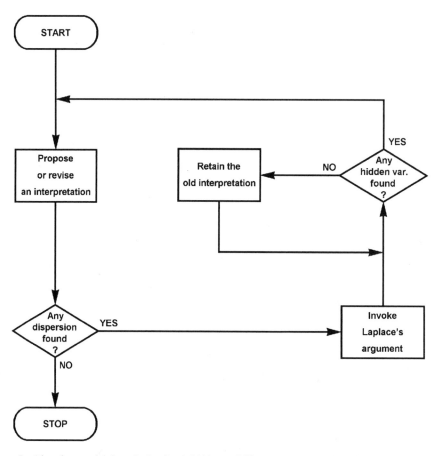

Figure 5. Flowchart explaining why Laplace's hidden variable argument cannot be refuted. The flowchart shows two loops in which the Laplace's hidden variable argument can be invoked indefinitely as long as there is dispersion. Each time the outer loop is traversed, a major advance about our understanding of dispersion is made. The inner loop is traversed repeatedly until a new hidden variable is found. Exhaustion of existing dispersion is only tentative because improvement measurement methodology may uncover new dispersion. Thus, Laplace's argument can neither be proved nor disproved See text for further discussion.

It is important to realize that it is impossible to refute the "hidden variable" argument originally enunciated by Laplace. An examination of the flowchart shown in Fig. 5 will explain why. Given the observation of a new phenomenon, the first step is to establish a viable interpretation (mechanism). If there is any dispersion present in the measured value of a parameter for which the interpretation provides a physical basis, one can invoke the "hidden variable" argument of Laplace to dismiss the dispersion. One then engages in the search of a hidden variable that can describe and explain the dispersion. If such a variable is not found, the old interpretation is retained but the search for the hidden variable continues. If a hidden variable is found, one must then revise the interpretation to include an explanation of the former dispersion. One then re-examines the phenomenon to see if there is any additional dispersion left. If so, one invokes Laplace's argument again. When no more dispersion is found, the search stops. However, if new dispersion is found after improvement of the measurement technique, the entire process starts all over again.

There are two loops in the flowchart. When a hidden variable is found, a major advance in our understanding is made (outer loop). If a new hidden variable is not found, then one traverses in the inner loop indefinitely. The fact that a new hidden variable cannot be found after a finite period of exhaustive search does not mean it does not exist. Thus, Laplace's argument cannot be refuted because the number of times one can traverse the two loops is indefinite. On the other hand, Laplace's argument cannot be proved either because it would require an infinite number of successful examples and which would cover time till eternity; the possibility that new dispersion will appear in the future as a consequence of improved instrumental resolution cannot be excluded ahead of time.

The true significance of Laplace's argument is the role it plays as the devil's advocate. It is like a carrot hung in front of a mule at a fixed but unreachable distance: in striving to reach the impossible goal the mule always makes "advances." Likewise, in our attempt to debunk Laplace's claim, we inevitably fail but we may gain additional insight about bio-computing.

CONFLICT BETWEEN FREE WILL AND CLASSICAL DETERMINISM

From the above discussion, relative determinism, which occupies almost the entire gray scale of determinism except the two extremes, is not prohibited by epistemology. It provides an alternative way to re-examine the problem of conflict between free will and classical determinism.

Here, we must make it clear that the following discussion is not intended to prove the existence of free will. Rather, we think the existence of free will is impossible to prove *scientifically*. A scientific hypothesis is usually tested by means of evaluating either a time-average or an ensemble-average of a series of repeated experiments and controls. It is impossible to perform a *time-average* or an *ensemble-average* in the analysis of a behavioral experiment with regard to free will. This is because free will is so *history*-dependent and so *personality*-dependent that when there is only one life to live the concept of probability, as applied to a free will experiment, is inherently problematic (cf: Ruelle, 1991). But the conflict between free will and determinism can be examined at a scientific level.

The notion of time-reversal in the formulation of the conflict is not prohibited by the equations of motion in either Newtonian or quantum mechanics. But the formulation of the conflict also tacitly assumes the *one-to-one correspondence* between the present boundary conditions and the future boundary conditions, and equivalently, the one-to-one correspondence between the present boundary conditions and the past boundary conditions. If the last claim of correspondence is valid, then the existence of free will means one has the will power and capability of altering the past boundary conditions even if it occurred before one's birth. The absurdity of this inference constitutes the well known conflict (see Fig. 6A). If the one-to-one correspondence does not hold, the time reversal is not valid. What entails in relative determinism is the mapping of a sharply defined boundary condition to one with dispersion, which is essentially *one-to-many* mapping. This latter point is crucially involved in the dispute between compatibilists and incompatibilists, who claim the existence and non-existence of the conflict, respectively (e.g., Goldman, 1990).

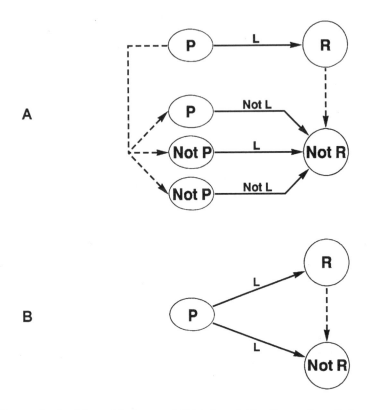

Figure 6. Diagram showing fallacy of the incompatibilists' view. A. The diagram shows the incompatibilist view. The past event P, together with natural laws L caused the result R to happen. Through free will, R was rendered false. This could happen only if (a) the natural laws could be altered, or (b) the past event P could be rendered false, or (c) both the natural laws could be altered and, at the same time, the past event P could be rendered false. Since none of the options (a), (b), and (c) are possible in reality, a paradox ensues if one insists that absolute determinism is valid and free will exists. B. The compatibilists claim that even though free will could rendered R false, not every past event P that led to R could be rendered false. Therefore, the compatibilists claim that there is no conflict between free will and absolute determinism. However, the diagram shows that the event P could cause two mutually exclusive outcome: R and not R. The causal relationship is not a one-to-one correspondence. Therefore, either the determinism is not absolute (the variance encompasses both either R or not R), or there is no cause-effect relationship between P and R to begin with. The compatibilists inadvertently invoked relative determinism, as defined in the present article, in their argument and strayed from the original problem.

If one assumes absolute determinism, then the mapping from a past boundary condition to a present one is one-to-one. Fig. 6A shows that the conflict does exist and incompatibilists are right. In contrast, compatibilists' escape from the incompatibilists' dilemma was the following argument: a person may be able to bring about a given state of affairs without being able to bring about *everything* entailed by it. In other words, for a certain event P, rendering R false does not necessarily render P false or alter natural laws L (Fig.

6B). Thus, as Fig. 6B implies, the event P can lead to multiple outcomes. This argument essentially ruins the assumed absolute determinism because the argument implicitly requires that the prior event P together with natural laws L lead to *multiple* but *mutually exclusive* outcomes — R and not R (one-to-many correspondence). The determinism implicitly held by compatibilists is a weaker form of determinism; it is better termed as *relative determinism* rather than *absolute determinism*. Thus, compatibilists have inadvertently altered the meaning of determinism implied in the original formulation of the conflict. Still, there is an alternative but trivial interpretation of Fig. 6B, that is, P and R are not causally related. But, apparently the latter is not what compatibilists mean.

The notion of time-reversal of the equations of motion together with the one-to-one correspondence is intimately related to the principle of *microscopic reversibility*. Macroscopically, time-reversal is not possible because the increase of entropy gives time a directionality as stipulated by the second law of thermodynamics.

Apparently, microscopic reversibility is not applicable at the mesoscopic scale where entropic changes are not considered exceptional (mesoscopic irreversibility?) (e.g., see Váró and Lanyi, 1991). In the mesoscopic events of light-induced proton transport in a photosynthetic membrane, validity of microscopic reversibility would require the following event to be observable: the event starts with reversed proton transport which leads to reverse electron transfer in the reaction center, and subsequently the reaction center emits a photon (or, in a green plant, the reaction centers emit two photons). The absurdity of such an event is well recognized. Perhaps one must go to an even smaller physical dimension to observe the postulated microscopic reversibility. But even at the scale of a small organic molecule, the principle of microscopic reversibility is not applicable. This becomes evident by considering the well established principle of photochemistry.

In a small organic dye molecule that fluoresces, the emitted photon always has a lower energy (longer wavelength) than the exciting photon. This is because an electron which has been excited by the incoming photon to a higher electronic orbital first settles to a lower orbital by vibronic relaxation before a photon is re-emitted; the energy difference of the two orbitals accounts for the loss of energy dissipated as heat and accounts for the difference in wavelengths between the exciting photon and the emitted photon. Microscopic reversibility as applied to this case is thus prohibited by the second law of thermodynamics. Reversal of the photophysical event even at this microscopic scale would exhibit an emitting photon that had a shorter wavelength than the exciting photon, and thus would betray the prohibited time-reversal. We suspect the principle of microscopic reversibility is an approximation at the microscopic scale as a consequence of negligible dispersion, which is nevertheless *nonzero*.

That the concept of microscopic reversibility is problematic can be seen by considering the spatial scale where (microscopic) reversibility meets (macroscopic) irreversibility: a point of abrupt transition or *discontinuity*. From the above discussion, the discontinuity apparently lies in a spatial scale that is much smaller than the mesoscopic scale. But where and how?

Alternatively, by abandoning absolute determinism, one can view microscopic reversibility as a mathematical idealization and approximation. This view offers an easy answer to the question about the point of transition between reversibility to irreversibility. This is because the transition can be viewed as the gradual breakdown of the mathematical approximation and gradual emergence of non-negligible dispersion. Thus the point of transition depends on how much error one can tolerate and is thus not sharply and uniquely defined — the transition is not an abrupt one.

Without a valid concept of microscopic reversibility, the argument leading to the conflict of free will and physical determinism would be seriously undermined: time-reversal invariance of equations of motion is not sufficient without the validity of one-to-one correspondence in the prescribed mapping.

In the above attempt to resolve the conflict of free will and determinism, we focus on the control laws *per se*: the conflict does not exist if the control laws are not absolutely deterministic. There is an alternative approach: focusing on the boundary conditions. Matsuno (1989) has pointed that control laws (equations of motion) alone is not sufficient to guarantee absolute determinism; uniqueness of boundary conditions should also be part of the argument. If the boundary conditions cannot be uniquely determined, then the strict one-to-one mapping stipulated by physical laws is not valid (Matsuno, 1989). Again, the principle of microscopic reversibility must then be considered an approximation: an excellent one at the microscopic scale.

DISCUSSION AND CONCLUSION

In this article, several examples of information processing at the membrane level have been examined. A two-dimensional network characteristic is apparently in bioenergetic systems. The networks are partly "hard-wired" and partly "dynamically connected." The fixed part of the network are maintained by short-range intermolecular forces which allows supramolecular complexes to be formed. The loosely connected part of the network are made possible by membrane fluidity. But the randomness entailed by diffusion-controlled encounters is substantially curtailed by short-range intermolecular forces.

The control laws governing each step of biocomputing may be highly deterministic when biochemical reactions are preferentially channeled along a certain desired pathway by an intricate control system that utilizes short-range interactions to activate key enzymes (switch-like control). Sometimes, the control laws may not be well defined but defined only probabilistically. However, one need not be alarmed: variance is not routinely propagated and amplified. Despite the presence of variance (error), biocomputing may still be accurate and reliable. *Analog* and *digital* modes of computing often alternate at various stages and various levels of biocomputing, and control laws swing from highly random to highly deterministic.

Consider the chain of events initiated by synaptic transmission at the neuromuscular junction of a skeletal muscle. The content of neurotransmitter acetylcholine (ACh) in a synaptic vesicle is somewhat fixed (quantal content), and the release of transmitter molecules is parceled in a quantal packet. Here, biocomputing acquires a digital characteristic. Subsequent diffusion of ACh across the synaptic cleft causes uncertainty in timing of its arrival at the postsynaptic membrane. While spontaneous release of ACh appears completely random (Katz and Miledi, 1972), neurally evoked release of ACh is synchronized by the arriving nerve impulse which precedes the release — a major switching event. Collectively, it gives rise to the end plate potential with a well defined time course. The initiation of a new action potential starts with an analog process of a graded change (depolarization) of membrane potential which in turn controls the opening and closing of Na^+ channels. While these channels open or close in a fixed quantal step, its timing is governed by a probabilistic control law and a great deal of timing uncertainty (or rather, indeterminacy) is introduced. Collectively, these channels give rise to a well behaved time course of Na^+ activation and inactivation. Once the threshold is reached, a new action potential is generated. The threshold phenomenon ensures that either a full-fledged action potential is generated, or not at all (*all-or-none principle*). Thus, biocomputing regains its digital mode of processing. However, the analog information of the intensity of neural activity is coded as the frequency of action potentials, giving rise to a graded response of muscle contraction, for example. Fine-tuning with various control processes via various neural reflexes at various key connections of the neural network guarantees a smooth and meaningful movement of any part of the body.

A nagging question is: why does Nature resort to probabilistic control laws in such crucial biocomputing steps as nerve impulse generation? This question is similar to a parallel question in genetic mechanisms: why did Nature not make gene duplication error-free? The answer is obvious for the latter question. Were it not for occasional errors in gene duplication, there would be no evolution. Likewise, were biocomputing error-free, there probably would not be creativity. Thus, the weaker form of determinism practiced by biocomputing is the price paid by a living organism for unleashing the internal dynamics of biocomputing which exhibits rich repertoire of elemental computation paradigms in carbon-based chemistry.

With regard to technological applications, exploiting carbon-based chemistry for its rich and diverse dynamics is the first step (e.g., Hong, 1994). But the inherent randomness of such dynamics must be brought under control. Nature achieves the "containment" of excessive randomness by means of a hierarchical architecture in which intermolecular forces are exploited (*controlled randomness*). In attempting to develop molecular electronic devices, investigators often find Nature's act hard to follow. Not only do all components form interlocking parts, the control laws are also interlocking in nature. Materials scientists must therefore confront a formidable task. Fabrication engineers also face an equally formidable task because the interlocking components with interlocking control laws must be put together in the correct sequence like putting together a three-dimensional jigsaw puzzle. In overcoming the technical difficulty, investigators often resort to various techniques of protein immobilization. But if the randomness is excessively contained, the "intelligence" inherent in the materials may not be fully unleashed. Future research should exploit the possibility of introducing structural variability into molecular devices. Lessons learned from developmental biology may shed light on future fabrication technology.

ACKNOWLEDGMENTS

The author thanks Don DeGracia, Koichiro Matsuno and Kaus-Peter Zauner for helpful discussions and Filbert Hong for reading and correcting several versions of the manuscript.

REFERENCES

Bezenilla, F., and Stefani, E., 1994, Voltage-dependent gating of ionic channels, *Annu. Rev. Biophys. Biomol. Struct.* 23:819.

Blank, M. (ed.), 1986, *Electrical Double Layers in Biology*, Plenum Press, New York.

Barber, J., 1983, Photosynthetic electron transport in relation to thylakoid membrane composition and organization, *Plant Cell Environ.* 6:311 (1983).

Bashford, C.L., 1997, A novel explanation for fluctuations of ion current through narrow pores: experiments in support of the hypothesis, *FEBS Advanced Course on Membrane Transport Processes and Signal Transduction*, August 24-31, 1997, Bucharest, Romania, Abstract, p. 1.

Conrad, M., 1984, Microscopic-macroscopic interface in biological information processing, *BioSystems* 16:345.

Conrad, M., 1985, On design principles for a molecular computer, *Comm. ACM* 28:464.

Conrad, M., 1989, The brain-machine disanalogy, *BioSystems* 22:197.

Conrad, M., 1990a, Molecular computing, in: *Advances in Computers, Vol. 31,* M.C. Yovits, ed., Academic Press, Boston, San Diego, New York, London, Sydney, Tokyo, Toronto, p. 235.

Conrad, M., 1990b, Quantum mechanics and cellular information processing: the self-assembly paradigm, *Biomed. Biochim. Acta* 49:743.

Conrad, M., 1992, Quantum molecular computing: the self-assembly model, *Int. J. Quantum Chem.: Quantum Biol. Symp.* 19:125.

Conrad, M., 1997, Origin of life and the underlying physics of the universe, *BioSystems* 42:177.

Cottingham, J. (ed.), 1996, *Western Philosophy: An Anthology*, Blackwell Publishers, Oxford and Cambridge, Massachusetts.

Goldman, A., 1990, Action and free will, in: *Visual Cognition and Action: An Invitation to Cognitive Science, Vol. 2*, D.N. Osherson, S.M. Kosslyn, and J.M. Hollerback, eds., MIT Press, Cambridge and London, p. 315.

Hodgkin, A.L. and Huxley, A.F., 1952, Currents carried by sodium and potassium ions through the membrane of the giant axon of *Loligo*, *J. Physiol.* 116:449.

Hong, F.T., 1987, Internal electric fields generated by surface charges and induced by visible light in bacteriorhodopsin membranes, in: *Mechanistic Approaches to Interaction of Electric and Electromagnetic Fields with Living Systems*, M. Blank and E. Findl, eds., Plenum Press, New York, p. 161.

Hong, F.T., 1989, Relevance of light-induced charge displacements in molecular electronics: design principles at the supramolecular level, *J. Molec. Electronics* 5:163.

Hong, F.T., 1994, Molecular electronics: science and technology for the future, *IEEE Engg. Med. Biol. Mag.* 13(1):25.

Katz, B. and Miledi, R., 1972, The statistical nature of the acetylcholine potential and its molecular components, *J. Physiol.* 224:665.

Kell, D.B., 1979, On the functional proton current pathway of electron transport phosphorylation: an electrodic view, *Biochim. Biophys. Acta* 549:55.

Lee, C.P. (ed.), 1987, *Structure, Biogenesis, and Assembly of Energy Transducing Enzyme Systems, Current Topics in Bioenergitics, Vol. 15*, Academic Press, San Diego, New York, Berkeley, Boston, London, Sydney, Tokyo and Toronto.

Lee, C.P. (ed.), 1991, *Photosynthesis, Current Topics in Bioenergitics, Vol. 16*, Academic Press, San Diego, New York, Berkeley, Boston, London, Sydney, Tokyo and Toronto.

Lev, A.A., Korchev, Y.E., Rostovtseva, T.K. Bashford, C.L., Edmonds, D.T., and Pasternak, C.A., 1993, Rapid switching of ion currents in narrow pores: implications for biological channels, *Proc. R. Soc. B* 252:187.

Matsuno, K., 1989, *Protobiology: Physical Basis of Biology*, CRC Press, Boca Raton, Florida.

Mitchell, P., 1966, Chemiosmotic coupling in oxidative and photosynthetic phosphorylation, *Biol. Rev.* 41:445.

Myers, J., 1971, Enhancement studies of photosynthesis, *Plant Physiol.* 22:289.

Newton, A.C., 1993, Interaction of proteins with lipid headgroups: lessons from protein kinase C, *Annu. Rev. Biophys. Biomol. Struct.* 22:1.

Pasternak, C.A., 1997, A novel explanation for fluctuations of ion current through narrow pores: experiments leading up to the hypothesis, *FEBS Advanced Course on Membrane Transport Processes and Signal Transduction*, August 24-31, 1997, Bucharest, Romania, Abstract, p. 19.

Ruelle, D., 1991, *Chance and Chaos*, Princeton University Press, Princeton, New Jersey.

Schrödinger, E., 1936, Indeterminism and free will, *Nature (Lond.)* July 4, p. 13.

Sigworth, F.J. and Neher, E., 1980, Single Na^+ channel currents observed in cultured rat muscle cells, *Nature (Lond.)* 287:447.

Smith, H.T., Ahmed, A.J., and Millett, F., 1981, Electrostatic interaction of cytochrome c with cytochrome c_1 and cytochrome oxidase, *J. Biol. Chem.* 256:4984.

Teissié, J., and Gabriel, B., 1994, Lateral proton conduction along lipid monolayers and its relevance to energy transduction, in: *Proceedings of 12th School on Biophysics of Membrane Transport, Vol. II*, S. Przestalski, J. Kuczera, and H. Kleszczynska, eds., Agricultural University of Wroclaw, Wroclaw, Poland, p. 143.

Váró, Gy., and Lanyi, J.K., 1991, Thermodynamics and energy coupling in the bacteriorhodopsin photocycle, *Biochem.* 30:5016.

Verwey, E.J.W. and Overbeek, J.Th.G., 1948, *Theory of the Stability of Lyophobic Colloids*, Elsevier, New York, Amsterdam, London, and Brussels.

Williams, R.J.P., 1978, The history and the hypotheses concerning ATP-formation by energized protons, *FEBS Lett.* 85:9.

A COMPUTER SYSTEM THAT LINKS GENE EXPRESSION TO SPATIAL ORGANIZATION OF CAENORHABDITIS ELEGANS

Tsuguchika Kaminuma,[1] Takako Igarashi,[1] Tatsuya Nakano[1] and Johji Miwa[2]

[1] Division of Chem-Bio Informatics
National Institute of Health Sciences
1-18-1, Kami-Yoga, Setagaya-ku, Tokyo 158, JAPAN

[2] Research & Development Group and Fundamental Research Laboratories
NEC Corporation
34 Miyukigaoka, Tsukuba 305, JAPAN

1. INTRODUCTION

Development of multicellular organisms has long been one of the most mysterious and fascinating biological phenomena (Slack,1991), and it has become one of the most attractive biological processes to study in various fields, including computer science, from both theoretical and engineering points of view (Kaminuma and Matsumoto, 1991). For the past decade, we have been gaining increasingly sophisticated knowledge about the basic molecular mechanisms that relate genes to the three-dimensional body plan of animal development, such as those for origins of the body axes of *Drosophila* embryos (Lawrence, 1992), neurulation (neural tube formation) of *Xenopus* and chickens (Lumsden and Krumlauf, 1996; Tanabe and Jessell, 1996; Tessier-Lavigne and Goodman, 1996), body axial formation and gastrulation of *C. elegans*, and neural plasticity of *Aplysia*, mice, and monkeys. Yet, we can not conceive of any emerging model that unifies animal development.

Among the many popular animals used for developmental studies, *C. elegans* is unique because each of its small number (about 1000) of somatic cells is lineally identified (Deppe et al., 1978; Sulston et al., 1983; Sulston and Horvitz, 1977; Kimble and Hirsh, 1979), the approximately 8800 connections fabricated by the total of 302 neurons are also worked out (White et al., 1986), and its entire genome will have been sequenced within a couple of years (Information is available from A C. elegans Database (ftp://lirmm.lirmm.fr, ftp://cele.mrc-lmb.cam.ac.uk, and ftp://ncbi.nlm.nih.gov)). The entire developmental process of *C. elegans* from a single-celled zygote to adulthood can be observed in the living state under Nomarski optics. Relatively easy isolation of mutants and the capability of performing genetic crosses provide a good genetic system from which one can gain general information on gene functions of interest (Brenner, 1974; Miwa et al.,1980). In addition, we demonstrated a computer graphics model that well simulates the three-dimensional organization of cells in the developing *C. elegans* embryo (Kaminuma et al., 1986). Therefore, we can reasonably expect that *C. elegans* will be the very first animal in which the genetic imple-

Information Processing in Cells and Tissues
Edited by Holcombe and Paton, Plenum Press, New York, 1998

mentation of the body plan is understood in molecular terms.

One of the ultimate goals in developmental biology is to trace entire cell lineages and identify all the genes that control the division, differentiation, shaping, and death of these cells. The number of genes that control the early embryonic development of animals has been estimated to be not very large (perhaps 20 to 30), and several of these genes have already been identified. These early expressed genes and some initial spatial factors, which are mostly still unknown, control the dynamic spatial organizations of embryonic cell aggregates during early animal development.

Several research groups including us have previously made tools by reconstructed cellular graphics which can represent early embryonic development in *C. elegans* (Kaminuma et al., 1986). In the present study, we have directed our effort to link the gene-controlling scheme, which represents genetic information, one-dimensionally stored in DNA and parallelly expressed, to the three-dimensional embryonic cell structure. The Virtual Reality Modelling Language (VRML) is very suitable for this purpose. We have thus developed a computer system using VRML and other Internet-related technologies that link the knowledge of genes controlling development to the knowledge of three-dimensional configurations of cells in developing animals.

In this paper, we describe the conceptual framework of our model and its initial implementation.

2. SYSTEM DESIGN

2.1 Basic Design Concept

The eventual goal of our system is to describe how genes control the three-dimensional spatial organization of cells in the embryo. However, genes themselves do not directly control cellular organization. Cells must first divide, differentiate, and sometimes die to form proper aggregate structures, and it is proteins, the products encoded by genes, that control these cellular events. Genetic information is linearly stored in the nucleic acid sequence. Division, differentiation, and eventual death of cell occur also linearly, whereas an entire cell lineage is represented by a binary tree. However, the formation of three dimensional cell aggregate structure requires parallelly occurring events.

Thus one of the most interesting points in building a model system to represent embryonic development is how to relate linear events of the gene expression to the three-dimensional parallelly processed events of cellular arrangements. This relation is not one-directional but multi-directional, and often circuitly, because the three-dimensional spatial distribution of cells and the products of their genes influence gene expressions of cells at later stages. In fact uneven spatial distributions of maternal gene factors in zygote is the cause of the first uneven gene expressions among cells at some early stage, which eventually trigger the first cell differentiation. So, the logic is circular from the very beginning of development. (In reality, development is not circular but quasi-circular because it repeats itself only through different generations from parents to offspring. Thus the logic is helical along time rather than circular. This is, however, not a main topic of our present report.)

Thus, to describe developmental structure in our model, one must always go back and forth between the three layers of events of different characters : gene expression, birth and death of the cells, and spatial arrangements of cell aggregates. We call this a three-layer model for describing embryonic development (Figure 1).

2.2 Basic Model: Three layer model for development

The first layer of this model is called the Genomic Space, where molecular interactions within and between cells, such as those involving gene expression or signal transduction, play important roles. This layer includes not only expressed genes and their proteins

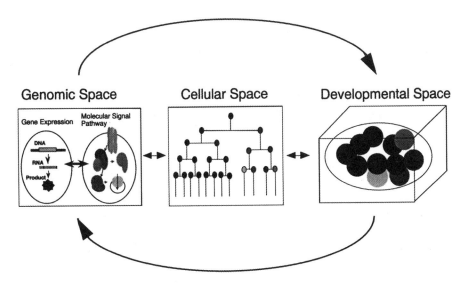

Figure 1. Our model consists of three different spaces which are causally related each other.

but also RNAs of various kinds and, ideally, small molecules of no genetic origin.

The second layer is called the Cellular Space, which consists of individual cells generated and differentiated through development. Such description is only possible for organisms whose constituent cells are identified throughout development or during certain specific stages such as early development. *C. elegans* is the only organism all of whose cells are identified throughout development. Cell lineage of early development was studied extensively for some other organisms other than *C. elegans*.

The third layer is the Developmental Space, which represents an organism itself as a three-dimensional aggregate of individual cells. In *C. elegans* we can determine three dimensional cell positions of an intact embryo by computerized Nomarski optics, but such determination used to be difficult for other organisms. Recently, however, because of advances in various computerized microscope techniques, spatial arrangements of cells that relate to important developmental events such as body axial formation, gastrulation, and neurulation of some organisms such as *Drosophila,* Zebra fish, *Xenopus,* and chickens have been studied in detail. Thus similar models must be applicable to these cases.

MENU	
PHENOMENA	GENE

I. Gametogenesis
II. Embryogenesis
   - Early
      - Polarity Determination
      - Gastrulation
   - Late
      - Neurulation
      - Other Organogenesis
III. Postembryogenesis
IV. Topical Developmental Events

Figure 2. The Top Menu from which one can look into specific phenomena.

In familiar biological terminology, the first layer corresponds closely to the genotype which determines the genetic characteristic of the organisms and the third to the phenotype that relates to shapes and behaviors of the organisms.

2.3 Advantage of *C. elegans* as a model organism

The eventual goal of our model is to describe how genes controlling information in the first layer organize the spatial arrangement of cells in the third layer and vice versa. The three layer model is particularly relevant to *C. elegans*, because all cells are identified and their lineage is invariant. The spatial positions of cells were identified by E. Shierenberg and his colleagues (Schierenberg et al., 1984) and by T. Kaminuma et al. (Kaminuma et al., 1986) up to about 100-cell stages. Later, J. White and R. Schanabel determined the cell positions at much later stages (Thomas et al., 1996).

Already at present, several embryonically interesting genes, such as par-1 & -2 (Boyd, 1996), pie-1 (Mello et al., 1996), glp-1, (Hutter and Schnabel, 1994; Evans et al., 1994), and emb-5 (Nishiwaki et al., 1993), have been cloned and characterized. Many more genes are currently under intensive study, and their properties will be identified in the near future. This information together with the complete genomic sequencing data should provide a sound knowledge base for the first layer of our model.

3. SYSTEM IMPLEMENTATION

We have implemented a computer system conceptually based on the three-layer model. Our system was implemented on Silicon Graphics Indigo running on a UNIX operating system.

Software was developed modularly. It consists of a top module and submodule systems each of which corresponds to the three-layered space world. These modules are hyperlinked by internal WWW mechanisms.

3.1 The Top Module : Main menu

We assume all users of our system enter from their (either internal or external) browsers. Thus it provides the home page and the main menu (Figure 2). From the main menu one can select either phenomena or genes. Phenomena are classified into different stages of development, i.e. gametogenesis, embryo genesis, postembryogenesis, and other topical developmental events such as neurulation. From the gene menu one can investigate certain genes that relate to the development.

3.2 The Genomic Space Module

The first layer, the Genomic Space, is represented by two basic systems ACEDB and CSNDB (Figure3). ACEDB (Durbin and Thierry-Mieg, 1991) is an object-oriented database originally developed for the *C. elegans* genome project, but it has become one of the most popular database management systems for genomic data of many other organisms. It contains a wide variety of the *C. elegans* genomic data, including DNA base sequences, genetic and physical maps, and contigs. (More precisely the database for the *C. elegans* genome is called ACeDB, while an empty data management system is called ACEDB.)

CSNDB is a data and knowledge base for cell signaling networks developed by our group (Igarashi and Kaminuma, 1997; Igarashi et al., 1996). It was constructed on ACEDB as the base (data management system). We plan to extend CSNDB to include transcription factor databases, such as TRANSFAC (Wingender et al., 1997), so that it can compile data and knowledge on gene expression during development.

With these basic facilities one can describe genes and their associated data and knowledge.

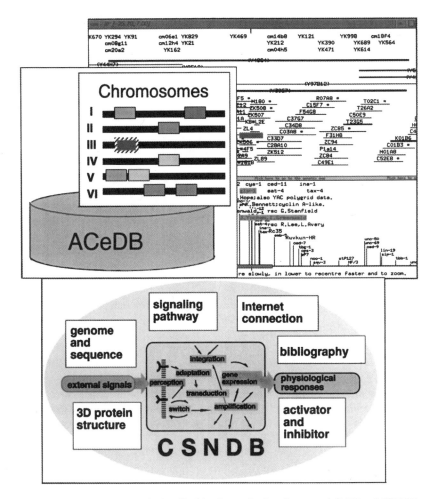

Figure 3. The genomic space is described by the two basic subsystems ACeDB and CSNDB.

3.3 Cell Dictionary and Cell Lineage

The Cellular Space for the second layer is represented by a Cell Dictionary that contains such information as cell identifications (cell names), cell positions, cell shapes, names of sister cells, and other cell characteristics (Figure 4). We used the nomenclature of the cells given by J. Sulston (Sulston et al., 1983). The Cell Dictionary is managed by relational databases such as ACCESS or Sybase.

3.4 Development as a Series of Spatial Events

About a decade ago, we developed a computerized system that could describe the spatial positions of dividing cells during early embryonic development. The system also successfully produced reconstructed images and animation of developing embryos by computer graphics.

We thus took advantage of various computer graphics technologies to represent the third layer of the Developmental Space. In our model, each cell is represented by a ball, which looks like a molecule, so that a popular molecular graphics software such as Rasmol (ftp://colonsay.dcs.ed.ac.uk) or Chemscape (http://www.mdl.com/chemscape/) can be used

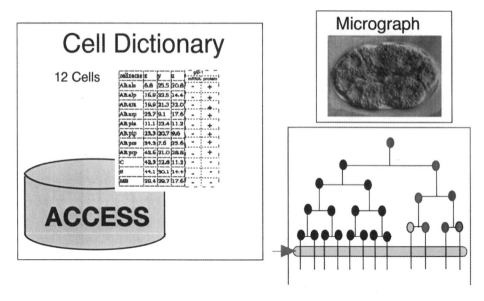

Figure 4. All cells that are generated during development are described by cell lineage (low right) and Cell Dictionary (left).

(Figures 5,A and C). Thus, well known molecular graphics display modes, such as wire frame, ball-stick, and space-filling models, are also applicable to describe objects in the third layer. Cellular connections are defined between two sister cells that are derived from a common parental cell. This connection can be traced back to the original zygote as a single cell, and all of these connections are represented by binary trees. Conversely, one can generate an animated image of a growing binary tree from cell coordinates and cell relation tables, which corresponds to what we call a Connection Table of Molecular Graphics. One can produce such animation by Java programming (Figure 5,B).

The hyper-links between the Cell Dictionary and the three-dimensional reconstructed embryo are straightforward if we use VRML (Figure 6) models. In VRML, we can easily hyper-link each cell in the reconstructed embryo to a unit object in the Cell Dictionary and the data and knowledge associated to it.

3.5 Linking Different Spaces

The basic linking mechanism between the first and the second layers is much more complicated than that for the second and the third layers, for we must consider gene expression at each cell separately. If cells, particularly their spatial positions, are different, their modes of gene expressions are expected to be different. The situation is very much like different members in an orchestra playing different parts of the same but an extremely complicated score. The variously resulting music (synthesized proteins) thus triggers cells to differentiate variously.

It is, however, impractical to provide each cell with a different mode of gene expression and with a different cell signaling network, because the number of cells is too large to accumulate the detailed data and knowledge per cell needed for practical and meaningful analysis. Alternatively, we have decided to assume only one universal cell signaling network database and gene expression database at the first layer and to focus on those signaling molecules or genes that characterize the cells at the second layer.

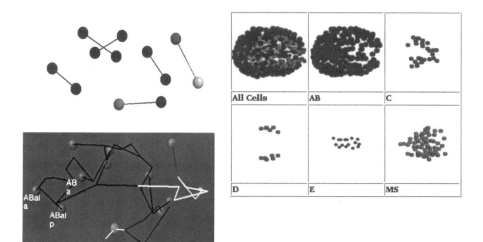

Figure 5. Various Computer graphics designs that represent three-dimensional reconstructed images of C. elegans embryo, (A) RasMol, (B) Java, and (C) Chemscape. Picture (A) and (B) represent developing cell trees. Picture (C) shows different cell groups in a same embryo.

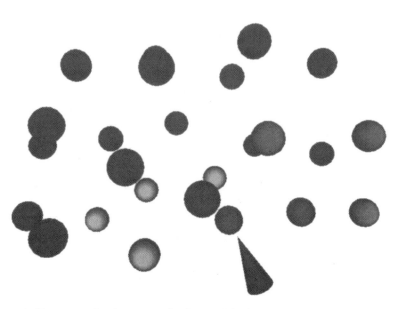

Figure 6. Three dimensional reconstruction image of C. elegans embryo using VRML. E cells, which initiate the gastrulation by moving towards inside of the embryo, are highlighted.

4. DISCUSSION

Although the implemented system is still a prototype, it reveals an essential feature of our model system. The heart of the system is the hyper-link. It connects genomic data in ACEDB which is a object-oriented type database to the Cell Dictionary (Cellular Space) based on a relational database, and it also connects the Cell Dictionary to the reconstructed image of the embryo represented by the VRML model. This type of connection was very difficult, if not impossible, to achieve before the advent of the new Internet-related technologies, such as WWW (HTML), multimedia WWW browser, and VRML as well as database interface (CGI, Common Gateway Interface) programs between WWW and databases. These new tools now enabled us to link different sets of data and knowledge of biological systems to that of others.

With *C. elegans* as a model organism, we are particularly interested in the reconstruction of its embryonic development up to a stage of about 50 cells, since technically speaking it is manageable to reconstruct the embryonic development up to this stage, and scientifically speaking, before the 50-cell stage we can observe most dynamic events in morphogenesis, such as the determinations of the three body axes, the birth of all the founder cells, and the initial phase of gastrulation, which is one of the most essential events in animal development.

A quite similar approach may also be applied to the early embryonic development of *Drosophila*. We also think that with some modification our model system can be applied to describe neural development of higher organisms such as Zebra fish, *Xenopus*, chickens, rodents, and even humans. In these cases, however, we may have to give up the idea of applying the model system to the level of a complete set of cells but rather be satisfied with applying it to only those of cells that are identified to play important roles in neural processes.

All these efforts of implementing developmental events at different levels for different organisms, when all put together, would give a more comprehensive and hopefully some unified view of animal development. Then, our system will be useful for both experimental biologist and theoretical researchers. For biologist the system may be used as a good reference with flexible viewers and a tool to map experimental data. For theoretician it may provide basic tools for representing causal relation between linear gene information and three-dimensional structural change of embryonic cells.

We are planning to make this system available on the Internet soon. We hope and expect the system to evolve and improve further at all three layers of development as we plan to keep up with and include future findings related to this subject, and as we would like all of you, too, to participate in evolving and improving the system. Finally it should be added that such a system may also be useful for educational purposes for both undergraduate and graduate students.

5. CONCLUSION

We have developed a computerized-system that aims to represent embryonic development as the dynamic organization of a structure, as formed by a cellular aggregate. The system also allows one to link intracellular phenomena, such as gene expression, to a member of the embryonic cell aggregate at a given stage. The system is based on a three layer model: the molecular level (Genomic Space that relates to gene expression, signal transduction pathways, etc.), cellular level (Cellular Space), and organismal level (Developmental Space of three dimensional cell aggregates).

Although conceptually general in nature, the system was specifically implemented to represent the embryonic development of *C. elegans*. Various new software techniques developed for the Internet, such as WWW, hyper link, VRML, Java, Web/database connection software like CGI, were used together with more conventional methods such as window, object oriented (-like) database, relational database, and 3D molecular graphics.

The system will soon be available on the Internet so that it may be used by a wide range of people from students to advanced researchers.

Acknowledgments

We thank Mr. T. Futaba, T. Umehara, and T. Murakoshi for most of the programming work for this project as part of their undergraduate theses, Drs. R. Durbin and J. Thierry-Mieg for *C. elegans* positional data stored in ACeDB. We also thank Ms. S. Miwa for suggestions and wording.

REFERNCES

Boyd, L., Guo, S., Levitan, D., Stinchcomb, D.T., and Kemphues, D.T., 1996, PAR-2 is asymmetrically distributed and promotes association of P granules and PAR-1 with the cortex in C. elegans embryos, *Development* 122: 3075.

Brenner, S., 1974, The genetics of Caenorhabditis elegans,*Genetics* 77: 71.

Deppe, U., Shierenberg, E., Cole,T., Krieg C., Schmitt, D., Yoder, B., and Ehrenstein, von G.,1978, Cell lineages of the embryo of the nematode Caenorhabditis elegans, *Proc. Natl. Acad. Sci.* 75: 376.

Durbin, R. and Thierry-Mieg, J., 1991, A C. elegans Database. I. Users' Guide, II. Installation Guide, and III. Configuration Guide. Code and data are available from anonymous FTP servers at lirmm.lirmm.fr, cele.mrc-lmb.cam.ac.uk andncbi.nlm.nih. gov.

Evans, T.C., Crittenden, S.L., Kodoyianni, V., and Kimble, J., 1994, Translational control of maternal glp-1 mRNA establishes an asymmetry in the C. elegans embryo, *Cell* 77: 183.

Hutter, H. and Schnabel, R., 1994, glp-1 and inductions establishing embryonic axes in C. elegans, *Development* 120: 2051.

Igarashi, T. and Kaminuma,T., 1997, Development of a cell signaling networks database, in: *Pacific Symposium on Biocomputing '97*, R.B. Altman, A.K. Dunker, L. Hunter, and T.E. Klein, ed., World Scientific, Singapore.

Igarashi, T., Kaminuma, T., and Nadaoka, Y., 1996, A data and knowledge base for cell signaling networks, in: *Computation in Cellular and Molecular Biological Systems*, R. Cuthbertson. M. Holcombe, and R. Paton, ed., World Scientific,Singapore.

Kaminuma, T. and Matsumoto, G., eds.,1991, *Biocomputers*, Chapman and Hall, London.

Kaminuma, T., Minamikawa, R., and Suzuki, I., 1986, 3D reconstruction of spatio-temporal series of optical pictures, in: *Pattern Recognition in Practice*, E.S. Gelsema and L.N. Kamal, ed., Elsevier Science Publishers B. V., North-Holland.

Kimble, J. and Hirsh, D., 1979, The post-embryonic cell lineages of the hermaphrodite and male gonads in Caenorhabditis elegans, *Developmental Biology* 96: 189.

Lawrence, P. A., 1992, *The Making of a Fly*, Blackwell Scientific Publications Ltd..

Lumsden, A. and Krumlauf, R., 1996, Patterning the vertebrate neuraxis, *Science* 274: 1109.

Mello, C.C., Schubert, C., Draper, B., Zhang, W., Lobel, R., and Priess, J.R., 1996, The PIE-1 protein and germline specification in C. elegans embryos, *Nature* 382: 710.

Miwa, J., Shierenberg, E., Miwa, S., and Ehrenstein, von G., 1980, Genetics and mode of expression of temperature-sensitive mutations arresting embryonic development in Caenorhabditis elegans, *Developmental Biology* 76: 160.

Nishiwaki, K.,Sano, T., and Miwa J., 1993, emb-5, a gene required for the correct timing of gut precursor cell division during gastrulation in Caenorhabditis elegans, encodes a protein similar to the yeast nuclear protein SPT6, *Mol. Gen. Genet.* 239: 313.

Schierenberg,E., Carlson, C., and Sidio,W.,1984, Cell development of a nematode: 3-D computer reconstruction of living embryos, *Roux's Achieves of Developmental Biology* 194: 61.

Slack, J. M.W, 1991, *From Egg to Embryo*, Cambridge University Press, Cambridge.

Sulston, J. E., Schierenberg, E., White, J.G., and Thomson, J.N., 1983, The embryonic cell lineage of the nematode Caenorhabditis elegans, *Developmental Biology* 100: 64.

Sulston, J. E. and Horvitz, H.R., 1977, Post-embryonic cell lineages of the nematode Caenorhabditis elegans, *Developmental Biology* 56: 110.

Tanabe, Y. and Jessell, T.M., 1996, Diversity and pattern in the developing spinal cord, *Science* 274: 1115.

Tessier-Lavigne, M. and Goodman, C. S., 1996, The molecular biology of axonguidance, *Science* 274: 1123.

Thomas, C., DeVries, P., Hardin, J., and White, J., 1996, Four-dimensional imaging: com puter visualization of 3D movements in living specimens, *Science* 273: 603.

White, J., Southgate, E., Thomson, J.N., and Brenner, S., 1986, The structure of the nervous system of Caenorhabditis elegans, Philos. Trans. R. Soc. Lond. B Biol. Sci. .

Wingender, E., Kel, A.E., Kel, O.V., Karas, H., Heinemeyer, T., Dietze, P., Knueppel, R., Romaschenko, A.G., and Kolchanov, N.A., 1997, TRANSFAC, TRRD and COM-PEL: Towards a federated database system on transcriptional regulation, *Nucleic Acids Res.* 25: 265.

VERBS, GLUE AND CATEGORIES IN

THE CELLULAR ECONOMY

Ray Paton[1] and Koichiro Matsuno[2]

[1] Department of Computer Science
The University of Liverpool
Liverpool L69 3BX, UK

[2] Department of BioEngineering
Nagaoka University of Technology
Nagaoka 940-21, Japan

INTRODUCTION

> *"If we try to squeeze science into a single viewpoint ... we are like Procustes*
> *chopping off the feet of his guests when they do not fit on the bed".*
> Freeman Dyson, 1995

Biology is faced with major challenges as it evolves towards the twenty first century. The diversification of the subject matter, the paucity of theoretical frameworks and the dramatic increase in knowledge beg for integrative tools of thought. For some, certain areas of the subject still seem little more than natural history. It is good at collecting bits and pieces of information but not so good at finding how these bits come to be interrelated. The Procrustean damage may have already been done. There can be many ways to decompose a system into its elements but putting the parts back together again remains a major challenge.

The purpose of this paper is to examine some ways of extending integrative ideas. To be fair there are a number of existing integrative approaches such as the varied forms of General Systems Theory. Our approach captures some of the strong isomorphic emphases of GST but in a way that is initially metaphorical in a verbal (literary) rather than mathematical sense. We shall demonstrate this by examining some issues associated with the nature of the enzyme concept.

There is a strong non-essentialist dimension to biological thought. One aspect of this non-essentialism is to apply ideas that deal with relation (e.g., Rosen, 1991) and process

Information Processing in Cells and Tissues
Edited by Holcombe and Paton, Plenum Press, New York, 1998

(e.g., Sattler, 1985; Kampis, 1994). In this sense concepts and ideas like species evolve and so knowledge about the biological world is not fixed. Biological qualities, like the concepts which refer to them are part of an evolving ecology of knowledge. In addition, many biological concepts are fundamentally multidimensional in nature and this is manifest in very simple issues such as: the lack of necessary and sufficient conditions for classifying dicotyledonous plants; conceptual relations between synonyms of the blood system; differences in view of whether blood is part of a circulatory system; the ontological status of the species concept. Do such questions really matter? Some people will think this kind of discussion is no more than *mere* semantics; an activity biologists should avoid and leave to philosophers. This is a serious mistake. In each case the multidimensionality means that simple yes/no answers will not suffice. Contrast this with the classical physical notions of mass, time, distance and (maybe) momentum which are easier to define - although such ideas as free energy, latent heat and electric potential are less clear unless their status is merely as mathematical constructs with no existence in the real world. In the first place, scientific thinking requires models to be constructed. These are not just mathematical but also verbal, descriptive, qualitative, diagrammatic, iconic and analogical. The prevalence of diagrams in biology textbooks illustrates the importance of visual models to the subject. Secondly, the effective communication of scientific ideas should be meaningful. However, when no thought is given to the semantics, the presuppositions underlying a model are never examined, the syntax of modelling dominates and the subject as a whole remains fragmented or governed by monoliths (e.g., the classical physics of molecular biology). One apparent problem with multidimensionality is that concepts may often seem fuzzy or imprecise. This is because they are multidimensional.

What is an enzyme? Is it the same thing *in vivo* as it is *in vitro*? Classically it can be summarised as being both a 'biological catalyst' (function) and a protein (structure). As we shall see, the concept is far from as straightforward as this. Firstly, enzymes are more than catalysts. They are sophisticated thermodynamic machines (Welch and Kell, 1986), switches or information processors (e.g., Bray, 1995; Marijuan, 1994). This is especially the case with enzymes in signalling networks but also with allosteric enzymes in general. More broadly they could be viewed as 'cognitive' agents (e.g., Conrad, 1992; Paton *et al*, 1996). Putting these three features together they could be described as smart thermodynamic machines. Added to this is the status of components of multi-enzyme complexes or multifunctional enzymes. Secondly, some biological catalysts are not simply/strictly enzymes. Many proteins other than enzymes have catalytic properties. For example, haemoglobin demonstrates enzymatic activities. This multifunctional protein not only transports oxygen, it acts as a molecular heat transducer through oxygenation-deoxygenation cycle, modulates erythrocyte metabolism, is linked with genetic resistance to malaria and is a source of certain metabolites (for more details see Giardina *et al*, 1995).

Function is insufficient in defining what an enzyme is. Even simple enzymes are multifunctional - they are more than catalysts. The problem is we tend to focus on a reaction or on one particular level of cellular organisation. Structure is equally as problematic not least because protein structure classifications can tend to use arbitrary classifying criteria. If we accept that the concepts we use **can** influence the science we undertake then we must expand our appreciation of the enzyme concept. To help us do this we introduce a number of interrelated themes.

THE ECONOMY OF THE CELLULAR ECOLOGY

We shall argue that a unique insight into cellular processes is the dichotomy between the cellular economy and its ecology. The economy presumes economic actors standing upon a local perspective, while the ecology assumes its own global perspective. The contrast between the local and the global with the cellular processes urges us to identify how such a contrast materialises in reality. The local actors in the cellular economy are both energy suppliers and consumers, and there are no prior co-ordinations between the two. The absence of prior co-ordination comes to imply that each acts asynchronously with the others. In contrast, the cellular ecology is a consequence of a global integration of those local activities, otherwise there would be no ecology. This relationship between the local and the global would become most vivid if one takes another contrast between information and thermodynamics to be applied to the cellular processes.

Cellular energy transduction is both informational and thermodynamic. Information presumes signalling of a local character, whereas thermodynamics takes its co-ordination on a global scale for granted as embodied in the notion of thermodynamic states. Cellular processes thus integrate both the local and the global phenomena in a coherent manner because thermodynamics could eventually specify what the global cellular ecology would look like. Information from a local perspective would have to come to be integrated into thermodynamics perceived from a global perspective.

Underlying the coherent co-ordination between information and thermodynamics is time (Matsuno, 1996, 1997). Communication of a signal in a cell proceeds in locally asynchronous time because of the absence of a means for global synchronisation right in the middle of communication, while thermodynamic states identify themselves in globally synchronous time because of the presence of a global perspective. Locally asynchronous time precipitating a globally synchronous counterpart serves as a factor for synthesising information and thermodynamics in a consistent fashion.

The global synchronisation underlying cellular energy transduction can be structurized in the network of metabolic pathways as demonstrating how locally asynchronous time in process comes to generate globally synchronous time in products. Accordingly, information precedes thermodynamics in time. This precedence of information makes the resulting thermodynamics open to time in the sense that thermodynamics is a derivative of the activities of precipitating a globally synchronous time from a locally asynchronous one. Thermodynamics is of itself silent on how the global comes to be precipitated from the local. In fact, thermodynamic openness of the energy transduction in cells resides in the incompleteness of the global synchronisation while constantly generating locally asynchronous time to be synchronised subsequently. This is due to dynamics intrinsic to information. In view of the fact that thermodynamics is a global consequence of information activities of a local character, there always remain some local activities that cannot be integrated into the consequential thermodynamics. That is locally asynchronous time left out of the then available thermodynamics perceived in a globally synchronous time.

The present contrast between locally asynchronous and globally synchronous time makes the occurrence of energy suppliers and consumers inevitable in cells. As far as the completed energy transduction is concerned, all the energy supplied has to be consumed. However, such a complete balance between energy supply and consumption is feasible only in the record operating in a globally synchronous time. As far as the completed energy transactions are concerned, time referred to there is global in the sense that it serves as a marker distinguishing what was simultaneous from what was sequential in the record.

Once one pays attention to the process of energy transduction in locally asynchronous time, no such balance between energy supply and consumption could yet be conceivable. Energy supply and consumption without prior co-ordination for the balance between the two necessitates the occurrence of energy suppliers and consumers operating in locally asynchronous time. The energy suppliers and consumers functioning in cells are the internal observers in the sense that the measurement proceeding internally is accompanied with realisation of quantities to be supplied or consumed (Matsuno, 1989). That measurable quantities could be realised without interventions from the external agents such as human scientists comes to imply that both the measurement and the realisation are internal to the cells. As a matter of fact, since the quantities eventually come to be shared by both parties, each of the energy suppliers and consumers extend the gluing activity towards the other. The glue acting between the energy suppliers and consumers derives from the transference from locally asynchronous to globally synchronous time.

The economy of the cellular ecology is in fact a material structurization of the global synchronization of time. Time underlies the relational activities in progress in enzymes in the sense that information in locally asynchronous time comes to be integrated into thermodynamics in globally synchronous time. That is, the cellular ecology derives from the cellular economy.

An interesting set of analogies now begin to emerge. We can begin to open up parallels between information, thermodynamics and cellular systems by looking at verbal attributes. Consider the following comment by Kelly:

> *"A distributed decentralised network is more a process than a thing. In the logic of the Net there is a shift from nouns to verbs. Economists now reckon that commercial products are best treated as though they were services ... It is not what something is, it is what it is connected to, what it does."*

Kelly (1994), p27

We shall use this idea of the "verbs of the Net" to explore enzyme processes.

ENZYMES AS VERBS

In this section we consider similarities between enzyme and verb and so promote the key issues of relation and process. The insightful comment from Kelly (above) is used as a integrating notion namely, the process-based thinking associated with a distributed decentralised network like the Internet. In so doing it is important to make a crucial comment about reasoning within this metaphorical scheme. The metaphor or simile of enzyme as (i.e., is like) verb, is a very general framework. As we shall see there are a number of similarities. However, we do **not** seek to cash out the metaphor in terms of a detailed analogy. Rather we are using metaphors to provide a general frame of reference for describing a relational view of the cell (this use of metaphor will be visited again when we consider 'glue' and 'category' in the following sections).

Enzymes have Cases

In natural languages case systems emphasise the central role of verbs in sentences (Fillmore, 1968). Apart from the thematic object, each case also has an associated preposition viz: agent(by), coagent(with), source(from), destination(to), duration(for). If enzyme is treated as a noun it could be the thematic object or the agent. We shall not explore this here (for more details see Paton and Matsuno, 1997).

A typical view of an enzyme as verb might consider substrate, product, regulator(s), location, co-agent(s) to be its cases. A mesoscopic description of the glycolytic enzyme phosphofructokinase (PFK) would take the following cases:

INPUT - the substrate materials involved in the catalytic reaction in this case, fructose-6-phosphate (F6P) and ATP.

OUTPUT - the product materials involved in the catalytic reaction namely, fructose-1,6-bisphosphate (F1,6BP) and ADP.

REGULATORY INFORMATION - information which affects allosteric sites. In this case there is stimulatory information from F6P, F1,6BP and F2,6BP and inhibitory information from citrate.

BINDING - how the enzyme binds to and interacts with other molecules. In this case binding to intracellular membranes.

LOCATION - the habitat or context of the enzyme e.g., membrane, bulk phase or microtrabecullar lattice. The membrane-bound enzyme exhibits a normal saturation curve for F6P whereas the free enzyme has sigmoidal kinetics (Uyeda, 1992).

In this example, we see that the specification of the enzyme shifts from input-output alone to include important networking features. This emphasises the relational aspects of enzymes beyond their catalytic roles and also introduces the importance of appreciating an enzyme's context.

Enzymes are Context-sensitive

In natural languages verbs are far less common than nouns. They carry out their role because they are sensitive to context. In a similar way enzymes demonstrate this characteristic. For example, CaM Kinase II. The extent of its context-sensitivity can be related to its molecular architecture. It is a very common and very large multimeric enzyme with subunits derived from 4 genes. Four domains are recognised:

- catalytic,
- regulatory - it has both inhibitory and CaM binding regions,
- association - with other subunits,
- variable - targetting and localisation.

These functional domains are highly conserved (except for the variable). This multifunctional enzyme acts on upwards of <u>49</u> substrates.

Another example concerns one family of enzymes, the hexokinases. There are several isozymes of hexokinase which (in mammals) are variously distributed in the tissues. The tissue type of a particular hexokinase is reflected in its abundance which, for some tissues also varies during development. Hence, the type of hexokinase found (including liver glucokinase) is context-dependent (i.e., tissue, age). However, we also find that hexokinase can be context-sensitive within a cell. Wilson (1980) introduced the term 'ambiquitous' enzyme to account for the way brain hexokinase (BHK) exhibits rapid and reversible changes between soluble and particulate forms. He proposed that the distribution of BHK forms is regulated by intracellular concentrations of metabolites such as ATP, glucose-6-phosphate and P_i. The particulate form is more active than the soluble form. BHK can select its location for catalysis. In making a selection it shows information processing activity.

Conrad (1992) proposes a 'seed germination' model for enzyme action (specifically enzyme pattern recognition) to help account for context-sensitivity. The analogy is with the multi-factor complex associated with seed germination in plants. It illustrates the value of using the ecological notion of a factor complex to represent a biochemical agent (Paton, 1996). Conrad argues that the common electronic analogy of enzyme-as-switch breaks down because enzyme behaviour is context dependent. This idea is also implicit within fuzzy models of enzymes (Paton *et al*, 1996).

Enzymes have Voices and Moods

Enzymes effect change in reversible reactions although because of the biochemical context in which they operate, they can be typified by one particular voice. For example, the active voice in glycolysis would include the reaction F6P --> F 1,6 BP whereas the reverse reaction would be passive in the same context. It could be argued that *in vivo* the active voice for most enzymes is the prerequisite for metabolic functioning. Welch notes:

> *"if left alone (i.e., not linked to other processes), the enzyme will, in time just equilibrate the substrate-product concentration pools, whereby higher order computational functions (e.g., metabolic pathways) require the cell to couple whole enzyme reactions per se as the individual logic elements"*
>
> Welch (1994) p43.

We could say that metabolism is characterised as a far from equilibrium process in which it is essential that enzymes keep the same 'voice'.

Enzyme modality may depend on a large number of factors including location and interaction with other macromolecules, small molecules, ions and electric fields (Welch & Kell, 1986). It could be argued that these features, coupled with the fuzzy nature of enzyme action (Paton et al, 1996) reflects a subjunctive rather than indicative mood. It could be argued that this type of view would also be consistent with the notion of a 'fluctuating' enzyme which could also address issues of why enzyme molecules are so big and their roles in signalling systems. In summary it is easy to see why enzymes are 'smart' molecules. It is also valuable to view them as verbs, combining or gluing together metabolism.

ENZYMES AND 'GLUE'

Enzymes play the central role of verbs in the cellular metabolic and information processing system. Several examples have already been discussed in this paper (e.g., CaM Kinase). In this case we see how the interaction of catalytic, regulatory, association and targeting domains interact to provide a high degree of context sensitivity. Another example of a context sensitive multifunctional enzyme is 6-Phospho-Fructo-2-Kinase/Fructose-2,6-BisPhosphatase (6PF2K/F2,6BP). This enzyme catalyses two opposing reactions:

$$F6P + ATP ----> F2,6BP + ADP$$
$$F2,6BP ----> F6P + ATP$$

As Pilkis *et al* (1995) note, 6PF2K/F2,6BP has, in addition to a catalytic role, a key function in intracellular signalling.

Enzymes are not only verbs they are also 'glue' (indeed, verbs are also glue). The notion of 'glue' is metaphorical in basis but is used here to capture ideas about adhesion, cohesion and coherence (for more details see Paton,1997). For the present purposes of placing enzymes within the cellular economy/ecology we note that glues integrate networks and hierarchies. The gluing function of enzymes is both informational and thermodynamic because it can adhere on a local scale and cohere globally.

Energy suppliers and consumers residing in cells are the major players presiding the cellular economy there. This aspect naturally raises the question of how each player could be influential towards the other. In principle, there could be three possibilities; supplier domination, consumer domination, and both on a par. Among the three possibilites, what is actually significant is consumer domination over supplier (Matsuno, 1995). This is because suppliers who could not find their customers cannot survive indefinitely in the economy. Only those suppliers that can meet their customers, that are consumers, can maintain themselves in the economy. That is consumer domination over supplier. However, consumers cannot survive unless supplemented by suppliers. The consumers exert an affinity towards the suppliers, while the latter require the former for their own sustenance.

The present mutual affinity underlies the occurrence of gluing between the two.

So far we have seen how ideas about economy, verb and glue can be used to enrich our understanding and appreciation of the 'society of enzymes' from a relational point of view. In the final section of this paper we look at a the application of the mathematical notion of a category to the

SOME CATEGORY THEORETIC IMPLICATIONS

Category theory in its mathematical formulation is highly abstract. We use the notions of category and morphism as a mathematical metaphor to help explain the relational nature of the enzymes to the cellular economy and ecology. The application of category theory to relational thinking in biology has a long history (see e.g., Rosen, 1991). A biosystem can be modelled as a category whose objects are its components and whose arrows (morphisms) are the interrelations such as: neurons in a network, pools in a metabolism, neuronal paths in a network, and cells in an organism. The gluing relation discussed in the previous section and which remained implicit in the previous two sections can now be defined in category theoretic terms. Thus, Ehresmann and Vanbremersch note:

> " ... an object in a category has such a complex structure when it is composed of a family of more elementary objects, 'glued' together, the gluing depending on some specified links between the components"
> Ehresmann and Vanbremersch, 1987, p17.

Clearly, there is much more to the detailed workings out than what is presented here but it should be clear that a separation is made between objects and arrows and in this sense we have compared enzymes to arrows rather than objects.

We now conclude with a general statement on enzyme functionality based on the previous arguments. Both energy suppliers and consumers residing in cells are internal observers that are capable of making distinction. The capacity of making this distinction presumes the underlying categories. What is unique to the cellular internal observers is that their categories can be variable in a non-preprogrammable manner. Alternation of the internal observers can facilitate reshuffling of the associated categories. The cohesiveness between energy suppliers and consumers remains invulnerable to alternation of these players. Precisely for this reason, the gluing activity of enzymes in cells applies to whatever categories may come up.

A process view, if legitimately describable, must assume the presence of some invariance, otherwise there could be no such view to descriptively be identified. The activity of gluing just happens to be a case of such an invariance. Gluing is a functional invariance that can be applied to any cellular economy, while it is not structurally invariant because enzymes as the agents of gluing can be multifarious depending upon the cellular context. The activity of gluing makes enzymes both a material manifestation of functional invariance in time and a representation of structural versatility in space. Transference from locally asynchronous time to globally synchronous one is ubiquitous in the cellular processes, whereas each expression of the functional ubiquity depends upon the cellular context under which the transference actually takes place.

SO WHAT ?

We have spent a lot of time looking at some textual or literary perspectives on the cell. Some people question the value of this kind of approach. What does it give us? The answer to this question is very simple. The information processing perspective on biological systems is highly pervasive - from molecules and brains to cells and evolution. The danger is that critical issues such as the nature of bioinformation are under-examined so that ideas like "the cell's software", "the brain's hardware", etc. persist but are often little more than empty rhetoric. We hope the study provided in this paper will if nothing else deepen our understanding of some of the fundamental issues.

REFERENCES

Bray, D. 1995, Protein molecules as computational elements in living cells", *Nature*, 376: 307-312.

Conrad, M. 1992, The seed germination model of enzyme catalysis", *BioSystems*, 27: 223-233.

Dyson, F. (1995), The scientist as rebel, in: *Nature's Imagination The Frontiers of Scientific Vision*, Cornwell, J. (ed) Oxford University Press, Oxford, pp 1-11.

Ehresmann, A.C. & Vanbremeersch, J-P. 1987, Hierarchical evolutive systems: mathematical model for complex systems, *Bull. Math. Biol.*, 49: 13-50.

Giardina, B., Messana, I., Scatena, R. & Castagnola, M. 1995, The multiple functions of haemoglobin, *Critical Reviews in Biochemistry and Molecular Biology*, 30: 165-196.

Kampis, G. 1994, Biological evolution as a process viewed internally, in: *Inside versus Outside*, Synergetics Series 63, Atmanspacher, H. & Dalenoort, G. (eds), Springer-Verlag, Berlin pp85-110.

Kelly, K. 1994, *Out of Control The New Biology of Machines*, Fourth Estate, London.

Matsuno, K. 1989, *Protobiology: Physical Basis of Biology*, CRC Press, Boca Raton, Florida.

Matsuno, K. 1995, Consumer power as the major evolutionary force, *J. theor. Biol.*, 173: 137-145.

Matsuno, K. 1996, Symmetry in synchronous time and information in asynchronous time, *Symmetry: Cult. & Sci.*: 7: 295-305.

Matsuno, K. 1997, Information: resurrection of the Cartesian physics, *World Futures*, 223-237.

Marijuan, P.C. 1991, "Enzymes and theoretical biology: sketch of an informational perspective of the cell, *BioSystems*, 25: 259-273.

Miller, G.A. & Fellbaum, C. 1991, Semantic networks in english, *Cognition*, 41: 197-229.

Paton, R.C. 1997, Glue, verb and text metaphors in biology, *Acta Biotheoretica*: 45: 1-15.

Paton, R.C., Staniford, G. & Kendall, G. 1996, Specifying logical agents in cellular hierarchies, in: *Computation in Cellular and Molecular Biological Systems*, Cuthbertson, R., Holcombe, M. & Paton, R. (eds), World Scientific: Singapore, 105-119.

Paton, R.C. & Matsuno, K. 1997, Some common themes for enzymes and verbs, in preparation.

Pilkis, S.J., Claus, T.H., Kurland, I.J. & Lange, A.J. 1995, 6-Phosphofructo-2-kinase / Fructose-2,6-Bisphosphatase: A metabolic signalling enzyme, *Annu. Rev. Biochem.*, 64: 799-835.

Rosen, R. 1991, *Life Itself*, Columbia University Press: New York.

Sattler, R. 1985, *Biophilosophy*, Springer-Verlag: Berlin.

Uyeda, K. 1992, Interactions of glycolytic enzymes with cellular membranes, *Curr. Topics in Cellular Regulation*, 33: 31-46.

Welch, G.R. 1994, The computational machinery of the living cell, in: *Computing with Biological Metaphors*, Paton, R.C. (ed), Chapman & Hall, London, Chapter 4.

Welch, G.R. & Kell, D.B. 1986, Not just catalysts - molecular machines in bioenergetics, in: *The Fluctuating Enzyme*, Welch, G.R. (ed), John Wiley, New York, 451-492.

Wilson, J.E. 1980, Brain hexokinase, the prototype ambiquitous enzyme, *Current Topics in Cell Regulation*, 16: 1-54.

A COMPUTATIONAL ARCHITECTURE BASED ON CELLULAR PROCESSING

Mark Shackleton, and Chris Winter

Systems Research
BT Laboratories
Martlesham Heath
Ipswich IP5 3RE
UK

INTRODUCTION

There has been considerable interest in recent years in applying biological metaphors to develop new approaches to information processing (Paton, 1994; Banzhaf et al., 1996; Berry and Boudol, 1992; Fontana, 1991) and in exploiting biochemical processes for computing. At the same time, computational models of cells and cellular systems have attracted increasing attention (Paton, 1993) within both biological and computational spheres. The field of molecular computing (Conrad, 1990; Conrad 1992) in particular is providing many new approaches to difficult computational problems, whilst Adleman(1994,1995) and Lipton(1994) have shown how the massive parallelism of molecular systems can be exploited to solve combinatorial computations. Although many of these areas are still in relative infancy in comparison with traditional computing techniques and systems their offer of new algorithms and vast computational speedup seems likely to ensure their future.

Within the computing industry, the von Neumann architecture (Pohl and Shaw, 1981) has been almost universally adopted and has proved extremely useful. It is well suited for many common processing applications, particularly those which are inherently serial or which involve straight forward arithmetic tasks. However, it is not well suited to implementation of those algorithms which are inherently parallel and asynchronous or to tasks which involve a significant pattern matching element. These more difficult problems have been tackled by other approaches such as neural networks (Rummelhart and McClelland, 1986; Wasserman, 1989) inspired by models of biological neurons, and "soft computing" techniques including fuzzy logic. In part due to the difficulty of designing and programming these systems, new approaches to programming based on evolutionary techniques such as genetic algorithms (Holland, 1992) and genetic programming (Koza, 1992) have been developed. Many of these more recent techniques as well as some optimisation techniques such as simulated annealing have been inspired at least in part by biology, physics and chemistry, and have proved remarkably successful. In this paper a

Information Processing in Cells and Tissues
Edited by Holcombe and Paton, Plenum Press, New York, 1998

model for information processing is described which is inspired by molecular processing within biological cells.

A common view of processing within cells is that of an intricate series of production lines which assemble increasingly complex molecules from simpler components, with the reactions which transform molecules into the required products being controlled by means of enzymes which selectively accelerate the necessary chemical reactions by many orders of magnitude. Probably the two most important features of enzymes are that they exhibit *specificity* for particular molecular substrates and that they carry out particular *functions* which transform substrates into products, via the reactions they catalyse. An enzyme's *binding sites* to which substrate(s) can bind determine an enzyme's specificity, whilst an enzyme's *active site* plays the key part in enabling and accelerating a particular reaction and thus provides the enzyme's function. An enzyme's shape, including the positions of its constituent amino acids determines the enzyme's specificity and function. The shape itself arises from the process of protein folding (Darnell et al., 1995); we discuss a simple analogue for the folding process later.

The architecture proposed in this paper contains many processing elements which are analogous to enzymes in that they have specificity, via pattern matching, for their operands (data as opposed to molecules) and cause the data to be transformed in some way. The emergent properties of the system taken as a whole, including the enzyme-data interactions and the dynamics (kinetics), comprise the information processing system.

The following section describes the information processing architecture which is based on a cellular processing analogy. Some results of the information processing architecture applied to a specific processing task are then presented; in particular, the time evolution of the processing system is described. Subsequently we describe in outline how a system could be automatically configured to perform a specific information processing task using evolutionary techniques rather than being explicitly programmed. The last two sections include a discussion of the concepts and results presented, and our conclusions.

THE INFORMATION PROCESSING ARCHITECTURE

We would like to better understand the processing capabilities of biological cells in order to use similar principles to design new or improved information processing systems and algorithms. It is likely that such an understanding will also be useful in the reverse direction of computationally motivated biology, for instance by gaining a better understanding of the sort of processing tasks these architectures are well suited to perform. An analogy is drawn between molecules and data, and between enzymes and processing units; thus cells can be viewed as processing systems which accept simple building blocks as input and progressively transform these to build up increasingly complex structures. As cells have solved many complex problems in the interests of survival in hostile and unpredictable environments and have clearly proved to be evolvable, there should be important principles which can be used to design systems which demonstrate similar attributes but which carry out specified information processing tasks.

Attributes of biological systems which are of particular interest include robustness against perturbations (homeostasis), processing dynamics (i.e. enzyme kinetics), pattern matching of enzymes leading to specificity for substrates, and evolvability (or programmability) of such systems. An architecture designed using these principles is substantially different from the traditional von Neumann computer architecture so we wish to create a simulation system which allows us to explore the importance of these attributes.

The proposed information processing architecture is illustrated in figure 1. The system consists of an enclosure termed a sac (rather like a very simple cell) which contains processing "enzymes", data items and, potentially, constituent sacs corresponding to intracellular vesicles. Each enzyme has a pattern matching method to select data items thus providing enzyme *specificity,* and has an associated operation which provides the enzyme *functionality.* An enzyme can thus consistently identify the classes of data it should operate upon from the data pool, based upon data types and values. The data values can be used to evaluate conditions associated with an enzyme which further determine whether the function will be evaluated; the conditions are analogous to the thermodynamics which determine when a molecular enzyme operates. Enzyme operations will typically combine data items to form a more complex data structure, or manipulate the data field values of a data item. Unlike traditional computer architectures there are no explicit fixed "data paths" linking processing elements together; instead the flow of data is governed by the specificity of an enzyme for a data item of a particular type and/or value. This is not to say that processing pathways cannot be formed at all, but rather that they are not fixed *a priori* in the architecture described.

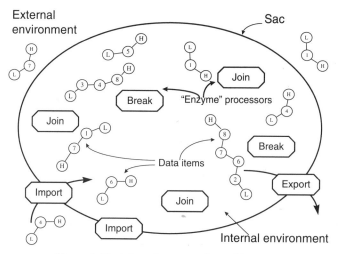

Figure 1. The information processing architecture.

The specificity and functionality of a biological enzyme are a result of its shape (tertiary structure) and surface characteristics; in a subsequent section a mechanism is described by which these enzyme attributes can be determined from shape via a simple analogue of protein folding. Note that figure 1 shows some processing enzymes which reside on the sac's membrane which are assigned the special role of selectively importing data items into the sac from the external environment or exporting data items from the sac to the external environment. In the case of a subsidiary sac these import/export enzymes provide a means of supplying the sac with its input and output data, whilst it carries out a sub-task of the main processing task.

Although real enzymes catalyse a set of pre-existing chemical reactions rather than carrying out the reactions themselves, the enzymes here are viewed as active processors carrying out a particular transformation on the data. This is equivalent to assuming that a corresponding reaction (operation) exists and that the enzyme (processing unit) is capable of catalysing it.

The architecture described currently does not include multiple binding effects known for molecular enzymes, such as cooperativity and various forms of inhibition. These have been omitted initially for simplicity but we expect to include these and other mechanisms in the future so as not to preclude the important effects of regulation of enzymes on the processing dynamics.

So far a general information processing architecture has been described. To actually apply the architecture to solve a particular task it must be configured appropriately by defining a set of system attributes; this is effectively how the system is programmed. The processing carried out by the system is emergent from the interaction of the parts however, rather than being a direct result of an explicit sequential program as would usually be the case with a von Neumann architecture. It is necessary to define a data representation for data items, a pattern matching method for each enzyme which identifies (binds) the data items to operate upon, and the function each enzyme will perform. In addition a system of dynamics must be specified which plays an important part in the processing system, as it governs the rate at which enzymes operate. Its role is to provide an analogue for enzyme kinetics, although it need not necessarily be as complex as the systems of equations typically used to model real systems. By providing such a system of dynamics a mechanism exists to smoothly control rates of processing based on "concentrations" of data items.

Clearly for all but relatively simple information processing tasks the system described is quite difficult to "program" as the processing is emergent rather than explicit. It is our intention that such systems will ultimately be programmed to perform a task automatically by using evolutionary techniques such as Genetic Algorithms (Holland, 1992) rather than being manually programmed by defining the above details explicitly. An outline description for how this could be achieved is given later. First we show how a system can be manually configured to carry out a specific information processing task.

AN EXAMPLE INFORMATION PROCESSING APPLICATION

This section describes a simulation of the architecture described above applied to a simple information processing task: sorting a set of unique numbers into an ordered list. This provides an insight into how the system processes information and an examination of the time evolution of the system as it does so. In order to apply the architecture to a specific task we must define certain details; for the number sorting application we define the following:

- a data representation comprising a list with low and high end terminator labels.
- two types of enzyme called "join" and "break.
- a system of dynamics governing the rate of operation of enzymes.

A "join" enzyme is responsible for assembling lists of increasing lengths by concatenating two shorter lists together. (See figure 2a). The specificity of this enzyme is such that it simultaneously binds two data lists using the high-end terminator of one and the low-end terminator of another. It then concatenates the two lists after removing the bound terminators, but only if the appropriate condition is met: the values of the elements adjacent to the bound terminators must satisfy $v_h < v_l$. This action results in lists which are monotonically increasing but which may contain large gaps of missing values.

A "break" enzyme is responsible for splitting a list into two sub-lists when it contains gaps of missing values (versus the fully sorted list). The specificity is defined such that the enzyme simultaneously binds a list containing exactly one value element and a list of length greater than one. (See figure 2b.) The longer list is split into two sub-lists provided that it does not already contain v_0 and that the following condition is satisfied: $v_l < v_0 < v_h$.

The join and break enzyme operations are defined such that at all times the data lists will be ordered lists; i.e. the list data type itself does not preclude unsorted lists, but unsorted lists will never arise through the specified enzyme operations.

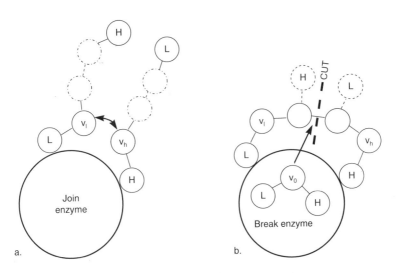

Figure 2. (a) join and (b) break enzyme operations.

The system of dynamics, together with the concentrations of enzymes and data items, governs the rate at which enzymes operate and thus affects the processing carried out by the system. In the current example concentrations of enzymes and data items are explicitly represented by multiple instances of a corresponding object. The objects are placed on a toroidal surface with random initial positions and the data items are given random initial velocities. Whenever a data item comes within a fixed distance of an enzyme it may become bound to the enzyme for a short period according to the enzyme's specificity. When the appropriate conditions for a join or break enzyme are met the enzyme operates on the data lists.

The system is initialised by creating a single sac containing multiple unit length data lists and multiple enzymes of both types. Each value to be sorted is represented multiple times by a corresponding unit length list. Subsequent to initialisation, the system carries out processing iterations on the data lists within the framework described above. Whenever the number of data items drops due to list concatenations, new unit length data lists are added to "top up" the system. Note that for this simple simulation, data is introduced directly into the processing sac rather than via the import enzymes shown in figure 1.

Figure 3 shows the time evolution of the system during processing. The sac initially contained 100 unit length lists comprising multiple instances of data lists for each of the 8 unique values to be sorted. The sac was topped up with further unit length lists whenever the number of data lists present dropped below 100. The system succeeds in creating fully sorted lists as expected, with concentrations of intermediate length data lists rising and falling as they are initially created then used up.

The graphs of frequency (concentration) of particular data items over time are rather reminiscent of plots of concentrations of molecular species present in systems of enzyme kinetics. Rather than using the system of dynamics described above we could in fact have used a set of dynamic equations derived from those used for modelling enzyme kinetics (Murray, 1993) to describe the concentrations of particular data items and enzymes present.

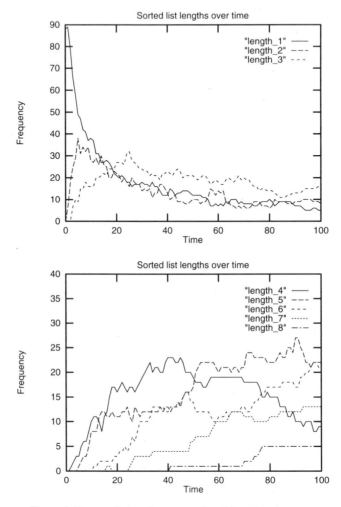

Figure 3. Time evolution of system sorting eight unique integers.

It is interesting to consider some of the attributes of the system described. Processing is highly parallel, depends on "concentrations" of data items and reaction dynamics, and the best solution available can be ascertained at any time by examining the longest data list present in the data pool. The system should be relatively stable to small perturbations such as fluctuations in concentration caused by the removal of a small quantity of some particular data list; this would simply alter the reaction rates slightly. Processing is "softer" than that of traditional systems where a small alteration would typically cause a program to fail.

It was observed when running the simulation that single valued data lists are quickly concatenated to form longer lists with the consequence that there are few available to allow the break enzymes to operate. In an attempt to compensate the break:join enzyme ratio was increased from 1:1 to 2:1. In fact this had virtually no effect on the time evolution of the system, suggesting that the system is very stable to this sort of variation in concentrations. An alternative approach to tackling this problem was then tried: the simulation was modified in such a way that the binding of a longer data list to a join enzyme inhibited the *subsequent* binding of single valued data lists to that enzyme for a finite period. This had the effect that as longer data lists were formed, single valued lists would be preserved longer in the system, thereby allowing the break enzymes an increased chance of operating. In this case the yield of solutions increased approximately threefold over the system without inhibition, although the time taken to find the first solution was relatively unaffected.

The yield of solutions may not seem important for such systems, it being sufficient to find just a single solution. However, if such systems are to be realised as real implementations in the future then yield could become much more important. In this case it is likely that a sufficient yield will be required so that the solution can be extracted or detected at all. The results reported so far are for a system in which the final products are not removed, thus these products tend to "clog up" the system. We therefore further modified the system such that in addition to injecting single valued data items to maintain the number of data items present, the solution data items were removed as they were produced by the system. Figure 4 shows the time evolution for this system, which also includes the inhibition described above. It can be seen that the system reaches a steady state in which the concentrations of intermediate products remain relatively constant whilst solution product continues to be produced.

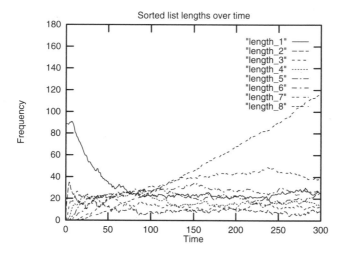

Figure 4. Time evolution of system with removal of final product.

EVOLVING A PROCESSING SYSTEM

It is interesting to consider how the architecture of figure 1 might be configured to carry out an information processing application without manual programming. It seems likely that systems employing the architecture described will prove be more evolvable than are algorithms implemented using traditional programming languages. The processing system's behaviour depends on a system of dynamics, concentrations, and pattern matching, all of which can be smoothly varied usually without catastrophic results, whereas a single change to a program written in a typical computer programming language such as "C" is almost always catastrophic to the functioning of the program. Evolutionary search systems are far more likely to succeed in finding solutions where the fitness landscape of potential solutions (systems or programs) is relatively smooth rather than where it is extremely sparse and rugged.

A Genetic Algorithm (Holland, 1992) can be used to evolve an information processing system based on the architecture of figure 1. The genetic algorithm requires that a mapping from a genotype (bit string) to the corresponding phenotype (system) is defined together with an objective function to evaluate the fitness of a system. The genotype to phenotype mapping proposed here is based on the biochemical mapping from a DNA gene to the corresponding enzyme. As each amino acid in an enzyme is chosen from 20 available amino acids by a DNA *codon* having 2^6 possible variants, consecutive strings of 6 bits are extracted from the genotype bit string and used together with a lookup table which gives the amino acid unit corresponding to each possible codon value. The sequence of units (the primary structure) can be used to determine a shape (tertiary structure) for the corresponding enzyme via a simple analogue of protein folding; the enzyme's data specificity and functionality can then be determined from this shape (see below). The genotype bit string can be partitioned into a series of sub-strings to specify a population of enzymes for use in the processing system. The objective function which is required by the genetic algorithm to determine the fitness of a phenotype (processing system) depends upon the particular task the system is to perform, for example the number sorting task might assign higher fitness to systems which succeed in producing greater numbers of longer sorted lists over a given number of processing iterations.

The shape associated with an enzyme can be ascribed by a process analogous to real protein folding. The exact details of the folding process employed are probably not very important for our purposes although the *general properties* of the process may prove to be important to evolvability of the system as discussed later. A protein folding model such as the protein toy model (Stillinger et al., 1993) or an alternative (Fernández-Villacañas, 1997) could be used to determine either a 2D or 3D shape.

Substrate specificity of real enzymes arises from the complementary lock and key fit between the enzyme surface and the substrate molecule. Just as a shape can be assigned to an enzyme, a similar process could be used for data items; enzymes and data items can then be inspected for complementary regions. This pattern matching process allows enzymes, within the evolutionary framework described, to develop specificity (a binding site) for particular data attributes. A mechanism for associating a function with an enzyme is to define a set of allowable functions appropriate to the data and then to provide a lookup table which maps particular amino acid configurations to corresponding functions drawn from the set of allowable functions. Spatially local clusters of the enzyme's constituent units can then be considered and the equivalent function (if any) looked up. This process is similar to the indirect mapping from a real enzyme's primary structure, via its tertiary structure, to the formation of its active site consisting of a particular local spatial configuration of amino acids.

DISCUSSION

Conrad(1992) notes that although some computational problems are addressed simply by increases of processing power and miniaturisation with time, other problems "such as pattern and object recognition, effective use of parallelism, and learning are not likely to succumb to downsizing alone. New computer structures and computing modes are necessary." The conceptual architecture which is described in this paper departs significantly from the traditional von Neumann computing architecture which is highly ordered and executes rigid steps sequentially and synchronously. Instead, the cellular architecture (figure 1) relies heavily on pattern matching as a primitive operation and processes data in a highly parallel, asynchronous and unstructured manner.

Traditional architectures usually have a fixed topology connecting the processing units together which determines the route data takes through the system; in contrast the route data takes in the cellular architecture is determined solely by the enzyme's data specificity. The processing performed is emergent from the interactions of its many parts under control of a system of dynamics rather like enzyme kinetics. Pattern matching, "concentrations" of data, and the system of dynamics (kinetics) can all be smoothly adjusted to affect the processing, which opens up new and rather unusual possibilities for controlling and programming information processing systems. There is little experience of implementing programs using these systems although Adleman(1994,1995) has illustrated that certain problems can be encoded in a suitable manner for processing by molecular computing systems. It seems likely that there will be mutual benefit in understanding the processing capabilities of such systems as computing will gain new architectures and algorithms whilst biological models of cellular processes can be better understood in light of computational analyses of these designs.

Although the emergent nature of the information processing makes traditional programming difficult when compared with a von Neumann machine and typical computer languages, the ease of programming of the latter contributes to its brittleness (Conrad, 1990). In contrast, the smooth control of processing in the cellular architecture should mean slight changes in conditions are not catastrophic which leads to improved robustness of the system including error resilience and coping with noisy data.

The function and substrate specificity of real enzymes derive from the indirect mapping from the sequence of amino acids encoded by the order of bases in DNA (primary structure) via protein folding to the enzyme's shape (tertiary structure). It is likely that the properties of this mapping are important to the evolvability of a system. In biological systems many base mutations in DNA do not change the enzyme's shape (and thus its properties) significantly, whereas a few mutations *do* effect a significant change. The distribution of significance of mutations is likely to be an important factor in the evolvability of the system and we hope to investigate this issue further. The main reason for choosing to adopt the analogue of the primary to tertiary structure mapping, and structure to enzyme specificity and function, is to allow us to explore these issues.

In the architecture described there is a clear distinction and asymmetry between processing enzymes and the data they operate on. This is not the case in biological cells where an enzyme can be operated upon by another in which case it takes the role of "data", or it can itself operate on a substrate in which case it is acting as processor. Hofstadter(1979) provides an interesting discussion of the levels of interaction and different roles that are played by enzymes and DNA within cells. These issues become more important when enzyme and gene regulation are being considered. We intend to extend the simulations to include enzyme regulation in the future.

The processing carried out by a system of the type described in this paper depends on the emergent behaviour of a network of enzyme processing elements and the interim products passed between them. Kauffman(1993) has investigated the global properties of similar networks parameterised by their degree of interconnectivity. His results suggest that the interconnectivity will have a significant effect on the dynamics and stability of the processing system. We hope to investigate these issues within the context of the architecture described.

The architecture shown in figure 1 is a conceptual architecture which we have so far simulated on a traditional computer. How such an architecture could best be implemented in practice is an open question, however work by Adleman(1994) suggests that there is a reasonable prospect of implementing such systems in the future using chemical and biochemical techniques.

CONCLUSIONS

In this paper we have presented a conceptual architecture for information processing based on an analogy to the processing of molecules by enzymes within biological cells. The processing carried out is emergent from the interaction of many processing "enzymes" under the control of a system of dynamics analogous to enzyme kinetics. The operation of the enzymes depends fundamentally on a pattern matching process which allows them to selectively match the data "substrates" which they should operate upon, drawn from a data pool.

The architecture has been used to configure a simulation system to carry out a simple information processing task, namely that of sorting a set of integers into an ordered list. Results of the time evolution of the processing system performing this task have been presented.

Finally, we have outlined a method by which the architecture could be automatically configured (programmed) to perform a specific task via an evolutionary technique. This method employs an analogy to protein folding in order to determine an enzyme's specificity and function.

REFERENCES

Adleman, L.M., 1994, Molecular computation of solutions to combinatorial problems, *Science* 226:1021.

Adleman, L.M., 1995, On constructing a molecular computer, in: *DNA Based Computers*, E.B. Baum and R.J. Lipton, ed., *DIMACS*, 27.

Banzhaf, W., Dittrich, P. and Rauhe, H., 1996, Emergent computation by catalytic reactions, *Nanotechnology*, 7:307-314.

Berry, G. and Boudol, G., 1992, The chemical abstract machine, *Theoretical Computer Science*, 96:217.

Conrad, M., 1990, Molecular computing, in: *Advances in Computers*, M.C. Yovits, ed., 31:236-324, Academic Press.

Conrad, M., 1992, Guest editor's introduction: molecular computing paradigms, *IEEE Computer*, 25,11.

Darnell, J.E., Lodish, H. and Baltimore, D., 1995, *Molecular Cell Biology*, W. H. Freeman.

Fernández-Villacañas, J.L., Fatah, J.M., and Amin, S., 1997, Evolution of protein interactions, *BCEC'97: Bio-Computing and Emergent Behaviour*, Skövde, Sweden.

Fontana, W., 1991, Algorithmic chemistry, in: *Artificial Life II: Proc. 2nd ALife Workshop*, C.G. Langton, ed., p159-.

Hofstadter, D.R., 1979, *Gödel, Escher, Bach: an eternal golden braid*, Harvester Press.

Holland, J.H., 1992, *Adaptation in Natural and Artificial Systems*, MIT Press.

Kauffman, S.A., 1993, *The Origins of Order: self-organization and selection in evolution*, Oxford University Press.

Koza, J.R., 1992, Genetic programming: on the programming of computers by means of natural selection, MIT Press.

Lipton, R.J., 1994, Speeding up computations via molecular biology, *Unpublished manuscript, internet URL*: ftp://ftp.cs.princeton.edu/pub/people/rjl/bio.ps

Murray, J.D., 1993, *Mathematical Biology,* 2nd ed., pp. 109-139, Springer.

Paton, R., 1994, ed., *Computing with Biological Metaphors*, Chapman and Hall.

Paton, R., 1993, Some computational models at the cellular level, *BioSystems*, 29:63-75.

Pohl, I., and Shaw, A., 1981, *The Nature of Computation: an introduction to computer science*, Computer Science Press.

Rummelhart, D.E., and McClelland, J.L., 1986, *Parallel Distributed Processing*, MIT Press.

Stillinger, F.H., Head-Gordon, T. and Hirsfield, C.L., 1993, Toy model for protein folding, *Physical Review E*, 48, 2:1469.

Wasserman, P.D., 1989, *Neural Computing: theory and practice*, Van Nostrand Reinhold.

DISTRIBUTIVITY, A GENERAL INFORMATION THEORETIC NETWORK MEASURE, OR WHY THE WHOLE IS MORE THAN THE SUM OF ITS PARTS

Roland Somogyi and Stefanie Fuhrman

Molecular Physiology of CNS Development
LNP/NINDS/NIH
Building 36, Room 2C02
Bethesda, MD 20892, USA
Phone: 301-402-1407
FAX: 301-402-1565
Email: rolands@helix.nih.gov (R. S.), sfuhrman@codon.nih.gov (S. F.)
WWW: http://rsb.info.nih.gov/mol-physiol/homepage.html

INTRODUCTION

Molecular life sciences are rapidly uncovering the elementary building blocks underlying complex biological functions. A common feature encountered at all levels is the organization of basic constituents into distributed, high-dimensional networks. A major challenge lies in understanding the abstract principles that govern the logic of these networks, enabling them to execute complex functions while retaining stability and adaptability (Somogyi and Sniegoski, 1996). Boolean networks provide a model framework that captures these features, serving as a testing ground for establishing novel techniques in computational network analysis.

A feature of primary concern is how particular "functions" of the network are distributed across its elements. For example, in a *genetic network*, we would like to understand to which degree individual gene products and their complexes regulate the expression of sets of downstream genes. Is a single gene product (biomolecule) generally functionally sufficient, or do we require minimal combinations of several biomolecules, or is a function absolutely dependent on one particular combination of effectors? An analysis of these dependencies at a high level of abstraction may provide us with generalizable measures. These could be applied to the study of e.g., genetic, metabolic and neural networks, and eventually to engineered systems.

Our approach is to build a fundamental analysis based on the principles of information theory, using such measures as *Shannon entropy* and *mutual information*. We extend these measures to the analysis of dependencies beyond input-output pairs, to interactions encompassing several input elements. A particular process (output; e.g., gene activity, secretion of signaling factor, etc.) is commonly influenced by several control signals (inputs; e.g., transcriptional regulators in a genetic network, protein phosphorylation and second messengers in intracellular signaling). Therefore, the information contained within the dynamic changes of the inputs, essentially their *Shannon entropies*, is transmitted to fluctuations of the output (Fig. 1). The question is how much of the information of each particular input is effectively transmitted to the output. We will show that combinations of inputs can make unique information contributions, only predictable by knowledge of all participating input states. In the ideal case of independent, maximum entropy inputs, and

Information Processing in Cells and Tissues
Edited by Holcombe and Paton, Plenum Press, New York, 1998

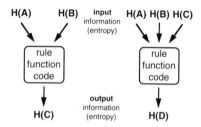

Figure 1. Information flow in distributed systems. Information (H) stemming from several inputs is transmitted to the output using a code determined by the biological signal integration function.

no bias of the output to particular inputs, the most efficient code should distribute the input information contributions evenly across each incoming channel and all of their combinations. This will be explored using the *distributivity* measure.

SHANNON ENTROPY

Information theory provides us with a quantitative information measure, the *Shannon entropy*, H. It should be stressed that information here is used as a technical term, more akin to "uncertainty" and "diversity", that should not be confused with "meaning". The Shannon entropy is defined in terms of the probability of observing a particular symbol or event, p_i, within a given sequence (Shannon and Weaver, 1963),

$$H = - \Sigma \, p_i \log p_i.$$

The maximum entropy, H_{max}, occurs when all states are equiprobable. Accordingly (Shannon and Weaver, 1963),

$$H_{max} = \log(n).$$

Entropies are commonly defined in units of "bits" (binary digits), when using the logarithm on base 2. Therefore, $H_{max} = 1$ for n=2, and $H_{max} = \log(3) \approx 1.585$ for n=3.

The *relative entropy* is a measure of the extent to which the system is using the full information potential of each possible state, we define (Shannon and Weaver, 1963),

$$H_{rel} = H/H_{max}.$$

The relative entropy essentially corresponds to the maximal information compression of a sequence while adhering to the same number of possible states (Shannon and Weaver, 1963). Conversely, we define the *redundancy* , R, of a system as,

$$R = 1 - H_{rel}.$$

It is also intuitively clear that low relative entropy must correspond to high redundancy; a sequence containing redundant states is not maximally compressed with respect to information content.

MUTUAL INFORMATION

Our aim is to compare different sequences using information measures to establish functional relationships between elements of a network. In a system of 2 binary elements, X (index i) and Y (index j), the individual and combined Shannon entropies are defined essentially as above:

$$H(X) = - \Sigma \, p_i \log p_i \, , \; H(Y) = - \Sigma \, p_j \log p_j \, , \; \text{and} \; H(X, Y) = - \Sigma \, p_{i,j} \log p_{i,j}$$

There are 2 conditional entropies which capture the relationship between X and Y, H(X|Y) and H(Y|X), which are related as follows (Shannon and Weaver, 1963):

$$H(X,Y) = H(Y|X) + H(X) = H(X|Y) + H(Y) .$$

In words, the uncertainty of X and the remaining uncertainty of Y given knowledge of X, H(Y|X), i.e. the information contained in Y that is not shared with X, sum to the entropy of the combination of X and Y.

We can now find an expression for the shared or "mutual information", M(X, Y), also referred to as "rate of transmission" between an input/output channel pair (Shannon and Weaver, 1963):

$$M(X, Y) = H(Y) - H(Y|X) = H(X) - H(X|Y).$$

The shared information between X and Y corresponds to the remaining information of X if we remove the information of X that is not shared with Y. Using the above equations, mutual information can be defined directly in terms of the original entropies; this formulation will be important for the considerations below:

$$M(X,Y) = H(X) + H(Y) - H(X,Y).$$

BOOLEAN NETWORKS

A Boolean network is simply a network of binary elements that can be freely connected to one another (wiring) , their interactions determined by Boolean or logical rules (for a more detailed description, please see Somogyi and Sniegoski, 1996; Wuensche, 1993 and Kauffman, 1993). The architecture of the network is defined in terms of the wiring and Boolean rules for each element. The rules strictly define how the state of the inputs of a particular element at time=t determine the output's state at time =t+1. Therefore, the state of the network at time=t precisely maps to the following state at time=t+1. A succession of states in time is referred to as a *trajectory*. It is standard to refer to the number of elements of the network as N (total number of states of the network = 2^N) and to the number of inputs an element receives as k. Given that there are 2^k different combinations of binary input states, and that each input state set must be mapped to a binary output state in a particular rule table, there are $2^{(2^k)}$ different Boolean rules for k inputs.

ANALYSIS OF K=2 RULES

We would now like to know which information contribution each input element, A or B, makes to the output element, C, for a particular k=2 rule. Boolean rules are numbered according to the output pattern corresponding to the table of all input possibilities; the output pattern is a binary number which can also be referred to as its decimal. This is illustrated for the *exclusive or* (no. 6), *only if A* (no. 12), and *or* (no. 14) rules (Fig. 2).

A	B	C		A	B	C		A	B	C
0	0	0		0	0	0		0	0	0
0	1	1		0	1	0		0	1	1
1	0	1		1	0	1		1	0	1
1	1	0		1	1	1		1	1	1

"xor"   "if A"   "or"

no. 6   no. 12   no. 14

Figure 2. Selected rule tables for k=2. The states of A and B correspond to the input at time=t, which determines the state of the output, C, at time=t+1.

Generally, we assume that what A and B tells us about C, M([A,B],C), can be decomposed into what A and B tell us about C individually, M(A,C) and M(B,C), minus the redundant part of what A and B tell us about C, i.e. what they also tell about each other, M(A,B,C),

$$M([A,B],C)=M(A,C)+M(B,C)-M(A,B,C) \ .$$

According to the above definitions,

$$M([A,B],C)=H(A,B)+H(C)-H(A,B,C),$$

$$M(A,C) = H(A) + H(C) - H(A,C), \text{ and } M(B, C) = H(B) + H(C) - H(B,C),$$

therefore, we can solve for M(A,B,C):

$$M(A,B,C)=H(A)+H(B)+H(C)-H(A,B)-H(B,C)-H(A,C)+H(A,B,C)$$

M(A,B,C) can only be positive in case of non-independent inputs (M(A,B)>0). However, for complete Boolean rule tables, the inputs are completely independent of one another (M(A,B)=0), since each conceivable input state is represented exactly once in the table. Here, M(A,B,C)<0 if combinations of input channels exclusively code for information on the third "virtual" channel that is not shared between the input channels. We would like to refer to this as the *unique mutual information* U([A,B],C), and define it as follows:

$$U([A,B],C)=-M(A,B,C).$$

Because M(A,B,C) is obviously symmetrical,

$$U([A,B],C)=U([A,C],B)=U([B,C],A),$$

so it should be simply referred to as

$$U(A,B,C).$$

Note that this does not apply to M([A,B],C), i.e. M([A,B],C) does not necessarily equal M(A,B,C). For the individual input/output pairs, the unique mutual information simply corresponds to the mutual information,

$$U(A,C)=M(A,C), \text{ and } U(B,C)=M(B,C).$$

We can now decompose M([A,B],C) in terms of the unique information contributions of the individual channels and their combination:

$$M([A,B],C)=U(A,C)+U(B,C)+U(A,B,C).$$

These are shown for all k=2 rules in Table 1.

Now we are beginning to get a feel for how distributed each rule is. Rule 12, *if A*, is trivial, all the information defining C is contained in A, therefore U(B,C) and U(A,B,C) are both 0. In this case, the effective number of inputs (k_{eff}) is 1, meaning that this rule is redundant with regard to the number of inputs and should be replaced by a k=1 rule. Generally if either U(B,C)=0 and U(A,B,C)=0, or U(A,C)=0 and U(A,B,C)=0, k_{eff}=1. There are four k=2 rules for which k_{eff}=1 (see Table 1). If U(A,C)=0, U(B,C)=0 and U(A,B,C)=0, then k_{eff}=0; this applies to rules 0 and 15, and generally to all rules for which the output is either always *off* or always *on*, i.e. rules 0 and $2^{(2^k)}-1$ for all k. We conclude that the first step in examining or constructing a Boolean network should be the determination of wiring and rule combinations leading to gross "overwiring", and replacement of such with minimal wiring and rules.

Decomposition of the *exclusive or* and *or* rules provides us with some interesting insights. In case of rule no. 6, *exclusive or*, neither A nor B contribute ANY information to C by themselves, i.e. U(A, C)=0 and U(B, C)=0. The information for determining C is solely contained in the combination of A and B, i.e. U([A, B], C)=H(C). Essentially, the *exclusive or* rule no less affirms that, with respect to predicting C, the *whole is more than the sum of its parts*, i.e. M([A, B], C) > U(A, C) + U(B, C). Lastly, we shall consider the more complex case of the *or* rule (no. 14). Here the unique mutual information with respect to C is distributed optimally across A, B and the combination A, B.

At this point, we introduce the relative or *proportional unique mutual information*, P, with respect to input(s) X and output Y,

$$P(X, Y)=U(X, Y)/H(Y).$$

For the *or* rule, P(A,C)=0.38, P(B,C)=0.38, and P([A,B],C)=0.23; of course, P(A,C)+P(B,C)+P([A,B]C)=1. The choice of P, reminiscent of probability, as a symbol for proportional mutual information is not coincidental. The probability of an event is defined as the limit of the ratio (proportion) of the number of occurrences of a particular event over all events encountered in a series of trials as the number of trials goes to infinity.

Considering the similarity between proportional mutual information and a probability, we will take the liberty of using P(X, C) as a probability as used in the definition of the Shannon entropy. With this in mind, we can construct the Shannon entropy, H_P, of the relative contributions of the combinations of input elements in determining C,

$$H_P = -\Sigma P(X_j, C)*logP(X_j, C).$$

For the *or* rule, taking $P(X_1, C)=P(A, C)$, $P(X_2, C)=P(B, C)$, and $P(X_3, C)=P([A, B], C)$, $H_P =1.55$. H_P for the other k=2 rules is shown in Table 1.

Redundancy was defined above. We have calculated this with respect to the output C, and refer to it as the *output redundancy* of a rule, reflecting how efficiently the output states are employed in coding (see Table 1),

$$R_O = 1 - H_O / H_{Omax} .$$

Analogously, we may consider the *input redundancy* of a rule,

$$R_I = 1 - H_P / H_{Pmax} ,$$

which gives an indication of how optimally each input state is used in coding for the output (see Table 1). Finally, we would like to determine the "effective" sources of information generated by the inputs, which we would like to refer to as *distributivity* or Δ:

$$\Delta = 2^{H_P} .$$

For k=2, $\Delta_{max} = 3$, since there are maximally three information sources determining C, i.e. A, B and the combination, A,B (discussed above). Indeed for the *or* rule, Δ = 2.93, coming very close to its theoretical maximum. In case of *exclusive or*, Δ = 1; one may wonder whether this implies inefficient coding (note: the input redundancy is 1), i.e. why use 2 inputs to code for an output if they effectively behave like a "single" information source?

Lastly, even given a high distributivity, not much information may be transmitted by each channel and combinations if the relative output entropy is low. A high distributivity value may be misleading in such a case. Here we introduce the *effective distributivity*, Δ_{eff}, as the product of the distributivity and relative entropy:

$$\Delta_{eff} = \Delta \cdot H_{Orel} = \Delta \cdot H_O / H_{Omax} .$$

Table 1. Information analysis of k=2 rules. The table shows all calculations for determining the rule redundancies and distributivities. We also show the number of canalyzing inputs, which are in close agreement with distributivity for k=2.

rule							"xor"		"and"				"if A"		"or"	
rule no.	0	1	2	3	4	5	6	7	8	9	10	11	12	13	14	15
Hs, Ms, and Us																
H(A)	1.00	1.00	1.00	1.00	1.00	1.00	1.00	1.00	1.00	1.00	1.00	1.00	1.00	1.00	1.00	1.00
H(B)	1.00	1.00	1.00	1.00	1.00	1.00	1.00	1.00	1.00	1.00	1.00	1.00	1.00	1.00	1.00	1.00
H(C)	0.00	0.81	0.81	1.00	0.81	1.00	1.00	0.81	0.81	1.00	1.00	0.81	1.00	0.81	0.81	0.00
H(A,B)	2.00	2.00	2.00	2.00	2.00	2.00	2.00	2.00	2.00	2.00	2.00	2.00	2.00	2.00	2.00	2.00
H(A,C)	1.00	1.50	1.50	1.00	1.50	2.00	2.00	1.50	1.50	2.00	2.00	1.50	1.00	1.50	1.50	1.00
H(B,C)	1.00	1.50	1.50	2.00	1.50	1.00	2.00	1.50	1.50	2.00	1.00	1.50	2.00	1.50	1.50	1.00
H(A,B,C)	2.00	2.00	2.00	2.00	2.00	2.00	2.00	2.00	2.00	2.00	2.00	2.00	2.00	2.00	2.00	2.00
M(A,B)	0.00	0.00	0.00	0.00	0.00	0.00	0.00	0.00	0.00	0.00	0.00	0.00	0.00	0.00	0.00	0.00
U(A,C)=M(A,C)	0.00	0.31	0.31	1.00	0.31	0.00	0.00	0.31	0.31	0.00	0.00	0.31	1.00	0.31	0.31	0.00
U(B,C)=M(B,C)	0.00	0.31	0.31	0.00	0.31	1.00	0.00	0.31	0.31	0.00	1.00	0.31	0.00	0.31	0.31	0.00
U(A,B,C)= -M(A,B,C)	0.00	0.19	0.19	0.00	0.19	0.00	1.00	0.19	0.19	1.00	0.00	0.19	0.00	0.19	0.19	0.00
M([A,B],C)=sum of Us	0.00	0.81	0.81	1.00	0.81	1.00	1.00	0.81	0.81	1.00	1.00	0.81	1.00	0.81	0.81	0.00
relative contribution of Us to output																
P(A,C)= U(A,C)/H(C)	0.00	0.38	0.38	1.00	0.38	0.00	0.00	0.38	0.38	0.00	0.00	0.38	1.00	0.38	0.38	0.00
P(B,C)= U(B,C)/H(C)	0.00	0.38	0.38	0.00	0.38	1.00	0.00	0.38	0.38	0.00	1.00	0.38	0.00	0.38	0.38	0.00
P([A,B],C)= U(A,B,C)/H(C)	0.00	0.23	0.23	0.00	0.23	0.00	1.00	0.23	0.23	1.00	0.00	0.23	0.00	0.23	0.23	0.00
Redundancy																
input redundancy: 1-Hp/Hpmax	1.00	0.02	0.02	1.00	0.02	1.00	1.00	0.02	0.02	1.00	1.00	0.02	1.00	0.02	0.02	1.00
relative Houtput: H(C)/H(C)max	0.00	0.81	0.81	1.00	0.81	1.00	1.00	0.81	0.81	1.00	1.00	0.81	1.00	0.81	0.81	0.00
output redundancy: 1-H(C)/H(C)max	1.00	0.19	0.19	0.00	0.19	0.00	0.00	0.19	0.19	0.00	0.00	0.19	0.00	0.19	0.19	1.00
effective k of rule	0	2	2	1	2	1	2	2	2	2	1	2	1	2	2	0
Distributivity																
Hp= -sum(Pi*logPi)	0.00	1.55	1.55	0.00	1.55	0.00	0.00	1.55	1.55	0.00	0.00	1.55	0.00	1.55	1.55	0.00
Δ	1.00	2.93	2.93	1.00	2.93	1.00	1.00	2.93	2.93	1.00	1.00	2.93	1.00	2.93	2.93	1.00
Δeff	0.00	2.38	2.38	1.00	2.38	1.00	1.00	2.38	2.38	1.00	1.00	2.38	1.00	2.38	2.38	0.00
no. of canalyzing inputs	0	2	2	1	2	1	0	2	2	0	1	2	1	2	2	0

ANALYSIS OF K=3 RULES

Let us now consider the case of k=3 rules in which the states of the inputs A, B and C determine the output, D. The goal again is to decompose the information contributions of the inputs and their combinations with respect to the output. For each k=3 rule, $M([A,B,C],D)=H(D)$, which can be confirmed according to the definition of $M([A,B,C],D)$ in terms of the contributing entropies:

$$M([A,B,C],D)=H(A,B,C)+H(D)-H(A,B,C,D).$$

By systematically substituting the definitions derived for the k=2 rules above, we find an expression for $M([A,B,C],D)$ in terms of the individual contributing mutual information measures (algebraic details of substitutions not shown):

$$M([A,B,C],D)=M([B,C],D)+M([A,B],D)+M([A,C],D)-M(A,D)-M(B,D)-M(C,D)+K$$

Let us define K as $M(A,B,C,D)$, and describe it in terms of Hs following another series of substitutions according to the above definitions of the Ms:

$$M(A,B,C,D)=H(A)+H(B)+H(C)+H(D)-H(A,B)-H(A,C)-H(A,D)-H(B,C)-H(B,D)-H(C,D)+H(A,B,C)+H(A,B,D)+H(A,C,D)+H(B,C,D)-H(A,B,C,D)$$

We can now formulate $M([A,B,C],D)$ completely in terms of contributing Ms:

$$M([A,B,C],D)=M(A,D)+M(B,D)+M(C,D)-M(A,B,D)-M(B,C,D)-M(A,C,D)+M(A,B,C,D)$$

Finally, we seek to define $M([A,B,C],D)$ in terms of the *unique mutual informations* (see above), $U(X,Y)=M(X,Y)$, $U(X,Y,Z)=-M(X,Y,Z)$ and $U(W,X,Y,Z)=M(W,X,Y,Z)$,

$$M([A,B,C],D)=U(A,D)+U(B,D)+U(C,D)+U(A,B,D)+U(B,C,D)+U(A,C,D)+U(A,B,C,D)$$

Note: While U(X,Y,Z)=-M(X,Y,Z), U(W,X,Y,Z)=M(W,X,Y,Z), i.e. there is no difference in sign between U(W,X,Y,Z) and M(W,X,Y,Z). Why? Consider complete redundancy, then M([A,B,C],D) = M(A,B,C,D). But also consider k=3 rule 150, for which all inputs have to be known to gain any knowledge of the output (Fig. 3). Then U(A,D)+U(B,D)+U(C,D)+U(A,B,D)+U(B,C,D)+U(A,C,D)=0, and therefore,

M([A,B,C],D)=M(A,B,C,D)=U(A,B,C,D).

First of all, and as above, using information measures we should be able to easily identify inputs that bear no influence on the output for k=3 rules. These must logically exist in rule tables which consider all possible combinations of inputs. Simply put, if the conditional entropy of X given Y equals the entropy of X, X tells us nothing about Y (X=inputs, Y=outputs). In case of k=3, we must determine if H(A|[B,C,D])=H(A), i.e. H(A)=H(A,B,C,D)-H(B,C,D), or if M([B,C,D],A)=0, for all input elements A, B, and C. If this holds true, then that particular input is redundant. For k=3 rules, there are 218 rules of k_{eff}=3, 30 of k_{eff}=2, 6 of k_{eff}=1 and 2 of k_{eff}=0. Having identified which rules of a particular k effectively depend on all of the k inputs, we can and should limit the construction of model networks to rules of an "effective k".

Since the complete k_{eff}=3 rule table consisting of 218 entries is too large to display, we shall discuss several interesting examples to illustrate their information parameters (Fig. 3). We will also consider the relationship between canalyzation and the information parameters. An input for a rule is canalyzing (Kauffman, 1984), if one of its states is always linked to one of the output states, allowing it to overrule the contributions of the other inputs.

For k=3, there are two analogies to the k=2 *exclusive or* rule, in terms of input configurations, or in terms of information content. For k=2, we found that the *exclusive or* rule required knowledge of all input elements to determine the output, i.e. the individual input Us were zero, resulting in a distributivity of unity. From an information standpoint, the equivalent rule for k=3 is rule 150 (Fig. 3, a). However, in this case in addition to the *exclusive or* input configurations coding for *on* in the output, the all *on* configuration also corresponds to the output being *on*. In case of the k=3 rule 22, in which the activity of strictly only one input leads to output activation, distributivity is 4.64, meaning that knowledge of individual or pair-wise combinations of the inputs gives us information about the output, as reflected in the constituent Us in Fig. 3, b. The differences between rules 150 and 22 are also reflected in their output and input redundancies.

A high distributivity may be expected for the *or* rule (no. 254; Fig. 3, c), since each input should make an equal contribution to the output. Indeed, Δ =5.23 for this rule. Note that while the input redundancy is low (due to high Δ), this rule exhibits a relatively high output redundancy of 0.46, therefore lowering the effective distributivity to 2.84. Here we see that high canalyzation on 3 inputs sacrifices effective distributivity.

One may ask how the "ideal" rule (if we can define such a thing) should be optimized with respect to input and output redundancies. Let us now consider the rule, *if at least two inputs are on*, essentially a "majority rule" (no. 232; Fig. 3, d). Of all k=3 rules, this rule (and its 7 relatives; not shown) exhibits the highest observed distributivity and effective distributivity of 6.66, a close approximation to $Δ_{max}$=7. Furthermore, this rule also shows minimal output and input redundancies. From an information coding standpoint, rules like no. 232 would appear favorable. However, this rule has no canalyzing inputs. One may ask which consideration is more important for biological rules, optimal information transmission as measured by the effective distributivity, or canalyzation?

Let us consider two rules that show some unusual properties. For the rule *if A or (B and C)*, no. 248 (Fig. 3, e), we observe an intermediate Δ=3.67, and low output redundancy. Note, however, that P(A,B,C)<0. We attribute this to the fact that in case of three elements, M or U may be negative depending on whether there is redundancy or unique information contributions by pair-wise combinations (see discussion above for k=2). Nevertheless, all the Ps add up to unity without altering any of their signs. We circumvent the problem of a negative P for the entropy calculation by allowing a negative sign before the log, and using the absolute value of P for the log,

H_P= - Σ P log (abs P) .

Rule # 150	input redundancy	1.00
	output redundancy	0.00
	Δ	1.00
	Δeff	1.00

A B C	D	canalizing inputs	0
0 0 0	0		
0 0 1	1	P(A)	0.00
0 1 0	1	P(B)	0.00
0 1 1	0	P(C)	0.00
1 0 0	1	P(AB)	0.00
1 0 1	0	P(AC)	0.00
1 1 0	0	P(BC)	0.00
1 1 1	1	P(ABC)	1.00

Rule # 22	input redundancy	0.21
	output redundancy	0.05
	Δ	4.64
	Δeff	4.43

A B C	D	canalizing inputs	0
0 0 0	0		
0 0 1	1	P(A)	0.05
0 1 0	1	P(B)	0.05
0 1 1	0	P(C)	0.05
1 0 0	1	P(AB)	0.11
1 0 1	0	P(AC)	0.11
1 1 0	0	P(BC)	0.11
1 1 1	0	P(ABC)	0.51

Rule # 254	input redundancy	0.15
	output redundancy	0.46
	Δ	5.23
	Δeff	2.84

A B C	D	canalizing inputs	3
0 0 0	0		
0 0 1	1	P(A)	0.25
0 1 0	1	P(B)	0.25
0 1 1	1	P(C)	0.25
1 0 0	1	P(AB)	0.03
1 0 1	1	P(AC)	0.03
1 1 0	1	P(BC)	0.03
1 1 1	1	P(ABC)	0.14

Rule # 232	input redundancy	0.03
	output redundancy	0.00
	Δ	6.66
	Δeff	6.66

A B C	D	canalizing inputs	0
0 0 0	0		
0 0 1	0	P(A)	0.19
0 1 0	0	P(B)	0.19
0 1 1	1	P(C)	0.19
1 0 0	0	P(AB)	0.12
1 0 1	1	P(AC)	0.12
1 1 0	1	P(BC)	0.12
1 1 1	1	P(ABC)	0.07

Rule # 248	input redundancy	0.33
	output redundancy	0.05
	Δ	3.67
	Δeff	3.51

A B C	D	canalizing inputs	1
0 0 0	0		
0 0 1	1	P(A)	0.57
0 1 0	0	P(B)	0.05
0 1 1	1	P(C)	0.05
1 0 0	1	P(AB)	0.11
1 0 1	1	P(AC)	0.11
1 1 0	1	P(BC)	0.11
1 1 1	1	P(ABC)	-0.01

Rule # 24	input redundancy	0.58
	output redundancy	0.19
	Δ	2.26
	Δeff	1.84

A B C	D	canalizing inputs	0
0 0 0	0		
0 0 1	1	P(A)	0.00
0 1 0	0	P(B)	0.00
0 1 1	1	P(C)	0.00
1 0 0	1	P(AB)	0.38
1 0 1	0	P(AC)	0.38
1 1 0	0	P(BC)	0.38
1 1 1	0	P(ABC)	-0.15

Figure 3. Information analysis for selected rule tables for k_{eff}=3. Elements A, B and C provide the inputs for determining the output, D. **a**, rule 150 (parity rule - no unique information contributions by individual and channel pairs); **b**, rule 22 (only one input); **c**, rule 254 (or); **d**, rule 232 (majority - at least 2 must be on); **e**, rule 248 (A or [B and C]); **f**, rule 24 (A exclusive or [B and C]).

This does not violate the consistency of our definitions.

Lastly, there is the rule *A exclusive or (B and C)*, which exhibits non-optimal output and input redundancies and a negative P(A,B,C). In this case (Fig. 3, f), Δ =2.26. Considering that this rule has an *exclusive or* component, which effectively results in a loss of combinatorial unique information sources (the Ps for individual elements are all zero; also discussed for k=2 above), only the remaining four information sources are left for coding. Indeed, these remaining information sources are optimized with respect to distributivity according to their respective Ps (Fig. 3, f).

In summary, for all k_{eff}=3 rules, we observe 9 distinct values of Δ (the number of occurrences of each shown in parentheses): 1.00 (2), 2.3 (8), 2.9 (24), 3.3 (48), 3.7 (48),

Figure 4. Example basin of attraction graph for n=15, k_{eff}=3 network of Δ=1. Attractor period, 2047 states; basin of attraction, 16376 states.

3.9 (48), 4.7 (16), 5.2 (16), 6.7 (8). Moreover, it is clear that distributivity and canalyzation do not go hand in hand. It remains to be determined what the respective roles of these properties are in the implementation of biological signal integration rules.

GLOBAL DYNAMICS OF BOOLEAN NETWORKS GIVEN $K_{EFF}=3$ RULES OF DEFINED Δ

We have defined *distributivity* as a measure of how optimally the information determining the output is spread across the input information sources, and provided an analysis of k=2 and k=3 rule tables. We are now concerned with the role distributivity may play in determining Boolean network "global dynamics", which we will examine using k=3 network examples. By global dynamics we refer to general features such as the structures of the networks' *trajectories* and *attractors*, and related statistical parameters (Wuensche, 1992; Wuensche, 1993).

Since a Boolean network has a limited number of possible states (2^N), and each state maps to exactly one resultant state, a *trajectory* (discussed above) must eventually lead to a repeating series of states, which is referred to as an *attractor* (Somogyi and Sniegoski, 1996; Kauffman, 1993; Wuensche, 1993). The number of states in the attractor itself corresponds to its cycle, which can be quite small in terms of the percentage of the total network states. However, an attractor may be reached from a very large number of different states, all of which constitute its *basin of attraction*. Basins of attraction can be displayed graphically (Figs. 4 and 5) by representing each state as a dot or vertex that is connected to its resultant state (or preceding state) by a line or edge. Each line leading from the perimeter of the graph to its center (the cycling attractor), represents a particular time series of states or *trajectory*. For each of the attractors shown, one state is highlighted as a bit-pattern.

We have constructed several randomly wired Boolean networks of n=15 elements (total no. of states: $2^{15}= 32768$), using rules of varying distributivities. Note the differences of the Basin of attraction structures shown in (Figs. 4 and 5). In case of $\Delta=1$ (Fig. 4), we observe an extremely long attractor period of 2047 states, which is typical of all $\Delta=1$

Figure 5. Example basin of attraction graph for n=15, $k_{eff}=3$ network of $\Delta=6.7$. Attractor period. 9 states: basin of attraction. 7651 states.

281

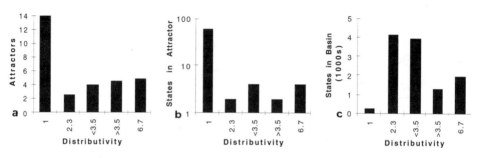

Figure 6. Analysis of ensembles (8 examples each) of n=15, k_{eff}=3 nets of varying distributivities. a, median number of attractors over each ensemble; **b**, median number attractor states or cycle; **c**, median number of states in basin of attraction.

examples we have studied (Fig. 6). Such networks would generally be characterized as "chaotic", even though the structure shown here is very symmetrical; we attribute this (somewhat misleading) symmetry to the fact that there are only two $\Delta=1$ rules for $k_{eff}=3$, which would suggest repetitive behavior. For $\Delta=6.7$, maximal distributivity, our example (Fig. 5) shows a basin of attraction consisting of 7651 states, leading to a much shorter 9 state attractor. Clearly, $\Delta=6.7$ corresponds to features much more closely analogous to what we infer about biological networks.

To establish general trends in the relationship between distributivity and network global dynamics, we compared 3 different measures (Fig. 6); a) the median number of attractors, b) median number of states in the attractor (cycle length), and c) median number of states in the basin of attraction, for 5 ensembles of 8 random $k_{eff}=3$, n=15 Boolean nets. Rules were chosen at random among $\Delta=1$, $\Delta=2.3$, $\Delta<3.5$, $\Delta>3.5$, and $\Delta=6.7$, respectively, for each ensemble.

Networks of $\Delta=1$ clearly stand out, showing maximal median number of attractors, attractor cycle lengths and the lowest median number of states in the basin of attraction. These extreme features are usually associated with "disordered" networks, unlike what we expect for biological networks (Kauffman, 1993). Regarding the median number of attractors, there is a steep drop on the transition of $\Delta=1$ to $\Delta=2.3$, and a steady upward trend from $\Delta=2.3$ to $\Delta=6.7$. It remains to be seen whether this upward trend is generalizable beyond the studied ensembles, perhaps even for networks of k>3. The median number of states in the attractor shows no trend beyond the strikingly high value for $\Delta=1$ networks. Although the *average* number of states in the basin of attraction for each ensemble of networks is strictly determined by the number of attractors, examination of the *median* number of basin states also depends on their distribution. Indeed, beyond the predictably low median number of states per basin for $\Delta<3.5$ (200; many attractors), we observe a significantly higher value for $\Delta=2.3$ and mixed $\Delta<3.5$, and an in-between value of 1000-2000 for $\Delta=6.7$ and mixed $\Delta>3.5$. Is there a phase transition at $\Delta=3.5$ for the distribution of basin of attraction size? It appears that the global dynamics of the studied $k_{eff}=3$ networks are to a degree determined by the distributivity of the chosen rules. Further analysis of additional stability parameters and networks of higher k (rule constraints play a more important role for networks of k>3; Kauffman, 1993) may provide deeper insights on the relationship between network stability and diversity.

DISCUSSION

The terms of *input* and *output redundancy* and *effective distributivity* suggest that these parameters could be targets for network optimization, at least in terms of information coding. We may ask how important efficient information coding is for "good" biological networks. Other rule parameters have been explored, canalyzation (Kauffman, 1984) and the P parameter (Kauffman, 1969), that are associated with network output patterns that display a degree of order similar to what we see in genetic networks. We have shown that canalyzation is correlated with optimal information transmission for only k=2; for k=3 high canalization sacrifices the relative output entropy and effective distributivity. However, the P parameter, which measures the amount that the probability of finding an element in the *off* or *on* state deviates from 0.5, is obviously closely related to the redundancy of each element. We must now determine which roles efficient information processing and canalyzation play in biological signaling networks, and how their somewhat contradictory relationship can be reconciled.

While we have applied our information measures to simple examples of discrete networks, our motivation is in part to establish analytical tools that can be applied generally. We are now using information decomposition in developing algorithms for *reverse engineering*, i.e. inferring Boolean network architectures solely from information on trajectories (Somogyi et al., 1996; Liang, Fuhrman and Somogyi, 1998). Of course, the principles of information theory are not restricted to discrete data sets, but find important applications for continuous data sources (Shannon and Weaver, 1963). While the transition from discrete to continuous states may be rather straightforward, the main challenge in the study of biological network data (e.g. large-scale gene expression time series) may lie in the definition of input vs. output. While input and output are clearly delimited by a single discrete time step in Boolean nets, functional time intervals vary across orders of magnitude in biological feedback networks. These problems need to be addressed satisfactorily before conducting an information analysis of biological data sets.

REFERENCES

Kauffman, S.A., 1969, Metabolic stability and epigenesis in randomly constructed genetic nets, *J. Theoretical Biol.* 22:437.

Kauffman, S.A., 1984, Emergent properties in random complex automata, *Physica D* 10:J45.

Kauffman, S.A., 1993, *The Origins of Order, Self-Organization and Selection in Evolution,* Oxford University Press.

Liang S, Fuhrman S, Somogyi R, 1998, REVEAL, A general reverse engineering algorithm for inference of genetic network architectures. *Proceedings of the Pacific Symposium on Biocomputing 1998, in press*

Shannon, C.E. and Weaver, W, 1963, *The Mathematical Theory of Communication,* University of Illinois Press.

Somogyi R, Sniegoski CA, 1996, Modeling the complexity of genetic networks: understanding multigenic and pleiotropic regulation, *Complexity* 1(6):45.

Somogyi R, Fuhrman S, Askenazi M, Wuensche A, 1996, The gene expression matrix: towards the extraction of genetic network architectures, *Proc. of the Second World Congress of Nonlinear Analysts (WCNA96).* Elsevier Science, *in press.*

Wuensche, A., Lesser, M.J., 1992, *The Global Dynamics of Cellular Automata,* SFI Studies in the Sciences of Complexity, volume 1, Addison Wesley

Wuensche, A., 1993, The ghost in the machine: basins of attraction in random boolean networks, in Artificial Life III, ed. Langton, C.G., 465-501.

Wuensche, A., 1995, Discrete Dynamics Lab (DDLAB), software and documentation at MIT Press Artificial Life Online (alife.santafe.edu/alife/software/ddlab.html).

GENET DATABASE AS A TOOL FOR ANALYSIS OF REGULATORY GENETIC NETWORKS

Alexander V. Spirov[1,2] and Maria G. Samsonova[1]

[1]Institute of High-Performance Computing and Data Bases
[2]The Sechenov Institute of Evolutionary Physiology and Biochemistry
St Petersburg, Russia

INTRODUCTION

The large-scale projects on human and several model organisms DNA sequencing lead to rapid growth of biological information. The well known *Genbank* and *SwissProt* databases contain essentially *structural* information. It is now necessary to design databases containing *functional* information that is focused on specific problems of molecular biology. These will contain a broad spectrum of information including pictures, schemes and movies. The distinctive feature of these databases will be their ability to serve not only as an information depository but also as tools for derivation of new knowledge by means of computer analysis.

One of such specialised data bases is the *GeNet* database located on site: *http://www.csa.ru/Inst/gorb_dep/inbios/genet/genet.htm*, which is designed in the Bio-information Systems Lab of our Institute.

GeNet contains the information on functional organisation of regulatory genetic networks acting at embryogenesis. The regulatory genes play a crucial role in embryo-genesis, controlling both activity of downstream regulatory genes (crossregulation), as well as their own activity (autoregulation). Most of genes of the network encode a transcription factors, which function is activation or repression of downstream target genes. In turn down-activated members of network switch on structural genes at the appropriate time and place. Thus the network of regulatory genes defines the genome activity during embryo development, and to reveal the mechanisms of embryogenesis it is necessary to understand the principles of regulatory genetic networks organisation.

The hypertext version of *GeNet* is build on the basis of comparative evolutionary approach. The information in the database is presented in several categories: genes entries, regulatory regions entries, gene interactions entries, bibliography entries and graphical representation of gene interactions. Each gene entry contains as mandatory such fields as definition, expression pattern, sequence, regulatory regions, regulatory connections (upstream and downstream genes), evolutionary homologues, links to other databases, bibliography. The expression pattern field involves the images of expression pattern of segmentation genes in fruit fly *Drosophila melanogaster*. The work on incorporation in

GeNet quantitative data on segmentation genes expression proceeds in collaboration with Dr. J.Reinitz (Brookdale Center of Molecular Biology, Mt.Sinai School of Medicine), whose group obtains gene expression data in experiments.

The regulatory element entry in *GeNet* contains such obligatory fields as definition, keywords, organism, bibliography, sequence and co-ordinates of sites for transcription factors binding. Gene interactions entries contain the target or effector gene name, mechanism of interaction, experimental proofs and bibliography. Graphical representation of gene interactions is accomplished in the form of flow diagrams, consisting of nodes and arrows, as well as in the form of the Java applets (e.g., *http://www.csa.ru/Inst/gorb_dep/inbios/genet/ Graph/genes.html*) that permits to emphasise the interacting genes in the network, to reflect the mode of genes action, to drag genes for better visualization of links between them, etc.

Using information collected in *GeNet* we present the results of analysis of genetic networks controlling early development in fruit fly *Drosophila*.

GENETIC NETWORKS CONTROLLING EARLY DEVELOPMENT IN DROSOPHILA

The information in *GeNet* provides better understanding of molecular mechanisms of regulation of genes activity at embryogenesis. We use this information for to reveal the principles of organisation of regulatory regions of genes controlling early morphogenesis stages in *D. melanogaster*, for to describe and analyse gap and pair-rule genetic networks, as well as for investigation of evolutionary conservation of these networks organisation.

In brief, initial steps of embryo development in every organism can be described as follows (Lawrence and Morata, 1994; Jackle et al., 1992). In fertilised zygote there are maternally predetermined concentration gradients of several morphogens, which exponentially decrease with distance. The global challenge in early development is a conversion of this analogue input into digital one. This conversion is fulfilled by the system of macromolecular devices which is formed by genes controlling morphogenesis. The central role in this system play the coupled complexes of general transcription factors with promoters as well as specific transcription factors with *cis*-regulatory regions (enhancers and silencers). This of a sort molecular probes co-operatively ("all - none") respond to exceeding threshold concentrations of transcription factors (Johnson and Krasnow, 1992). It is believed that such molecular probes can recognise at least two-fold change in transcription factor concentration (Schulz and Tautz, 1994). Essentially that the conversion of input is accomplished not by a single gene, but by the cascade of interacting genes.

Organisation of Regulatory Regions Controlling Early Morphogenesis

The main function of genes controlling early morphogenesis stages (See legends to Figures 1 and 2) is "reading" and translation of maternal morphogen gradients into the parasegmental organization of early embryo (Driever and Nusslein-Volhard, 1988; Driever et al., 1989). Reading of maternal morphogen gradients is accomplished by high sensitivity of transcription initiation complexes to morphogens concentration thresholds. In turn this sensitivity is based on a multitude of maternal morphogen binding sites in the *cis*-regulatory regions of each gene, as well as on their differential affinity (Driever et al., 1989: Jiang and Levine, 1993; Simpson-Brose et al., 1994; Stanojevic et al., 1989). As the result each region along embryo's anterior-posterior and dorsoventral axis is characterised by a specific set of expressed genes.

Structure of Enhancers, Reading the Morphogen Gradients. There are the gap and the pair-rule genes which read the gradients of maternal morphogens *BCD* and *HB* along anterior-posterior axis. The binding sites of these transcription factors are found in quantity in many enhancers of gap and pair-rule genes and are substantially differentiated with respect to affinity. Table 1 shows the properties of 12 the most thoroughly characterised cis-regulatory elements of 4 gap, 2 pair-rule and 3 selector genes. It appeared that each cis-regulatory region contains on the average 15 - 20 binding sites for 5 transcription factors. The density of binding sites distribution is very high as one site falls within the region of 20 - 100 bp.

Table 1. Enhancers of *Drosophila* segmentation genes.

Regulatory element	Length of element (in bp)	BCD		HB		KR		KNI		GT		CAD		TLL	
hb anterior	731	6	+	2	+	2	-								
hb posterior	>1400	p	-	5	±									8	+
Kr730	730	6	+	10	+			1	-	6	-			7	-
kni-enhancer	~1000	4	+	10	-	6	-			1	-	12	+	8	-
tll-enhancer	350	8	+												
eve stripe#2	~800	5	+	3	+	6	-			3	-				
eve stripe#3+7	508			11	-			5	-						
h stripe#5	302					4	-	p	+	p	-				
h stripe#6	205			p	-	7	-	8	+						
en 5'element	100			2	+	1	-								
Ubx: PBX	254			3	-										
Ubx: BRE	500			5	-									p	-

[1]p means "putative" binding sites, '+' activation, '-' repression.

The transcription factor *HB* controls the greatest number of regulatory elements - 11. Transcription factors *BCD* and *KR* regulate 7 elements each. Homeodomain protein *BCD* in 5 instances from 7 acts as activator, while *Kr* containing Zn-finger as a DNA-binding domain produces in all cases repression. *HB* protein which is highly homologous to *KR* acts just as activator so also repressor. Nuclear receptors *KNI* and *TLL* act in the same way. Homeoproteins (*BCD, FTZ* and *CAD*) appear more often to act as activators. The similarity of *Kr* and *kni* genes enhancers in the set of binding sites is evident.

In the whole regulatory elements presented in Table 1 contain 161 experimentally characterized binding sites for 9 transcription factors. More detailed analysis of regulatory elements sequences shows that 52 sites form 18 overlapping clusters. Among them 10 clusters consist of 2 overlapping sites, in 7 such cases one of sites binds activator, the other - repressor. Three clusters are formed by the overlap of 3 sites. The rest 5 clusters are composed from 4 or 5 partly overlapping sites.

Thus, summing up the data presented above, it is possible to conclude that regulatory elements of genes controlling early embryogenesis show striking similarity of their structure and function:

• they are compact, functionally autonomous regulatory elements, which fit the classical definition of an enhancer, as their regulatory action as a rule does not depend on location;

- these enhancers are about several hundreds base pairs in length;
- each segmentation gene enhancer contains binding sites both for activators and repressors;
- usually each element contains several copies of binding site for transcription factor given, which often differ in affinity;
- repressor binding sites are located adjacent or partly overlap with activator binding sites;
- the transcription factor may act for some enhancers as activator and as repressor for another.

Networks of Genes, Controlling Early Development in Drosophila

The networks of genes controlling embryo development in fruit fly *D.melanogaster* are among the most thoroughly studied (Jackle et al., 1992). Gentic cascades, each member of which controls expression of downstream target genes, are followed in Drosophila from maternal genes acting at oogenesis to genes controlling development of imago. These networks encompass up to hundred known at present genes, which code for transcription factors, transmembrane receptors and their ligands (Casares and Sanchez-Herrero 1995; Driever et al., 1989; Hoch et al., 1992; Margolis et al., 1995; Pignoni et al., 1992; Rivera-Pomar et al., 1995; Small et al., 1991; 1992; Pankratz et al., 1992). Despite of them, it is possible now to reconstruct the complete scheme of genetic network functional organization only for genes controlling early stages of development.

Gap Genetic Network. Gap genes act on the initial stage of the process of conversion of smooth exponential gradients of maternal morphogens along anterior-posterior and dorsoventral zygotic axes into discreet succession of embryonic para-segments, further pair-rule and selector genes take part in this process.

Gap genetic network (Figure 1) is formed by a small number of genes encoding transcription factors, which belong to Zn-finger, steroid-retinoid receptor and homeo-protein superfamilies. Gap genes activation and repression is triggered by exceeding the threshold concentrations of maternal morphogens, each gene being characterized by individual threshold values. These morphogens are transmembrane receptor *TOR*, which forms concentration gradients at the embryo termini, as well as *BCD* and *HB*, *CAD* and *NOS*, *DL* proteins, which gradients decrease correspondingly in the anterior-posterior, posterior-anterior and dorsoventral directions (Casanova and Struhl, 1993; Driever and Nusslein-Volhard, 1988; Pignoni et al., 1992; Simpson-Brose et al., 1994). It is noteworthy that practically all morphogens (*BCD*, *HB*, *DL*) may activate some genes and repress the other. It is also essential that several gap genes (*Kr* and *hb*) are autoregulated.

At the same time members of gap genetic network does not form closed activation circuits, which are characterised by interactivation of genes (eg. gene *A* activates gene *B*, *B* activates *C*, while *C* is the activator of *A*). It should be mentioned that such closed activation circuits are present in genetic networks controlling later stages of embryonic development.

Even cursory examination of gap genetic network scheme reveals that each gene is involved in many regulatory interactions and that negative regulatory links dominate (Table 1). At most thoroughly characterised genes *hb*, *Kr*, *kni* and *tll* regulate about from 7 to 10 other genes each, on average only two of them being activated. Such structure of genetic network leads to activation of no more than two gap genes in each particular region of sincytial blastoderm, which activity inhibits expression of other genes from this group.

We shall also underline now that terminal and trunk gap genes at blastoderm stage have usually two bands of expression, one of them located closer to the anterior, the other -

to the posterior end of the embryo. Gene *Kr* is an exception, as its band of expression divides early embryo approximately on anterior and posterior halves. In some genes (eg. *hb*) each expression band is regulated by autonomous enhancer, analysis of other gene regulatory regions up till now does not reveal such one-to-one correspondence (e.g., it seems that one regulatory element controls both bands of *kni* gene expression).

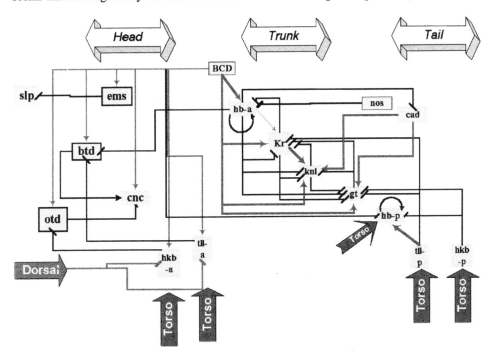

Figure 1. Genetic network defining gap gene expression domains in head, trunk and tail regions of the blastoderm embryo. Maternal morphogen *BCD* activates both head gap genes *ems, btd, cnc* and *otd*, and trunk gap genes *hb* (anterior element), *Kr, kni* and *gt*. Head gap gene *btd* activates, in turn, another member of this group *cnc*, while *otd* represses it. Maternal and zygotic *CAD* activates *kni* and *gt*. *hb* activates and represses *Kr* (in concentration-dependent manner) and represses the anterior limits of both *kni* and *gt* expression, while *Kr* activates *kni* and represses *gt*. *Torso* cascade activates terminal gap genes *tll* and *hkb*, as well as anterior element of the *hb* gene and head genes *otd* and *cnc*. Maternal morphogen *DL* represses *tll* and *hkb* in the anterior pole. Arrows indicate positive regulation; lines ending with a vertical bar designate negative inputs. "**-a**" marks regulatory elements, controlling anterior bands of gene expression; "**-p**" marks regulatory elements for posterior bands of expression.

Pair-rule Genetic Network. The next step in conversion of maternal morphogen gradients into segmental organisation of early embryo is accomplished by pair-rule genetic network (Reinitz and Sharp, 1995; Small et al., 1991; 1992). The primary pair-rule genes are activated firstly as the result of maternal morphogens and gap genes products action. The activity of primary pair-rule genes is necessary for activation and maintenance of secondary and tertiary pair-rule genes. Pair-rule genes express in a series of seven stripes along anterior-posterior axes of early embryo.

Several distinctive characteristics of gap genetic network are more distinctive in the case of pair-rule genetic network. As in the case of the gap genetic network the most thoroughly studied primary pair-rule genes *even-skipped (eve)* and *hairy (h)* are involved in many regulatory interactions with other genes. The activity of these genes is regulated by maternal morphogens, gap genes as well as by pair-rule genes themselves (Figure 2).

Altogether the total number of regulatory interactions may be as much as 12, with only 2 or 3 being activations.

As in the case of several gap genes each stripe of primary pair-rule genes expression is controlled by autonomous enhancer (Table 1). Each enhancer contains binding sites both for activators and repressors. A great number of regulatory interactions in which pair-rule genes are involved may be ascribed to the necessity of control of each of these genes expression in seven different embryos regions, which differ in concentrations of maternal morphogens and gap gene products.

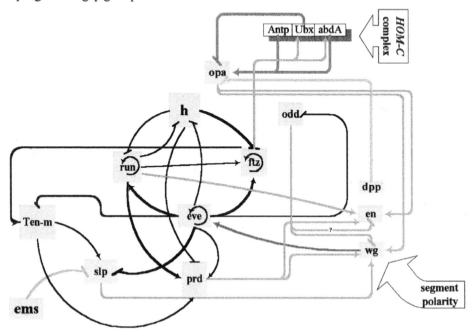

Figure 2. Functional relations of members of genetic network controlling pair-rule genes expression patterns in the blastoderm of early embryo. Initially the expression of primary pair-rule genes is controlled by maternal morphogen *BCD* an maternally and zygoticaly expressed *HB* and *CAD*. Pair-rule gene cross-regulation exhibits hierarchical structure prior to gastrulation. One pair-rule gene *eve* is not regulated by other pair-rule genes. Two others, *h* and *run*, are regulated by *eve*, but not by other pair-rule genes. These three are known as primary pair-rule genes. Another pair-rule gene, *ftz*, is regulated by *eve*, *h* and *run* but not by other pair-rule genes. *ftz* and *odd* are considered as secondary pair-rule genes. *prd* and *slp* are regulated by all other pair-rule genes and therefore are considered to be tertiary pair-rule genes. In turn pair-rule genes control expression of segment polarity (*wg*, *en*) and homeotic (*Antp*, *Ubx*, *AbdA*) genes.

Several secondary pair-rule genes (eg. *ftz*) show lack of multiple enhancers each controlling separate stripe of expression pattern and seems to be under control of more limited genes number.

It should be pointed out that autoregulation is typical of pair-rule genes. *eve* gene autoregulatory element is one of the most thoroughly studied among autoregulatory elements in Drosophila (Small et al., 1991).

Analysis of Gene Ensembles by Means of Boolean Networks Theory

The study of Boolean network models by S. Kauffman with co-workers over last thirty years leads to formulation of fundamental features of genetic networks global behaviour (Kauffman, 1993). On the other hand at present a wealth of experimental data permits us to understand in details the functioning of genetic networks controlling early development in *Drosophila*. Thus we can interpret these data in form of Boolean network models for comparison of behaviour of theoretical networks and the real biological ones.

A Boolean network is a system of interconnected binary elements and any element in the network can be connected to a series of other elements. Each individual element uses a logical (Boolean) rule to compute its value based on the values of the other elements it is connected to. The state of the system is defined by the pattern of ON/OFF states of all of its elements (For Introduction See Somogyi and Sniegoski, 1996).

One of the key features of the Boolean networks is that all states, i.e., ON/OFF pattern of its elements at a particular time point, lead to or are the part of an attractor. The attractor is a distributed structure, based on the state (a point attractor) or series of states (a dynamic attractor) which repeats on itself. All the states leading to or being a part of this attractor, constitute the basin of attraction.

Each network must reach one of several possible attractors depending on the initial conditions. Any state within a particular basin of attraction can be switched to any other state within this basin, without changing the global characteristics of the system. Despite the general resistance of attractors to point alterations in states, the boundaries of each basin of attraction must have at least one state, in which a single point alteration will determine which basin the system will fall into.

Let us illustrate the prospects of modelling of *Drosophila* segmentation networks in framework of Boolean model. Considering the data of *GeNet* database we formulate the Boolean rules for segmentation genetic networks as follows: action of at least one repressor switches off a target gene, while activation by at least one activator switches the target gene on, provided that a repressor is absent.

We perform the analysis of segmentation genetic networks by means of Discrete Dynamics Lab package developed by Dr. Wunsche.

Modelling of Head Segmentation Genetic network. The head segmentation genetic network consists of 3 genes encoding maternal morphogenes, *Hb*, which is both maternal and zygotic gene and is presented in network by its anterior enchancer hb_a, 5 genes encoding head gap genes, 2 anterior enchancers tll_a and hkb_a of terminal genes and trunk gene *Kr* (Figure 1).

It should be pointed out that correct modelling of expression pattern of head segmentation network in framework of Boolean model requires elimination of negative link from gene *Kr* to *slp*, as its presence leads to inactivation *slp* in all attractors. However, the inclusion *Kr* in head segmentation genetic network is mandatory, as elimination of this gene profoundly alters a whole picture of network dynamics as compared with experimental data.

The basin of attraction for all possible states of head segmentation genetic network consists of 9 point attractors (Table 2).

It turns out that some of these attractors may be correlated with definite region of embryo head on blastoderm stage. The 1st attractor "all off" reflects the trivial fact that in the absence of maternal morphogens inputs all elements of genetic network will be switched off. The other attractors are characterised by combined activity of head gap genes and morphogens.

Attractor III is formed by *cnc* and terminal gap genes, while attractor IX is characterized by activation *Hb* in addition to these genes. Earliest expression *cnc* in the late blastoderm is found in *labial* and *mandibular* parasegments (Grossniklaus et al., 1994; Mohler et al., 1995; Pignoni et al., 1992). Hence attractors III and IX may correspond to these parasegments.

ems alone and in combination with *DL* locally represses *slp* in two adjacent regions.

The so-called ventral repression depends on *DL* in conjunction with *ems*, while splitting of initially single band *slp* expression in two bands is under control *ems* only (Grossniklaus et al., 1994). As we can see, the attractor VII, characterised by activity *btd* and *ems*, but not *slp* could correspond to one of these two regions. On the other hand, attractor VI and VIII, formed by *otd*, *ems*, *cnc*, *Kr*, *BCD* and *DL* genes may correspond to ventral part of prospective *ocular* parasegment, where *slp* activity is repressed by *ems* and high *DL* concentration (Cf. Mohler et al., 1995; Walldorf and Gehring, 1992).

Attractor V, characterized by activation *slp*, *ems*, *btd*, *hkb* and *Kr* genes in presence *BCD* may correspond to the region of combined expression *slp*, *ems* and *btd* genes, which defines the morphogenesis of *intercalary* segment (Grossniklaus et al., 1994).

At last, the most anterior domains in a head region are marked by *otd* gene expression (Finkelstein and Perrimon, 1990). Attractor IV may mark this zone. Data of mutation analysis suggest that domain solely expressing *otd* does not correspond to any head segment.

Table 2. Basin of attraction field of *Drosophila* head-gap genetic network.

No	Switched-on genetic elements	per-cents from all possible states	Number of layers
I	-	12.5	4
II	DL	12.5	4
III	cnc, tll$_a$, hkb$_a$, TOR	10.9	2
IV	cnc, otd, DL, TOR	12.5	4
V	slp, ems, btd, hkb$_a$, Kr, BCD	12.5	5
VI	cnc, otd, ems, Kr, BCD, DL	12.5	3
VII	btd, ems, hkb$_a$, Kr, BCD, TOR	12.5	5
VIII	cnc, otd, ems, Kr, BCD, DL, TOR	12.5	3
IX	cnc, hb$_a$, tll$_a$, hkb$_a$, TOR	1.6	2

Thus, in spite of imperfect knowledge on mechanisms of *cis-*, *trans-* regulation of head gap genes activity, we get a reasonable correspondence of attractors to the regions of cell fate determination in the limits of at least four head segments.

S. Kauffman investigated the characteristics of Boolean nets that correlate with the appearance of spontaneous order, namely, the connectivity and the homogeneity. Order arises for networks with connectivity k=2 or in networks characterised by higher connectivity and the parameter P value, which measures the internal homogeneity, greater then some critical value P_c.

The evaluation of head segmentation genetic network parameters shows that

the average value of P is equal to 0.65,

each element of network has in average 2.5 functionally linked neighbours,

totally, the number of positive links is nearly equal to the number of negative links (17 and 13 correspondingly),

one more feature of the net is the low branching level for all attractors (maximum 5).

Moreover, clearly defined hierarchical structure is inherent to head segmentation genetic network: the initial input of 4 morphogens activates 9 target elements. These downstream elements are interconnected by a number of negative links.

Analysis of head segmentation genetic network stability towards removal of elements confirms the hierarchy of network organisation. The morphogens removal (i.e. *BCD*) decrease the attractor number to 5, while 3 of them are new. On the contrary the removal of downstream targets of morphogens action (the 2nd level of network organisation) does not substantially change attractors number, but modifies their structure. For example *Kr* gene exclusion leads to elimination of the 4 and appearance of the 7 new attractors, while 5 other attractors do not change. Thus the resistance of head segmentation genetic network to elimination of elements depends on weight of their input into network organisation.

In comparison with the head segmentation genetic network the random Boolean network, characterised by the same neighbourhood k=2.5 but P value equal to 0.5, reaches quite different attractors. Hence, this comparison shows the peculiar features of head segmentation network. The basis for these features is in the fact that the head segmentation genetic network was formed in the process of evolution over ten millions of years (Patel, 1994).

CONCLUSIONS

It becomes evident now that a cell function cannot be understood in terms of "one gene - one function" paradigm. For to solve this problem it is necessary to understand how genetic networks operate.

To this end it is necessary to perform large-scale measuring of gene expressions spectra, to design databases containing gene expression data and to develop program tools to infer the information on genetic networks functioning on the basis of expression databases data.

The *GeNet* database is intended as a data base of new type oriented on representation of results of analysis of genetic networks structure, function and evolution. Now it contains the images of expression pattern of segmentation genes in fruit fly *D.melanogaster*. The inclusion of these images is the first step towards design of quantitative atlas of *Drosophila* segmentation genes expression, containing both 3D and numerical data.

The information collected in *GeNet* can be used for derivation of new knowledge on genetic networks functioning by means of computer analysis.

The head segmentation network is relatively autonomous part of *Drosophila* embryo segmentation network. At present most of functionally essential regulatory links in these networks are revealed. Most of them consist in interaction of *trans*-acting proteins with *cis*-regulatory regions.

Good understanding of mechanisms of gene interactions permits to model the information processing in head segmentation network by means of Boolean network theory.

The unique feature of segmentation genetic networks is the presence of several maternal morphogens, which are outputs of other genetic networks. Thus in basin of attraction for all possible states all combination of morphogens can be switched on. However only attractors with natural combination of morphogens reflect the actual pattern of genes activation in networks.

Boolean models of segmentation networks substantially differ from random Boolean network. These networks have hierarchical structure. They are sensitive to elimination of elements. Moreover the degree of disturbance elicited by elements elimination depends on the weight of element in network organisation.

Acknowledgments

Supported by Russian Foundation for Basic Researches (Grant No 96-04-49350).

REFERENCES

Casanova, J. and Struhl, G., 1993, The torso receptor localizes as well as transduces the spatial signal specifying terminal body pattern in Drosophila, *Nature* 362:152.

Casares, F. and Sanchez-Herrero, E., 1995, Regulation of the infraabdominal regions of the bithorax complex of Drosophila by gap genes, *Development* 121:1855.

Driever, W. and Nьsslein-Volhard, C., 1988, A gradient of bicoid protein in Drosophila embryos, *Cell* 54:83.

Driever, W., Thoma, G. and Nusslein-Volhard, C., 1989, Determination of spatial domains of zygotic gene expression in the Drosophila embryo by the affinity of binding sites for the bicoid morphogen, *Nature* 340:363.

Finkelstein, R. and Perrimon, N., 1990, The orthodenticle gene is regulated by bicoid and torso and specifies Drosophila head development, *Nature* 346:485.

Grossniklaus, U., Cadigan, K.M., Gehring, W.J., 1994, Three maternal coordinate systems cooperate in the patterning of the Drosophila head, *Development* 120:3155.

Hoch, M., Gerwin, N., Taubert, H. and Jackle, H., 1992, Competition for overlapping sites in the regulatory region of the Drosophila gene Kruppel, *Science* 256:94.

Jiang, J., and Levine, M., 1993, Binding affinities and cooperative interactions with bHLH activators delimit threshold responses to the dorsal gradient morphogen, *Cell* 72:741.

Jackle, H,. Hoch, M., Pankratz, M.J., Gerwin, N., Sauer, F. and Bronner, G., 1992, Transcriptional control by Drosophila gap genes, *J Cell Sci* Suppl 16:39.

Johnson, F.B. and Krasnow, M.A., 1992, Differential regulation of transcription preinitiation complex assembly by activator and repressor homeo domain proteins, *Genes Dev* 6:2177.

Kauffman, S.A., 1993, *The Origins of Order, Self-Organization and Selection in Evolution*, Oxford University Press, New York.

Lawrence, P.A. and Morata, G., 1994, Homeobox Genes: Their Function in Drosophila Segmentation and Pattern Formation, *Cell* 78:181.

Mohler, J., Mahaffey, J.W., Deutsch, E., and Vani, K., 1995, Control of Drosophila head segment identity by the bZIP homeotic gene cnc, *Development* 121: 237.

Margolis, J. S., Borowsky, M. L., Steingrimsson, E., Shim, C. W., Lengyel, J. A. and Posakony, J. W., 1995, Posterior stripe expression of hunchback is driven from two promoters by a common enhancer element, *Development* 121: 3067.

Patel, N.H., 1994, Developmental evolution: Insights from studies of insect segmentation, *Science* 266:581.

Pignoni, F., Steingrimsson, E. and Lengyel, J.A., 1992, bicoid and the terminal system activate tailless expression in the early Drosophila embryo, *Development* 115:239.

Reinitz, J. and Sharp, D.H., 1995, Mechanism of eve stripe formation, *Mech Dev* 49:133.

Rivera-Pomar, R., Lu, X., Perrimon, N., Taubert, H. and Jakle, H., 1995, Activation of posterior gap gene expression in the Drosophila blastoderm, *Nature* 376:253.

Schulz, C. and Tautz, D., 1994, Autonomous concentration-dependent activation and repression of Kruppel by hunchback in the Drosophila embryo, *Development* 120:3043.

Simpson-Brose, M., Treisman, J. and Desplan, C., 1994, Synergy between the hunchback and bicoid morphogens is required for anterior patterning in Drosophila, *Cell* 78:855.

Small, S., Kraut, R., Hoey, T., Warrior, R. and Levine, M., 1991, Transcriptional regulation of a pair-rule stripe in Drosophila, *Genes Dev* 5:827.

Small, S., Blair, A. and Levine, M., 1992, Regulation of even-skipped stripe 2 in the Drosophila embryo, *EMBO J* 11:4047.

Somogyi, R. and Sniegorski, A., 1996, Modelling the complexity of genetic networks: understanding multigenic and pleiotropic regulation, *Complexity* 1:45.

Stanojevic, D., Hoey, T. and Levine, M., 1989, Sequence-specific DNA-binding activities of the gap proteins encoded by hunchback and Kruppel in Drosophila, *Nature* 341:331.

Pankratz, M. J., Busch, M., Hoch, M., Seifert, E. and Jackle, H., 1992, Spatial control of the gap gene knirps in the Drosophila embryo by posterior morphogen system, *Science* 255:986.

Walldorf, U. and Gehring, W. J., 1992, empty spiracles, a gap gene containing a homeobox involved in Drosophila head development, *EMBO J* 11:2247.

CELLULAR-AUTOMATA-LIKE SIMULATIONS
OF DYNAMIC NEURAL FIELDS

Jörg Wellner[1] and Andreas Schierwagen[2]

[1]Chemnitz University of Technology
Department of Computer Science
09107 Chemnitz, FRG

[2]University of Leipzig
Department of Computer Science
Augustusplatz 10/11
04109 Leipzig, FRG

1 INTRODUCTION

In modelling neuronal phenomena one can generally choose two different methods (Levine, 1991): One way is to setup a network of discrete model neurons and all their interconnection weights, which is usually done in the so-called PDP approach (Rumelhart and McClelland, 1986; McClelland and Rumelhart, 1986). The main interest lies on the learning capabilities of the established network, i. e. on the correct adaptations of the connection weights, in order to generate the correct output for a given input. Another approach to neuronal modelling is using continuous networks (so-called fields) with special attention to the spatio-temporal activities of the network. The number of neurons is unlimited in these models, and the connections between the neurons are handled in a general way (e. g. statistically) without having individually changing weights. Therefore, instead of being interested in the learning mechanisms of the network one investigates the various dynamics of the fields. It is a convenient way to describe the evolution equations of the field by integro-differential equations (IDEs).

There are three most important states of activity in biological neural networks: a state of rest, an excited state, and a refractory state. The latter one means that a neuron being in this state cannot immediately become excited even when the excitatory input is strong enought. In most of the common neuronal (discrete) networks of the PDP approach this fact is neglected.

One important class of continuous networks are dynamic neural fields (DNFs). DNFs were first introduced by (Amari, 1977). Similar formulations are made by (Amari and Arbib, 1977) and (Masterov et al., 1989). This paper is motivated by two aims. Firstly, we want to describe a method for simulations of two-dimensional DNFs and secondly, we show their use to two different problems in computational neuroscience.

Amari (Amari, 1977) investigated a model of DNFs as a set of IDEs and found solutions for the one-dimensional spatial case. However, most spatio-temporal computations of mammalian brains involve more than one spatial dimension. Therefore we present a method to simulate two-dimensional DNFs which is based on cellular automata (CA) simulations as used for reaction-diffusion systems. The simulations offer a powerful way to investigate qualitatively the dynamics of the spatio-temporal patterns described by the IDEs. Further they can be used in applications employing biologically motivated space-time computations.

In the next section we look shortly at the results of Amari's work. Then we give details of our own simulations for two-dimensional DNFs. In Section 4 we discuss the usefulness of our method. In Section 5 we exemplify the approach by applying it to two problems in computational neuroscience. The last section ends with some conclusions.

2 DYNAMIC NEURAL FIELDS

Usually one regards DNFs as spatially distributed populations of model neurons which are connected in a random manner (*cf.* (Amari, 1977)). Being more precise one can distinguish different layers of neurons of the same type. The average connectivity function is of the lateral inhibition type.

Although one wants to model a huge number of neurons, that would be desirable e. g. for models of cortices, one cannot consider complete models, due to the limited computing resources. Therefore, one includes properties of groups of neurons, i. e. one assigns the average membrane potential of a group of neurons to one model element.

Let us consider m two-dimensional neuronal layers fully connected. The average membrane potential of the group of neurons in the ith layer at position $\mathbf{x} = (x_1, x_2)$ and time t is represented by $u_i(\mathbf{x}, t)$. The general field equations (1) incorporate a sigmoid activation function f_i, a (usually negative) value h_i as a resting potential, an external independent stimulation function $s_i(\mathbf{x}, t)$, and a connectivity function $w_{ij}(\mathbf{x})$ which gives the strenght of the influence of neurons from layer j to the i-th layer at place $\mathbf{x}$:

$$\tau_i \frac{\partial u_i(\mathbf{x}, t)}{\partial t} = -u_i(\mathbf{x}, t) + \sum_{j=1}^{m} \iint_{R^2} w_{ij}(\mathbf{x}, \mathbf{x}^*) f_j[u_j(\mathbf{x}^*, t)] \, d\mathbf{x}^* + h_i + s_i(\mathbf{x}, t) . \qquad (1)$$

The linear term in the ith field equation defines the time scale τ_i. From a mathematical point of view it is clear that one cannot analyse this system of equations for general functions f, w and s. In a first step Amari made the following simplifications: The sigmoid functions f_i are simplified to a step function:

$$f[u] = \begin{cases} 1, & \text{if } u > 0 \\ 0, & \text{if } u \leq 0, \end{cases} \qquad (2)$$

which means, that the field at a certain place is only for $u > 0$ in the excited state[1].

The connectivity function w is supposed to be symmetrical and homogeneous, i. e. $w(\mathbf{x}, \mathbf{x}^*) = w(|\mathbf{x} - \mathbf{x}^*|)$. There are no further constraints on w. The stimulation function

---

[1]In (Kishimoto and Amari, 1979), Amari showed that the following results hold also for smooth sigmoid functions.

s is kept constant. Although these constraints simplify the investigations it is however still an difficult task to analyse the system due to the great complexity of (1) and different initial values of the fields u_i. The simplest case is an one-dimensional field consisting of one layer.

For the case of one neural layer,

$$\tau \frac{\partial u(x,t)}{\partial t} = -u + \int w(x,x^*) f[u(x^*)]\, dx^* + h + s(x,t), \tag{3}$$

Amari proved the existence of five types of pattern dynamics:

- monostable field in which all excitations will die out

- monostable field which is entirely excited

- (explosive type) bistable field in which localised excitations up to a certain range spread without limit over the entire field, but vanish if the range of the excitation area is to narrow

- bistable field in which initial excitations either become localised excitations of a definite length or die out; localised excitations move in direction to the maximum of the input s

- fields showing spatially periodic excitation patterns depending on the average stimulation level.

The type of dynamics of a field depends mainly on the connectivity function w. For a Mexican-hat type connectivity function, also known as difference of two gaussians (Fig. 1), one assumes that excitatory connections dominate for proximate neurons, and inhibitory connections dominate at greater distances. This type of network is also known as a lateral inhibition type (see e.g. (Levine, 1991)). In particularly, the positive range of $w(x)$ along the x-axis determines the length of excited ranges of $u(x)$.

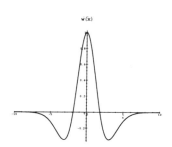

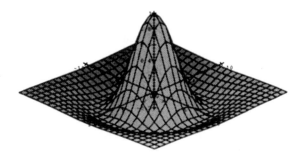

Figure 1: The one-dimensional connectivity function $w(x)$.

Figure 2: The connectivity function $w(\mathbf{x})$ used in single-layered DNFs of two dimensions.

The complexity of dynamics increases if one adds a second layer to the field u. Apart from the types mentioned for the case of one layer one can further detect oscillatory patterns and traveling waves. Amari simplified (1) in order to cope with the analysis of both layers in the following way: one layer, with a membrane potential u_1, is a layer consisting of only excitatory neurons, and the other layer, with membrane potential u_2, consists only of inhibitory neurons. Moreover, one can restrict the connections so that the inhibitory neurons only inhibit the excitatory neurons and the excitatory neurons have very narrow fan-out connections to the inhibitory neurons. The latter can mean that the excitatory neurons at place x activate the inhibitory neurons at place x only. Thus, the field equations turn out to be:

$$\tau_1 \frac{\partial u_1(x,t)}{\partial t} = -u_1(x) + \int w_1(x,x^*) f[u_1(x^*,t)]\, dx^*$$

$$- \int w_2(x,x^*) f[u_2(x^*,t)]\, dx^* + h_1 + s_1(x,t)$$

$$\tau_2 \frac{\partial u_2(x,t)}{\partial t} = -u_2(x) + w_3 f[u_1(x,t)] + h_2 + s_2(x,t). \tag{4}$$

Note that in Equations (4) the functions w_1 and w_2 are now of the gaussian type (realizing a Mexican-hat) and w_3 is a constant.

A travelling wave across the field can be established as follows: Suppose a localised excitation in u_1 at position x and an localised excitation in u_2 at position $x + \epsilon$. The value of ϵ depends on the significant range of w_1 and w_2. The excitation in u_2 prevents in u_1 the propagation of the excitation in both of the two directions[2]. Therefore, the excitation in u_1 moves in the opposite direction, compared to the placement of ϵ. If the maximum of the first excitation is moved far enought the excitation in u_2 will inhibit the previously excited regions in u_1. The excited region in u_2 will follow the travelling wave in u_1. This process without external stimuli s keeps on indefinitely.

Here we cannot discuss in greater detail the analysis made by Amari (but see also (Schierwagen and Werner, 1996)). However, for the two-dimensional case of Equations (3) and (4) there are so far no analytical results. There are two major ways to overcome this situation. On the one hand, one can calculate solutions of certain cases of (1) numerically in a direct fashion. On the other hand, one can simulate the evolutionary equations. We decided to take the second way, because direct numerical calculations may still be too complex or are not very suitable to indicate the temporal evolution of the dynamic patterns.

3 SIMULATIONS OF DNF'S IN TWO SPATIAL DIMENSIONS

If one looks for solutions of the DNF equations for the two-dimensional case one may suppose to find in general the same types of dynamics as Amari found for the one-dimensional fields. In this section we reproduce different dynamic patterns of two-dimensional DNFs. This is mainly done by displaying excitation patterns of neuronal layers at different discrete moments of time.

The Simulation Method

In our approach, we have adopted the way of CA-simulations for reaction-diffusion systems (see e. g. the work of (Gerhardt and Schuster, 1989; Gerhardt et al., 1990a; Gerhardt et al., 1990b; Gerhardt et al., 1990c; Gerhardt et al., 1991) and the papers by (Weimar et al., 1992a; Weimar et al., 1992b)). In these methods, time and space of the equations were discretised in order to find a limited number of possible states of the medium. In contrast to common cellular automata they had to choose larger neighbourhoods with different weights. This is more realistic because diffusion is not

---

[2]An excitation in the layer u_2 indicates that the field at this position is in the refractory state.

restricted to the nearest neighbours, and, on the other hand, the amount of diffusion decreases with greater distance.

A more straightforward approach is to rewrite the evolutionary equations as a set of difference equations. Then, one can use for the field variable u numerical values of a certain range. The convolution term $\int w(x)f[u(x)]dx$ will be treated as a sum: $\sum_{k=-r}^{k=r} w(k)f[u(x+k)]$. For example, the two-dimensional Equation (5) can be rewritten as a difference equation (6) in the above sense:

$$\tau \frac{\partial u(\mathbf{x}, t)}{\partial t} = -u(\mathbf{x}, t) + \iint_{R^2} w(\mathbf{x}, \mathbf{x}^*) f[u(\mathbf{x}^*, t)] \, d\mathbf{x}^* + h + s(\mathbf{x}, t) \tag{5}$$

$$u_{ij}^{t+1} = u_{ij}^t + \frac{-u_{ij}^t + \sum\limits_{k=-r}^{r} \sum\limits_{l=-r}^{r} \left(w_{kl} f[u_{i+k,j+l}^t] \right) + h + s_{ij}^t}{\Delta t \tau}. \tag{6}$$

Note that in Equation (6) the lower index indicates a position in space and the upper index is a discrete moment in time. Of great interest will be the value of r, which is the range of the neighbourhood. If r is relatively small, then the calculations will be faster compared to greater values of r. Ideally, r should be large (ranging over the entire field) to ensure, that the difference equations are as similar as possible to the differential equations. Of course, by doing the simulations one should find a good compromise: the value of r can be decreased as long as the qualitative behaviour of the dynamics doesn't change significantly. Equation (6) can be easily generalised to express the difference form of Equations (4). But one should be aware that the calculations for the neighbourhood for more than one layer are more time consuming.

Before we look in detail at the simulation results we address a last question in this respect: How one should deal with the boundary of a field. There are two major methods of traditional CA-simulations. Either one thinks of the field as a circle (in one dimension) or as a torus (in two dimensions). Or, on the other hand, one applies the zero-flux boundary condition, which states that there will be no influence to border cells from the outside. For all the following simulations we used the second point of view, because the constraints of this situation are biologically more plausible.

Single-Layered DNFs

Our field consists of 100x100 discrete points, stored in a matrix $\mathbf{U}$. The connectivity function is stored in another matrix $\mathbf{W}$, and the input in matrix $\mathbf{S}$. The values for the elements of $\mathbf{W}$ are generally calculated by the function $w(\mathbf{x}) = W(x_1, x_2)$:

$$\begin{aligned} W(x_1, x_2) &= a_1 * exp(-(x_1)^2/\sigma_1 - (x_2)^2/\sigma_1) - \\ &\quad a_2 * exp(-(x_1)^2/\sigma_2 - (x_2)^2/\sigma_2), \end{aligned} \tag{7}$$

which is a two-dimensional Mexican-hat (Fig. 2). The range for x_1 and x_2 depends on r (see Equation (6)).

In the following figures[3] we show some pattern formation processes of different dynamic types. Qualitatively, in the two-dimensional case of one-layered DNFs one finds the same types of pattern dynamics as in the one-dimensional case (*cf.* (Schierwagen and Werner, 1996)). In Fig. 3 we have two initial excitations of different diameter. The larger one dies out whereas the smaller one survives due to the nearly same sized

---

[3]For all pictures we have the following colour conventions: a medium gray shows the field in the resting state (about $u = h$), a darker gray indicates refractory regions, and sites of a light gray are in the excited state.

Figure 3: An example of an initial excitation which dies out (on top of the pictures) and one which becomes stable (on bottom of the pictures). The value of r is 12, the excited region of the lower activity is about 4 space units wide. The pictures are snapshots after $t = 0, 1, 3, 10$ time steps from the left to the right.

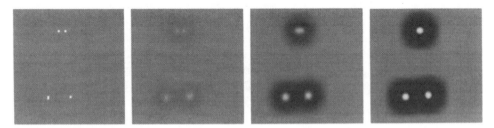

Figure 4: An example of converging (on top of the pictures) and coexisting (on bottom of the pictures) excitations. The value of r is 16, the radius of the positive region of the kernel w is about 7 space units wide. The pictures are snapshots after $t = 0, 1, 3, 33$ time steps.

positive (excited) region as determined by the connectivity function w. Note, the two excitations don't influence each other, because r is small. One interesting feature of DNFs is the support of converging and coexisting excitations (Fig. 4). This becomes important in modelling short term memory functions of the brain by layered neuronal tissue. A richer dynamic emerges if one gets rid of the homogenity of the field, e. g. by introducing asymmetric connectivity functions as shown in Section 5.

Two-Layered DNFs

As mentioned in Section 2, DNFs of two layers support travelling waves. Spiral waves are very stable examples of them and are of great interest in investigating active media (see e. g. (Mikhailov, 1990)). If one sets up the field by appropriate initial conditions (as described in Section 2) a spiral wave will emerge (Fig. 5).

Waves in two-dimensional excitable media ((Wiener and Rosenblueth, 1946) were the first who modelled excitable media by cellular automata) have some important properties, among them are the curvature and dispersion relation. Without analysing these relations in detail for our model we like to mention that our simulation model

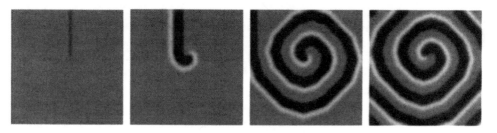

Figure 5: A spiral wave in a two-dimensional DNF with two layers. Only the u_1-layer (in Equations (4)) is shown in the pictures. The u_2-layer shows a similar dynamics with a small time lag. The value of r is 14. The pictures are snapshots after $t = 0, 4, 14, 34$ time units.

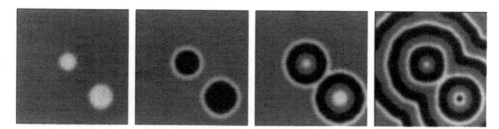

Figure 6: The curvature relation. The four shnapshots show a curvature effect ($t = 0, 2, 4, 14$). Here again only the u_1-layer is shown. The centre of each target wave is excited by a constant external stimulus.

supports both effects, which influence the propagation speed of a wave. The curvature relation states, that convex wave fronts travel slower than plane waves, and concave wave fronts travel faster than plane waves. The dispersion relation in excitable media holds, if the speed of waves increases with increasing wave length. In the pictures of Fig. 6 one can see the demonstration of the curvature relation, i. e. the part of the converging waves travels quicker (due to the convex wave front) as the rest of the wave.

4 SIMULATION RESULTS

One question one should ask is: How valueable are all the simulations of the IDEs? Answering this question rises some problems due to the missing analytical results, at least in the two-dimensional case. Of course, it is possible to compare the results of Amari for the one-dimensional DNFs with the results one gets in applying the simulation method to one-dimensional DNFs. In principle, one can say that qualitatively one finds the same types of pattern dynamics as Amari stated. However, a closer look at the quantitative level reveals some differences, some of them should be mentioned here. Because Amari assumed unlimited fields the simulations show sometimes unexpected patterns at the boundary. Amari determined certain conditions which produce the five types of dynamics. These conditions hold only for continuous fields. In simulations with discrete time and space we only can approximate these conditions. Nevertheless, we could produce the supposed behaviour of all types. A detailed analysis of this approach can be found in (Wellner, 1996).

Another way to find out something about the correctness of the simulations is to vary a few simulation parameters and then to compare the results of the simulations. If one can detect differences then the simulations are unstable and sensitive to simulation parameters, which would state that the results are of no great interest. As an example one can change the parameter r. From a certain value the enlarging of r should produce no significant difference. For example, in Fig. 1, the crucial part of the one-dimensional Mexican-hat function is between -7 to $+7$, i. e. the influence of cells which are placed further than 7 space units away can be neglected. For larger r the calculations need more time. Therefore, we have choosen always the smallest possible value for r. In the case of Fig. 1 it has been $r = 7$. Of course, a smaller r would narrow the range of inhibiting connections and the result of the experiments would be not meaningful.

There is one more crucial simulation parameter in our model. It is the product of the time constant τ and the time step Δt in Equation (6). In our simulations we assume $\Delta t = 1$ to be constant. It turned out that the denominator of Equation (6) should be greater than 1 in order to prevent the field from oscillating, i. e. $\tau > 1$. But, if one is interested in oscillations one should chose $\tau \leq 1$

The above discussion shows that the CA-like approach is worthwhile. Moreover, the method is of great advantage in studying the influence of different parameters whilst the simulation is running. A direct numerical analysing of DNFs with changing parameters or even with a time dependend input would be on the other hand very hard and time consuming, compared to our very simplyfied method.

5 APPLICATIONS

One of the neural organizations found in mammalian brains which are involved in spatio-temporal computations are so-called computational maps (CMs). CMs are a class of neural maps with certain functions. The computational properties of CMs vary with their spatial position. The position of a neuron in the map largely determines (1) which part of the input it receives, (2) how this input is processed, and (3) to what target the result eventually is transfered (see (Knudsen et al., 1987) for a discussion). CM discovered so far are mostly involved in processing sensory information and programming of movements (Knudsen et al., 1987). In motor maps, systematic variations of movement parameters (amplitude and direction) are represented topographically on the neural layer. The computational character of these maps is obvious: the topographically represented movement command must be transformed into spatio-temporal patterns of motoneuron activity, and the centre of activity on the map determines the features of the transformation.

An example of a CM is the motor map of the mammalian superior colliculus (SC), which major role is to control rapid gaze shifts (saccades), for a review see (Schierwagen, 1996). Recording studies of the cat's SC (Munoz et al., 1991) showed that during a saccade a hill of activity travels in the motor map from its initial location towards the fixation zone. The instantaneous hill location on the map specifies the remainig motor error. This dynamic can be modelled by using a two-dimensional DNF of one layer. A localiced excitation in that DNF can be thought of as a hill of activity. In order to realize a movement of such a hill we introduce asymmetric place-dependend connectivity functions w. Let $X_1 = X_1(\mathbf{p})$, $X_2 = X_2(\mathbf{p})$ with $\mathbf{p} = (p_1, p_2)$ being a position in space. The connectivity function (7) can be now written as:

$$w(x_1, x_2, X_1, X_2) = a_1 * exp(-(x_1 - X_1)^2/\sigma_1 - (x_2 - X_2)^2/\sigma_1) -$$
$$a_2 * exp(-(x_1 - X_1)^2/\sigma_2 - (x_2 - X_2)^2/\sigma_2). \qquad (8)$$

For $X_1 \neq 0$, $X_2 \neq 0$ the centre of the function (8) is replaced compared to the origin of the coordinate system (Fig. 2). We cannot discuss the model here in detail (for more information see (Schierwagen, 1996)), but the idea should be clear: A smooth variation of X_1 and X_2 in the model of the motor map would ensure, that every activity hill can travel in a predefined direction (here to the region representing the fovea).

So far, the connectivity function w has been kept fixed in time. The dynamic of excitation patterns will exhibit still more complex behaviour if we introduce a (global) time dependend connectivity function, i. e. $X_1 = X_1(t)$, $X_2 = X_2(t)$ in Equation (8). This change opens the way to another application, namely to model the process of dead reckoning. Dead reckoning, also termed path integration, is a navigation process which allows an animal to update its position (in relation to a point of reference) in an internal representation based on signals generated during locomotion (McNaughton et al., 1996). No further visual information which identifies landmarks is needed. The only necessary information is solely generated by movements. By means of this navigation process an animal is able to keep track of its position in the environment in relation to its starting point (or any other point of reference). The necessary information is stored in a geocentric coordinate system, i. e. the position of the animal and of the point of reference is kept in an earth-centered map. Thus, the animal is able to determine the direction of the starting point from its current point and can always move to it without any visual guidance.

Physiological experiments of the navigation system in rats have shown that certain neurons in the hippocampus fire when the rat is placed at a particular position in the environment, regardless of how it is oriented. Cells in the presubiculum fire when the rat's head is oriented in a specific direction regardless of where the rat is in space (McNaughton et al., 1996). This situation can be modelled by an two-dimensional dynamic neural field $u(\mathbf{x}, t)$ and a dynamic (place independend) connectivity function $w(x_1, x_2, X_1, X_2)$. The current position of the animal is indicated in u by a hill of activity (the exact position is the maximum of u, i. e. the top of the activity hill). If the animal moves in a specific direction with a particular velocity this hill will move also in the field in the same direction and a certain distance. These movements are specified by w, whereas w will be updated continuously as the animal moves on. Information of the speed and direction of the animal's movements has to be transformed to X_1 and X_2 in an appropriate manner.

6 CONCLUSION

The paper has described successful CA-like simulations of two-dimensional DNFs, consisting of one and two layers. The results were compared to the analytical investigations of Amari. It turned out, that the proposed method is of great use, if an analytical solution of the regarded IDEs is not available. Further we have argued, that our approach offers some advantages compared to direct numerical calculations, e. g. for the visual following of an ongoing simulation process with time-varying parameters. Particularly, we found the same types of pattern dynamics as Amari found for one-dimensional fields. Additionally, our results indicate that DNFs of two dimensions support target and spiral waves.

In the context of information processing in tissues we used the results to model two spatio-temporal transformation processes found in the mammalian brain. Further experiments, e. g. in application to autonomous mobile robots, have to elaborate and refine the models. An exhaustive search for different dynamic behaviours in two-dimensional DNFs has not been reported so far, at least not for more than one layer. Thus, simulations like those shown in this paper may help to analyse DNFs with greater complexity as Amari has regarded.

REFERENCES

Amari, S.-I., 1977, Dynamics of pattern formation in lateral-inhibition type neural fields. *Biological Cybernetics*, 27:77–87.

Amari, S.-I. and Arbib, M., 1977, Competition and cooperation in neural nets, in Metzler, J., ed., *Systems Neuroscience*, New York, Academic Press.

Gerhardt, M. and Schuster, H., 1989, *Physica D*, 36:209–221.

Gerhardt, M., Schuster, H., and Tyson, J., 1990a, *Science*, 247:1563–1566.

Gerhardt, M., Schuster, H., and Tyson, J., 1990b, *Physica D*, 46:392–415.

Gerhardt, M., Schuster, H., and Tyson, J., 1990c, *Physica D*, 46:416–426.

Gerhardt, M., Schuster, H., and Tyson, J., 1991, *Physica D*, 50:189–206.

Kishimoto, K. and Amari, S.-I., 1979, *J.Math.Biol.*, 7:303.

Knudsen, E. I., du Lac, S., and Esterly, S. D. ,1987, *Ann. Rev. Neurosci.*, 10:41–65.

Levine, D., 1991, *Introduction to Neural and Cognitive Modeling*, Lawrence Erlbaum Ass., Hillsdale, New Jersey.

Masterov, A. V., Tolkov, V. N., and Yakhno, V. G., 1989, Spatio-temporal structures in opto-electronic devices, in Gapanov-Grekhov, A. V., Rabinovich, M. I., and Engelbrecht, J., eds., *Nonlinear Waves 1: Dynamics and Evolution*, Berlin, Springer-Verlag.

McClelland, J. L. and Rumelhart, D. E., eds., 1986, *Parallel Distributed Processing, Vol II*, MIT Press, Cambridge.

McNaughton, B. L. et al., 1996, *Jour. Exp. Biology*, 199:173–185.

Mikhailov, A. S., 1990, *Foundations of Synergetics I*, Springer-Verlag, Berlin.

Munoz, D. P., Pelisson, D., and Guitton, D., 1991, *Science*, 251:1358–1360.

Rumelhart, D. E. and McClelland, J. L., eds., 1986, *Parallel Distributed Processing, Vol I*, MIT Press, Cambridge.

Schierwagen, A., 1996, Saccade control in active vision: Mapped neural field model of the collicular motor map, in Zangemeister, W. H., Stiehl, S., and Freska, C., eds., *Visual Attantion and Cognition*, pages 237–248, Amsterdam, Elsevier.

Schierwagen, A. and Werner, H., 1996, Analog computations with mapped neural fields, in Trappl, R., ed., *Cybernetics and Systems '96*, pages 1084–1089, Vienna, Austrian Society for Cybernetic Studies.

Weimar, J. R., Tyson, J. J., and Watson, L. T., 1992a, *Physica D*, 55:309–327.

Weimar, J. R., Tyson, J. J., and Watson, L. T., 1992b, *Physica D*, 55:328–339.

Wellner, J., 1996, *Distributed active media in analog neuronal information processing*, Department of Computer Science, University of Leipzig, Diploma Thesis (In german).

Wiener, N. and Rosenblueth, A., 1946, *Arch. Inst. Cardiol. Mexico*, 16:205–265.

TOWARDS A METABOLIC ROBOT CONTROL SYSTEM

Jens Ziegler, Peter Dittrich, and Wolfgang Banzhaf

Department of Computer Science
University of Dortmund
44221 Dortmund, Germany
email: ziegler, dittrich, banzhaf@ls11.informatik.uni-dortmund.de
http://ls11-www.informatik.uni-dortmund.de

INTRODUCTION

Bacteria must be able to detect rapid changes in their environment and to adapt their metabolism to external fluctuations. They monitor their surroundings with membrane-bound and intra-cellular sensors. The regulatory mechanisms of bacteria can be seen as a model for the design of robust control systems based on an artificial chemistry. The capability to process information which is needed to keep autonomous agents surviving in unknown environments is discussed.

Biological cells can be seen formally as information processing systems of high complexity [Lengeler, 1996]. They react to a large number of stimuli which are perceived by sensors. These mostly chemical stimuli are converted by *two-component-systems* into signals that are (i) collected, (ii) computed, (iii) integrated and (iv) transmitted through signal transduction pathways. Biochemical changes of proteins, protein-protein interactions and facilitated diffusion along ordered cell structures play a decisive role.

Cellular signal-transmitting systems are hierarchical and often require communication between hundreds of proteins. Membrane-bound two-component systems (see Fig. 1) provide the connection between the *sensory domain* and an attractant or repellant. This domain is connected via helical transmembrane protein structures with the *regulator domain*. Conformational changes in the sensory domain at the external surface induce changes in the relative positioning of the helical membrane-cytosol interface. For instance, binding of a ligand causes a change of the electrostatic density and hence a rotation that increases the distance between charged groups [Stock and Surette, 1996, Lengeler, 1995].

This reaction is transient, i.e. it ends even though the stimulus might continue. Formally, this adaption is achieved through release of a slow feedback reaction which inhibits the signal. Different pathways interlace and so a hierarchical cellular net-

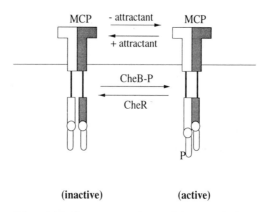

(inactive)   (active)

Figure 1. Ligand binding causes the modulation of the electrostatic properties of the helices and therefore the transition between **active** and **inactive** state.

work emerges with a high capacity to store information [Lengeler, 1995, Conrad, 1992]. The computational functions in such biochemical reaction networks have been analysed by [Arkin and Ross, 1994, Bray and Lay, 1994, Bray and Bourret, 1995] and are supposed to function akin to biochemical neuronal networks [Hjelmfelt et al., 1991, Hellingwerf et al., 1995, Okamoto et al., 1995].

Through encapsulation of correlated functions into compartments and the use of superimposed regulating mechanisms (see Fig. 2) the cell reduces the vast flow of information to a communication between a few clearly arranged functional blocks [Darnell et al., 1993, Maynard Smith and Szathmary, 1995].

The signal processing system of a cell is very robust in its dependence upon the observation of central metabolites and, on the other hand, on the hierarchical division into functional blocks. Therefore it can act as a model for the construction of robust, highly parallel and distributed control systems.

CELLS AS INFORMATION PROCESSING SYSTEMS

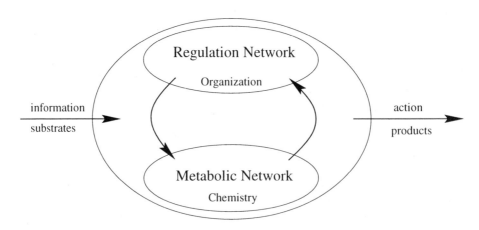

Figure 2. The metabolic and the regulatory network of a cell.

The metabolism of a cell can be divided into two levels: (i) the metabolic network, where all bio-chemical reactions occur and (ii) the regulatory network which controls the velocity of the reactions and supervises the transduction pathways (see Fig. 2).

All reaction pathways and all regulatory elements together form a kind of highly connected *molecular network*. Each element of this network is a collector of different stimuli (see Fig. 3).

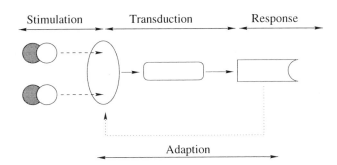

Figure 3. A node in the molecular network of the cell. Stimuli are collected and transmitted depending on the level of adaption to the specific stimulus. This node can be seen as a more general two-component system.

Signals are transmitted depending on the actual state of the adaptor. They are reduced to binary information: by molecules appearing in only two states: *activated* or *inactivated*. Usually these differ only in the absence or presence of a single atom or group at the molecular level. The regulatory network decides through slow or fast feedback or forward reactions which of the connections of the molecular network are activated or inhibited [Lengeler, 1995, Stanier et al., 1986].

One very interesting information processing system of a bacterium controls its movement. Thus, the rest of this section is devoted to a more detailed discussion of the information flow in the chemotaxis system. *Chemotaxis* is the ability of a bacterium to follow concentration gradients by modulation of tumbling frequency. (Fig. 4).

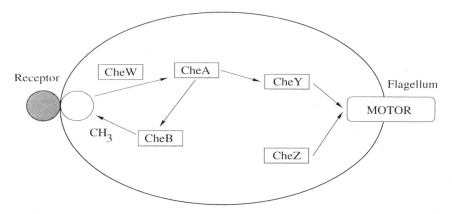

Figure 4. Information flow in the chemotaxis system of *Escherichia Coli*.

In *E. coli* there are six cytoplasmic transduction proteins: *CheA, CheB, CheR, CheW, CheY and CheZ* [Stock and Surette, 1996]. The increase of CheA, for example, indicates the presence of a carbon source (glucose, fructose, etc.). CheA phosphorylates CheY which binds to the switch complex of the flagellar motor and CheB which closes the feedback loop to reset the receptor. The binding of CheY causes clockwise rotation or, to be more specific, it interrupts counterclockwise rotation. Clockwise rotation of the majority of the flagellar motors causes a re-orientation (*tumble*). In this way, the frequency of tumbling is influenced by the concentration of the carbon source.

The cell thus executes a biased random walk, caused by concentration gradients, towards a more favourable environment. It is assumed [Stock and Surette, 1996, Boos, 1977, Alt, 1994] that chemotactic responses are mediated by a *temporal*, rather than by a *spatial*, sensing mechanism, for the following reasons: (i) the length of a bacterium is too small and it moves too quickly to allow for spatial sensing. (ii) Bacteria even respond to sudden changes in concentrations when the concentration is spatially homogeneous. The decision whether to change the movement direction or not depends on the state of some key-metabolites and constitutes an inherent memory of the cell. For a detailed description of bacterial chemotaxis see for example [Bray et al., 1993, Jones and Aizawa, 1991, Lengeler, 1995, Stock and Surette, 1996].

A SPATIAL MODEL OF THE CELL

Information processing in living systems means transduction of substances or molecular interactions. The reception of a stimulus causes the transmission of substances to inner regions of the cell. They arrive either unchanged via diffusion along ordered cell structures or through a cascade of sequential biochemical reactions. Thus, a changing concentration level of X_i can indicate the reception of Y_i. Since chemical reactions take place at very high velocity the delay between encountering a specific substance and a significant change in concentration of the signalling substance is small. Response to an actual stimulus is very fast.

Compartments

Although a procaryotic bacterium appears to lack a cytoskeletton it must not be seen as a "bag full of enzymes" but as a highly ordered system of interacting molecules which are interconnected and connected to the inner membrane.

A model of spatially continuously distributed substances can be based on nonlinear *partial* differential equations. Replacing the spatial derivatives with finite differences divides the system into subsystems (see Fig. 5) which can be treated as homogeneous regions and modelled by a system of coupled *ordinary* differential equations [Jetschke, 1989]. These subsystems are called *compartments*. The state of a compartment is represented by a concentration vector. A realistic model of a complex biochemical system requires 10^2 - 10^5 compartments. In order to keep the simulation of the system tractable the number of compartments should be reduced. The first experiments with a real robot are performed with only one compartment. A model system of this kind can be called a (formal, algorithmic) *reactor*.

Metadynamics

Describing the dynamics of the reactor leads to a system of coupled differential

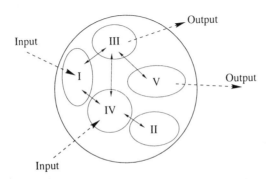

Figure 5. Spatial model of the cell. Homogeneous subsystems (*compartments*) are connected through diffusion flux. For further details see text.

equations. In order to avoid the number of reactions becoming infinite there must be a focusing on the most relevant ones. A real reactor only contains a finite number of substances and so a possible reaction can only take place when at least one single molecule is inside the reactor. Simulating the dynamics with differential equations, however, assumes a continuous model even with infinitesimal concentrations of substances.

Since a reactor in reality is not continuous but discrete there must be a concentration threshold representing the presence of a single molecule above which the substance is treated as a reactant. If its concentration be lower, it should not participate in the reactor's dynamics. Thus the system has two different time scales: a fast time scale of the concentration changes and a slower time scale for changing the number of participating substances. This change of the ODE system over time is called *metadynamics* [Bagley and Farmer, 1992]. The metadynamics is additionally influenced by diffusion fluxes from other compartments.

ARTIFICIAL REACTION SYSTEMS

Artificial Chemistry

In a real chemistry the kinetic parameters and efficiencies depend on thermodynamics, quantum mechanics and chemical composition of the molecules. For that reason it is impossible to calculate them exactly in practice. To circumvent the troublesome computation of approximations one can formulate an *artificial chemistry* (AC) [Bagley and Farmer, 1992, Fontana, 1992, Fontana, 1994]. An AC is not able to reproduce all properties of real chemistry but it can produce complex behaviour which can be studied instead. By changing the rules different chemistries can be instantiated and, adding layers of more realistic behaviours, it is possible to reproduce properties of the "real" chemistry.

A straightforward way for building a metabolic controller would be, first, to define an arbitrary AC and then use this chemistry to create a controller composed of certain substances. In the following sections we will define two artificial chemistries to show how the abilities of a metabolic controller are depending on its underlying AC. A more radical approach would be to set up a system with the metabolism emerging from the reactions [Banzhaf, 1994].

Example A: Artificial Polymer Chemistry

We begin our discussion with an artificial *polymer chemistry* which is an AC where only polymerisations reactions exists [Bagley and Farmer, 1992]. A substance is represented by a string $s = (s^1, s^2, \ldots s^n) \in A^n$ composed of characters from the alphabet A. The reaction mechanism is simply defined as the concatenation:

$$(s_i^1, s_i^2, \ldots s_i^{n_i}) + (s_j^1, s_j^2, \ldots s_j^{n_j}) \overset{s_k}{\rightleftharpoons} (s_i^1, s_i^2, \ldots s_i^{n_i}, s_j^1, s_j^2, \ldots s_j^{n_j}) \tag{1}$$

The reaction may be catalyzed by a string s_k with catalytic efficiency ν. The dynamic behaviour of the polymer chemistry should be demonstrated by the example shown in figure 6 with the alphabet $A = A, B$. The catalytic constants k_i' are set to $\nu = 10$. Polymerisation is ten times faster than de-polymerisation ($k_i = 1.0$, $k_{-i} = 0.1$).

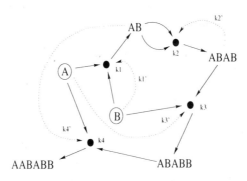

Figure 6. Example A: Reaction network for the polymer chemistry. The monomers A and B (food set) will be used as inputs. Dotted arrows indicate catalytic activity.

Details of the reaction mechanism and its system of ordinary differential equations (ODE) can be found in the Appendix, part A. The continuous influx flow $\Phi(A)$ and $\Phi(B)$ of the monomers A and B is defined as input. Fig. 7 shows the steady-state input-output relation of the example reaction network.

Example B: Artificial Enyzme-Substrate-Chemistry

The reaction network in Fig. 8 is an example based on enzyme-substrate kinetics. Reaction systems of this kind are referred to as chemical neuronal networks [Hjelmfelt et al., 1991, Okamoto et al., 1995]. The input is represented by the substances C and D, the output by T. The substances marked by "*" are held at a constant concentration level. The detailed reaction mechanism and its corresponding ODE model can be found in the Appendix, part B. Fig. 9 shows the steady-state concentrations of the output species T and several internal substances as a function of input C and D.

Comparison

The polymer chemistry shows a very smooth input-output surface in contrast to the sharp surface of the enzyme-substrate example. The smooth surface insinuates that

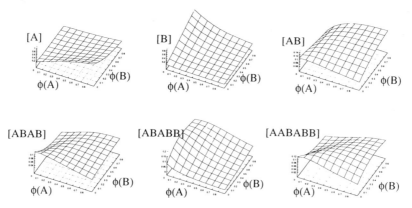

Figure 7. Example A: Steady-state input-output relation for the polymer chemistry (Fig. 6).

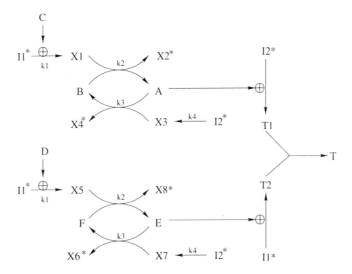

Figure 8. Example B: Reaction network based on enzyme-substrate kinetics. C and D are input species. T is the output. Substances marked by "*" are held constant.

input information is not processed. The state space which can be reached for different inputs is very large. The surface in state space can not be intuitively divided into different regions with different meanings. In the enzyme-substrate example the quasi-continous input information is condensed by mapping it on two regions in state space which can be clearly separated (Fig. 9). Given some concentrations of possible output-substances the input vector can be much more easily reconstructed than in example B. The robustness of the *classification* can be demonstrated by analysing the information passed from reaction network to reaction network. If these networks are considered to be identical the development of a small error in the first classification is shown in Fig. 10. Transfer functions like a) will reduce the classification to a simple *all-or-nothing*. A correct classification is achieved, even with a disturbed input signal. The course of concentrations in Fig. 8 is similar to a) and thus small fluctuations will be reduced. If the transfer function is of type b) or c) the error will be amplified and the

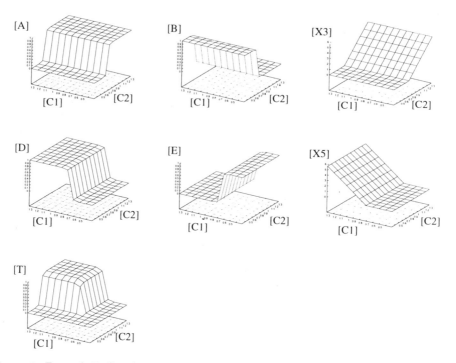

Figure 9. Example B: Steady-state input-output relation of the enzyme-substrate chemistry example (Fig. 8) which shows the function of an logic AND gate.

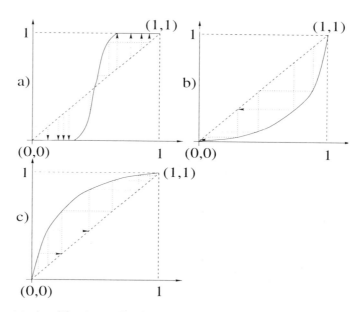

Figure 10. Amplification and reduction of a small error during repeated iteration.

original classification can not be restored. Some substances in Fig. 7 are of type b) or c) and thus the classification on incoming information is not robust. Additionally, the total information conservation of the polymer chemistry may lead to information destabilisation when the information should be stored in reaction loops, processed or transmitted. Small fluctuations will be amplified resulting in a quick loss of information.

THE ROBOT CONTROLLER

We shall now discuss the application of the above model to robot control. In order to make use of the computational capabilities of a metabolism it will be necessary to connect it to the environment. If it were simulated without any additional fluxes (e.g. sensor information) the system would run through a transient phase until it finally reached a steady-state.

If we connect the model of our reactor with the infrared sensors of a robot and the motor control with the concentration level of some metabolites we shall be able to use the emerging computational capabilities of the metabolism to control the robot's behaviour. The metabolism is driven away from equilibrium through the sensor information flux, so that all concentrations are influenced by the surroundings of the robot. The sensors are connected with different parts of the metabolism. A change of one sensor value may increase or decrease the concentration of a certain species. The resulting dynamics causes a representation of the environment inside the robot's metabolism.

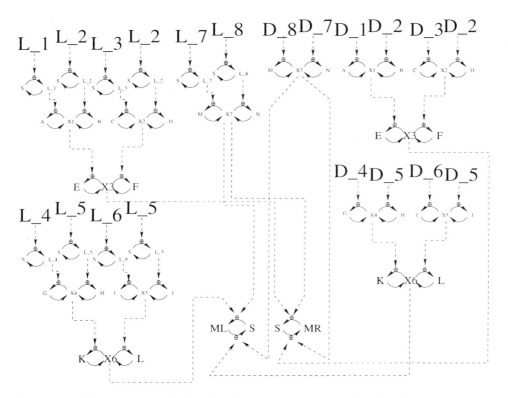

Figure 11. The robot's metabolism based on enzyme-substrate kinetics. L_i are the sensor substances emitted by the ambient light sensors. D_i are emitted by the distance sensors. Wheel motors are controlled by the substances ML and MR.

In Fig. 11 a schematic metabolism is shown based on an enzyme-substrate kinetics which has been used to control a real robot. The robot has 8 proximity and 8 ambient light sensors. Two basic behaviours are integrated into the metabolism: *obstacle avoidance* and *light seeking*. The part of the reaction network mainly concerned with the obstacle avoidance is processing the substances emitted by the proximity sensors. The light seeking behaviour gets its input from the ambient light sensors.

Experiments (Fig. 12) show, that the metabolism is able to control the robot in the desired way. It is also able to perform behaviour selection in critical situations where obstacle avoidance has to suppress light seeking. Furthermore, the controller is shown to be robust against disturbance of the reactor, which has been simulated by inserting small random amounts of substances. Figure 12 shows an example of an experiment in an artificial environment.

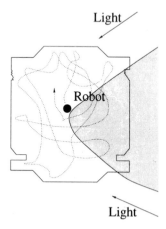

Figure 12. Example run of the robot. A typical trajectory of the robot's movement is shown in the environment. The robot tries to stay in the lighted area while simultaneously avoiding obstacles.

CONCLUDING REMARKS

The behavior of the robot emerges through chemical reactions between metabolites. It is possible for an autonomous robot to get information about its environment by sensors emitting substances into the robot's metabolism. The control system is parallel and distributed and the ability of navigation emerges only through computation with catalytic reactions [Banzhaf et al., 1996].

The increase and decrease of key-metabolites allows to achieve primitive goals such as obstacle avoidance or luminance gradient following (*phototaxis*). By adding different sensors it would be possible to solve more complicated tasks by stimulus-specific chemical reactions and alarmones. The current focus in our work lies in developing an algorithm which automatically evolves the metabolism to get a better stimulus-response mechanism, but this work is still in progress.

ACKNOWLEDGMENT

This project is supported by the DFG (Deutsche Forschungsgemeinschaft), grant Ba 1042/2-1.

APPENDIX

A: ODE System for Polymer Chemistry

The reaction mechanism for Example A:

$$A + B \quad \rightleftharpoons \quad AB, \qquad \omega_1 = k_1[A][B] - k_{-1}[AB]$$

$$AB + AB \quad \rightleftharpoons \quad ABAB, \qquad \omega_2 = k_2[AB][AB] - k_{-2}[ABAB]$$

$$ABAB + B \quad \rightleftharpoons \quad ABABB, \qquad \omega_3 = k_3[ABAB][B] - k_{-3}[ABABB]$$

$$A + ABABB \quad \rightleftharpoons \quad AABABB, \qquad \omega_4 = k_4[A][ABABB] - k_{-4}[AABABB]$$

$$A + B \quad \overset{B}{\rightleftharpoons} \quad AB, \qquad \omega_5 = k_5\,\nu[A][B][B] - k_{-5}\,\nu[AB][B]$$

$$AB + AB \quad \overset{ABAB}{\rightleftharpoons} \quad ABAB, \qquad \omega_6 = k_6\,\nu\,[AB][AB][ABAB] - k_{-6}\,\nu\,[ABAB][ABAB]$$

$$ABAB + B \quad \overset{A}{\rightleftharpoons} \quad ABABB, \qquad \omega_7 = k_7\,\nu[ABAB][B][A] - k_{-7}\,\nu[ABABB][A]$$

$$A + ABABB \quad \overset{AB}{\rightleftharpoons} \quad AABABB, \qquad \omega_8 = k_8\,\nu\,[A][ABABB][AB] - k_{-8}\,\nu[AABABB][AB]$$

Table 1. Kinetic rate constants of the polymer chemistry.

parameter	value
k_i	1.0 $\forall i$
k_{-i}	0.1 $\forall i$
ν	10

Resulting ODE system:

$$\frac{[A]}{dt} = -\omega_1 - \omega_4 - \omega_5 - \omega_8 + \varphi_A - [A]\frac{\varphi}{C}$$

$$\frac{[B]}{dt} = -\omega_1 - \omega_3 - \omega_5 - \omega_7 + \varphi_B - [B]\frac{\varphi}{C}$$

$$\frac{[AB]}{dt} = \omega_1 - \omega_2 + \omega_5 - \omega_6 - [AB]\frac{\varphi}{C}$$

$$\frac{[ABAB]}{dt} = \omega_2 - \omega_3 + \omega_6 - \omega_7 - [ABAB]\frac{\varphi}{C}$$

$$\frac{[ABABB]}{dt} = \omega_3 - \omega_4 + \omega_7 - \omega_8 - [ABABB]\frac{\varphi}{C}$$

$$\frac{[AABABB]}{dt} = \omega_4 + \omega_8 - [AABABB]\frac{\varphi}{C}, \tag{2}$$

with the size C of the reactor. The influx φ is defined as $\varphi = \Phi(A) + \Phi(B)$.

B: ODE System for Enzyme-Substrate Chemistry

The reaction mechanism for Example B:

$$I1^* + C \;\rightleftharpoons\; X1 + C, \qquad \omega_1 = k_1[C][I1^*] - k_{-1}[C][X]$$

$$X1 + B \;\rightleftharpoons\; X2^* + A, \qquad \omega_2 = k_2[X1][B] - k_{-2}[A][X2^+]$$

$$X3 + A \;\rightleftharpoons\; X4^* + B, \qquad \omega_3 = k_3[X3][A] - k_{-3}[B][X4^*]$$

$$X3 \;\rightleftharpoons\; I2^*, \qquad \omega_4 = k_4[X3] - k_{-4}[I2^*]$$

Table 2. Kinetic rate constants of the enzyme-substrate mechanism.

parameter	value	parameter	value
k_1, k_5	70	k_{-1}, k_{-5}	1
k_2, k_6	$5 \cdot 10^4$	k_{-2}, k_{-6}	1
k_3, k_7	$5 \cdot 10^4$	k_{-3}, k_{-7}	1
k_4, k_8	1	k_{-4}, k_{-8}	100

Resulting ODE system:

$$\frac{d[X1]}{dt} = \omega_1 - \omega_2$$

$$\frac{d[X3]}{dt} = \omega_3 - \omega_4$$

$$\frac{d[A]}{dt} = -\frac{[B]}{dt} = \omega_2 - \omega_3, \tag{3}$$

REFERENCES

[Alt, 1994] Alt, W. (1994). Cell motion and orientation. In *Frontiers in mathematical Biology*, pages 79 – 101. Springer.

[Arkin and Ross, 1994] Arkin, A. and Ross, J. (1994). Computational fuctions in biochemical reaction networks. *Biophysical Journal*, 67:560–578.

[Bagley and Farmer, 1992] Bagley, R. J. and Farmer, J. D. (1992). Spontaneous emergence of a metabolism. In Langton, C. G., editor, *Artificial Life II, Proceedings*, Cambridge, MA. MIT Press.

[Banzhaf, 1994] Banzhaf, W. (1994). Self-organization in a system of binary strings. In Brooks, R. A. and Maes, P., editors, *Artificial Life IV, Proceedings*, pages 108 – 119, Cambridge, MA. MIT Press.

[Banzhaf et al., 1996] Banzhaf, W., Dittrich, P., and Rauhe, H. (1996). Emergent computation by catalytic reactions. *Nanotechnology*, 7:307 – 314.

[Boos, 1977] Boos, W. (1977). *Intelligente Bakterien: Chemotaxis als primitives Mittel von Reizleitungssystemen.* Universitätsverlag Konstanz.

[Bray and Bourret, 1995] Bray, D. and Bourret, R. B. (1995). Computer analysis of the binding reactions leading to a transmembrane receptor-linked multiprotein complex involved in bacterial chemotaxis. *Molecular Biology of the Cell*, 6:1367 – 1380.

[Bray et al., 1993] Bray, D., Bourret, R. B., and Simon, M. I. (1993). Computer simulation of the phosphorylation cascade controlling bacterial chemotaxis. *Molecular Biology of the Cell*, 4:469 – 482.

[Bray and Lay, 1994] Bray, D. and Lay, S. (1994). Computer simulated evolution of a network of cell-signalling molecules. *Biophysical Journal*, 66:972 – 977.

[Conrad, 1992] Conrad, M. (1992). Molecular computing: The key-lock paradigm. *Computer*, 25:11–22.

[Darnell et al., 1993] Darnell, J., Lodish, H., and Baltimore, D. (1993). *Molekulare Zellbiologie.* de Gruyter.

[Fontana, 1992] Fontana, W. (1992). Algorithmic chemistry. In Langton, C. G., editor, *Artificial Life II, Proceedings*, Cambridge, MA. MIT Press.

[Fontana, 1994] Fontana, W. (1994). *Molekulare Semantik - Evolution zwischen Variation und Konstruktion.* Rowohlt Verlag.

[Hellingwerf et al., 1995] Hellingwerf, K. J., Postma, P. W., Tommassen, J., and Westerhoff, H. W. (1995). Signal transduction in bacteria: phospho-neural network(s) in escherichia coli? *Federation of Microbiological Societies. Microbiology Review*, 16:309 – 321.

[Hjelmfelt et al., 1991] Hjelmfelt, A., Weinberger, E. D., and Ross, J. (1991). Chemical implementation of neural networks and turing machines. *Proc. Natl. Acad. Sci. USA*, 88:10983–10987.

[Jetschke, 1989] Jetschke, G. (1989). *Mathematik der Selbstorganisation.* Vieweg Verlag.

[Jones and Aizawa, 1991] Jones, C. and Aizawa, S. (1991). The bacterial flagellum and flagellar motor: Structure, assembly and function. *Advances in microbial physiology*, 32:108 –171.

[Lengeler, 1995] Lengeler, J. W. (1995). Basic concepts of microbial physiology and metabolic control. Manuscipt.

[Lengeler, 1996] Lengeler, J. W. (1996). Die biologische Zelle als komplexes Informationssystem. *Biologie in unserer Zeit*, 426:65 – 70.

[Maynard Smith and Szathmary, 1995] Maynard Smith, J. and Szathmary, E. (1995). *The Major Transitions in Evolution.* W. H. Freeman, Oxford.

[Okamoto et al., 1995] Okamoto, M., Tanaka, K., Maki, Y., and Yoshida, S. (1995). Information processing of neural network system composed of 'biochemical neuron': Recognition of pattern similarity in time-variant external analog signals. In Paton, R., editor, *Information Processing in Cells and Tissues. Proceedings.*

[Stanier et al., 1986] Stanier, R. Y., Adelberg, E. A., and Ingraham, J. L. (1986). *The Microbial World.* Prentice-Hall.

[Stock and Surette, 1996] Stock, J. B. and Surette, M. G. (1996). Chemotaxis. In *Escherichia coli and Salmonella typhimurium*, pages 1103 – 1129. American Society for Microbiology, ASM Press.

EPILOGUE - CONCLUDING DISCUSSIONS

Mike Holcombe

Department of Computer Science
University of Sheffield
Sheffield S1 4DP

One of the main characteristics of the workshop is its interdisciplinary nature. The mix of biologists, computer scientists, mathematicians, biochemists and other scientists provides a very stimulating environment for discussion. The many viewpoints and experiences of trying to understand complex biological phenomena ensure a lively debate.

During the workshop three discussion groups were established, the intention being that the groups would focus on a number of important issues and report back at a plenary session. After some discussion the following themes were chosen.

(i) Multi-level Descriptions - Chair: David Friel;
(ii) Continuous/Discrete Systems - Chair: Mike Stannett;
(iii) Biological Information - Chair: Gareth Leng.

Later on there was a plenary session, chaired by Lev Belousov, when the groups reported back and a general discussion on all these issues took place. A brief review of the three discussions follows together with a record of the final discussion session. I am indebted to Alex Bell, Pedro Mendes, Mike Stannett, Sinead Scullion and Marie Willett for help in making this record of these stimulating debates.

GROUP SESSIONS

(i) Multi-level Descriptions - Chair: David Friel. Notes by Sinead Scullion

In biological systems there are multiple levels of complexity that lead to corresponding different levels of description. The important questions to be asked are: What is the relationship between the different levels in a biological system? How many variables are needed to accurately describe a system? The Gas Laws were used as a metaphor for these problems. In the case of the a single mole of gas, 3 times Avogadro's number of variables are needed to describe the position of every single molecule, yet the Gas Laws also completely describe the system in only 3 variables. Can we use this example to draw conclusions about how many variables are needed to accurately describe the behaviour of a biological system? For example can tissue behaviour be described using only a few variables or does the behaviour of each individual cell in the tissue have to be accounted for?

Some further conclusions:

1. Between different levels of description statistical averages are taken.
2. We can approach the modelling of a system from the bottom up or from the top down.
3. The level of description chosen depends on what can be measured experimentally.
4. The relative scales change between different levels of description (e.g. time/space).
5. These problems are not unique to biological systems. Accurate long term weather prediction is still almost impossible.

(ii) Continuous/Discrete Systems - Chair: Mike Stannett. Notes by Mike Stannett

Hierarchy is an important aspect of the need to move between discrete and continuous models. It is hard to deal with changing or resetting inputs and outputs and discontinuities with continuous models but it is more difficult to go from continuous to discontinuous accurately. We can simply reset continuous equations to deal with sudden change. We must make steps in a discrete model so small that the inaccuracies caused are insignificant, whatever that is - perhaps significant compared with e.g. noise. We can use discrete models which have continuous looking results by means of having more states. But this requires prior knowledge of the interesting features we want to preserve.

Time steps must reflect the speed of any changes in model behaviour. At the leading edge of an epidemic (often the key time) the data is discrete (only one or two cases), but for high numbers, we need a continuous approach. At low numbers we use a stochastic system. So we need two types of model tacked together. We can use both continuous and discrete modelling at the same time with hybrid models but they are low level and difficult to use. With very many variables, continuous models also become very hard to deal with, for example some systems involve ~ 100 enzyme molecules, so continuous models are no good. Working with both large and small numbers is difficult. We might use a large number as a continuous input to a discrete stochastic model for a small one which it affects.

So where is the boundary point between the regions where one or the other is suitable? Will there be a boundary region where both are appropriate? If there is such a region then we use the most convenient method. What about using automated discretisation techniques? These go from DEs to ODE lattice to Phase diagrams to automata models, but it gets hard as the number of variables increases.

Going from one regime to the other requires that we assume the system is predictable, which may not be true.

(iii) Biological Information - Chair: Gareth Leng. Notes by Gareth Leng

There are multiple views on how information may be described including:- Shannon Entropy; Thermodynamic Entropy; Kolmogorov complexity; Other measures of complexity; Algorithmic information content; Common usage as in newspaper.

A number of key issues were discussed, for example:- the fact that there are two views on the purpose of information; what is useful to you, the modeller; what is useful to the system.

Information only makes sense in a large array of [cells] and relationships between the parts e.g. not in a single neurone. Mutual Information has been discussed earlier in D'haeseleer et al and Somogyi and Fuhrman's contributions. Information only exsits when measured! Information is GLUE - part of the mechanism which unifies the subsystems and components.

Information Storage. This manifests itself in a variety of ways:- changes of conformation (e.g. enzyme, NN connections); cell states; information may be represented in bounded states and as processors: attractors; regions robust to perturbation.

Computers are designed explicitly by us as a computational device to solve a problem but in biology we see the expression of phenomena. These phenomena represent systems that also solve problems - e.g. an organism's survival in a niche, the subsystem's behaviour subject to constraints, the survival of the genome, etc.

Information theory is a useful tool in describing biological systems, but it does not provide a complete picture.

PLENARY SESSION - CHAIR: LEV BELOUSOV

The group panel chairs reported back and then a lively debate and discussion involving all the delegates ensued. Notes of the discussions were taken by Alex Bell and Mike Holcombe.

(i) Multi-Level Descriptions - Chair: David Friel

In biological and non-biological systems we need to understand them at different levels. What is relationship between these levels and how many variables are required at each level e.g. microscopic description of gas laws can be described by approximately 10^{23} variables, or by 3. By analogy: might there be similar things for tissues?

Conclusion: One way to reduce the number of variables is to use statistical averages, so our variables may be averages of cells. If you are interested in a `top down' approach you must describe the phenomena you want to explain before you look at the lower level thus preserving a link between the levels.

Sinead Scullion: Timescale is important in the choice of level of description since this implies the ability to use statistical measures.

Chair: Could we ever find differences between the levels in biological systems and non-biological systems?

Robin Callard: Is there a natural break in the levels of description at the genetic level? Above the genetic level the various levels interact, but is there any meaningful BIOLOGICAL information below it?

Pedro Mendes: Enzymes are expressed by the genetic information.

Mike Stannett: For example the Krebs cycle.

Roland Somogyi: Genetics also expresses dynamics.

Ray Paton: There are other factors at work which are just as important and "low level" as genes.

Chair: Genome has a very long "characteristic time" and could have the role of providing "parameters" but we need to know the equations describing the overall model as well.

Robin Callard: What I mean is, do the sub-molecular, quantum mechanical interactions really matter, for example does quanta effects in water molecules really affect cells??

Tim Hely: In computers there are different levels of modelling down to the silicon.

Mark Butler: In computers, aren't the levels "designed in" whereas biology is less clean cut?

Paul Leng: We need different techniques for different levels - e.g. hardware architecture - electrical circuit, etc. ....

David Friel: Are there different biological laws at the cellular/supercellular level? It would be useful to know the whole genome, but we need to know the other stuff too.

Patrick D'haeseleer: The higher the level the less precision.

Robin: Before the understanding of molecular biology there was a huge attempt to explain at the cellular level, and this didn't work. But the molecular level didn't solve it either, and we have to go back to a more holistic viewpoint. We must redevelop the cellular model based on a molecular foundation. Modelling is an iterative process.

A general conclusion was that some things can be expressed at one level and not another the key point is to relate things in a coherent way across the level boundaries.

(ii) Continuous/Discrete Systems - Chair: Mike Stannett

Key problem seems to be around sizes of populations. Large populations lend themselves to continuous models, but then can be misleading for small populations. Discrete models are better for small populations, but are hard to use for large ones. So we need to be able to patch the two together as in the paper by Arun Holden. And at what level do we do that, what is "significance". Difficulty is: we don't know what the interesting stuff is or where it is, until we have completed the model.

Problem: Is where we have one very small population, interacting with another very large one.

Problem: Where do we look to see if other people have thought about this stuff. Could we use a bulletin board?

Ray: Get on the IPCAT mailing list.

Patrick D'haeseleer: Reference "the importance of being discrete and spatial".

Patrick: Threshold effects can lead to discretisation.

Ray: Some other references: "Models in Biology", "Systems analysis in ecology".

There are some things you can do with particle models that simply cannot be done with differential equations.

Mark: Temporal effects in both paradigms are important.

Particle models can lead to artefacts like oscillations if the time scale is wrong.

David: We might be able to use discrete models if we know the system is predictable within certain perturbations, and use steps as appropriate.

Should we be obsessed with timesteps, would it be better to use other measures (e.g. fractions of total concentration).

(iii) Biological Information - Chair: Gareth Leng, Report by Paul Leng

What is information?
1. Something that passes through - a flow
2. A property (e.g. the state) of the system
3. Glue (holding it all together)

Can we really model things discretely and model it digitally. If we can, then our notions of information in computing are enough and then it is Turing computable. If not, we need a new paradigm.

Patrick: There are many definitions of information and people tend to talk past each other.

Roland: There is mutual information, correlations which are important.

Tim: One definition: "Information is what you are interested in". We may not know how, but in principle we should be able to model anything. But we don't have enough information.

Robin: The gene encodes a sequence of amino acids - we cannot predict from the gene the protein structure and its function.

Roland: The cell is hardware which is a deterministic computer that will compute the structure.

Mark: But we may need a quantum computer to do it.

Mike Stannett. It is NOT true that anything in nature can be modelled digitally e.g. wave equations. There is always divergence and there are problems which can be done in analogue, but not in digital. Also there are calculations which require more information than they actually contain. So information and predictability are very different.

Mark: Shannon said that information is a measure of uncertainty, the complement of redundancy. We need a universally acceptable definition of information.

Pedro: The rigorous definition of information is not much use for biology.

Yves: Information is biology - almost any expression of information is a result of natural selection!

Ray: This is a semantic issue. Is it all just semantics. Is information a conserved property?

A REFLECTION OF THE DISCUSSIONS BY SINEAD SCULLION

Hierarchies of Complexity and Multiple Levels of Description

Is there a natural break at the genetic level? At the genetic level do descriptions become more quantitative? Genes code, not only for proteins and structures but also for processes. So what is the importance of the genetic levels of description? Is gene expression important at all levels i.e. is it all pervading or is this not the case? Genomes have an exceptionally long characteristic time and exert their influence over many life cycles. Therefore genomes affect everything in biological systems (defining areas of parameter space). The dynamics of macromolecular interactions etc. are also important in understanding the complete picture. Genes contain all the information about protein sequence and structure, yet we can not predict secondary and tertiary protein structures or function from genetic code or amino acid sequences.

Are quantum changes important to cell functioning? Do quantum variations average out and so are unimportant at the cellular level or do they exert effects at a distance (in space or time)? If quantal variations and cell-cell interactions are important in describing cell behaviour then is it possible to develop a complete description of a cell at any one level and does that help our understanding? The links between the different levels of description are important in order to fully understand biological systems. Different modelling approaches are needed for modelling different levels (cf. computers, from structural model to electrical circuits).

Different levels of complexity usually have distinct spatial and time scales. The higher up the hierarchy you go the less predictive power the laws have about individuals in the system. Models of cellular interactions failed to explain all the complexities of the immune system, and observations at the molecular level have assisted our understanding. However both levels (molecular and cellular) have to be combined to gain a fuller understanding of the system as a whole.

There is a progression in model building where by models are hopefully predictive and are good models until something better comes along. There is no ideal level of complexity and description to choose when building a model and the levels chosen are defined by what we can measure.

Discrete versus Continuous Models of Biological Systems

This issue is of particular interest to those interested in the nature of models with respect to populations. In continuous models oscillatory behaviour can sometimes be seen and at the minimum value fractions of individuals may be involved (e.g. 0.0001 molecules). However continuous models have large scale behaviours that can be predicted by the model itself. Discrete models do not involve fractions of individuals (so useful when dealing with low numbers). It may often be beneficial to use both discrete and continuous models and piece them together. However, then you have to know how and where to do it. It depends on what is important in your experiment or model and where the boundaries are. There are several difficulties associated with this approach: What problems can it be used for? Where is information lost? Where do you change between models? What about the situation where you have very high and very low numbers of individuals together in the same system?

There are examples of things that may be achieved with particle models that cannot be done with continuous models and vice versa

What is Biological Information?

What is the nature of information in a biological system?
1. Information as something that passes through a system and affects it as it does so.
2. Information as a property (the state) of the system.
3. 'Glue' in the system - holds the elements of the system together.

Information is very closely connected to measurement - have to be able to measure things. To what extent can we represent these measurements discretely - are these things we can not?

Any kind of system, in principle, can be measured discretely and can be modelled digitally, therefore all properties of computational systems we know apply. Otherwise we need to establish new computational systems and approaches.

People have many different definitions of information but nevertheless individuals tend not to explain their definition when referring to information, assuming that all others will use the same definition. Information and mutual information (Shannon's entropy etc., a measure of the coherence between different parts of system, only one measure).

Different concepts of information: 'Information is what is useful' 'Information is a pool of resources'

A gene is a code for a sequence of amino acids. The process that controls the structure and interaction of the proteins is deterministic, yet we can not determine the processes that govern it. Processes are at a quantum level.

Not everything in the real world can be modelled on a digital computer. There are functions whose derivative can not be computed.

There is a difference between information and predictability.

INDEX

α-cell, 27
Adenylate cyclase, 111
Amplification, 125
an der Heiden, 111
Anisotropy, 76
Arrhythmia, 81
Artificial chemistry 261, 305
Artificial limb, 17
Artificial neural network, 17
Autophosphorylation, 125
Autopoiesis, 179

β-cell, 27, 30
B cell, 213
Bacteria, 39, 305
Baigent, 7
Banzhaf, 305
Basal ganglia, 95
Bell, 213
Beloussov, 165
Bezerianos, 75
Biocomputing, 227
Biological determinism, 227
Biophysical modelling, 27
Bogdan, 17
Boolean networks, 273, 285
Brown, 151
Brumen, 137
Butler, 177

Caenrhabditis elegans, 243
Caffeine, 47
Calbindin, 138
Calcium, 27, 47, 70, 95, 125, 137
Calcium-induced calcium release (CICR), 47
Calcium releasing channel, 27, 47, 95, 137
Calmodulin, 125, 137
CaM Kinase II, 125, 257
Cancer pagurus, 19
Carbon dioxide, 85
Category theory, 185, 259
Causality, 191, 197
Cell intercalation, 165
Cellular automata, 75, 295

Cellular graphics, 243
Chandler, 185
Chay, 27
Chay's store operated model, 27, 30
Chemoreceptor, 40, 305
Chemotaxis, 39
Clark, 39
Classification, 21
Clock, 35
Coding, 273
Complexity, 185, 273
Computation, 177, 197, 213, 261
Computational map, 302
Computational tissue, 177
Continuous/Discrete, 320
Control Law, 227
Cortical neuron, 57
Cottam, 197
Coupled map lattice, 75
Crosstalk, 95
Cyclic AMP, 95, 111
Cytokine, 107

δ-cell, 27
Database, 107, 243, 285
Development, 165, 243, 286
D'haeseleer, 203
Dissipative structure, 179
Distributivity, 273
Dittrich, 305
Dopamine, 95
Drosophila melanogaster, 286
Dynamic neural field, 295

Electrochemical potential, 12
Embryo, 8, 165, 243, 286
Embryomechanics, 166
Endoplasmic reticulum, 30, 137, 139
Energy, 229, 255
Enzyme, 232, 254, 263, 309
Ermentrout, 57
Escherichia coli, 39, 185, 307
Evolvability, 261